Professor Herbert Fröhlich

Energy Transfer Dynamics

Studies and Essays in Honor of Herbert Fröhlich on His Eightieth Birthday

Editors: T. W. Barrett and H. A. Pohl

With 87 Figures

Springer-Verlag Berlin Heidelberg NewYork
London Paris Tokyo

Terence W. Barrett, Ph. D.
W. J. Schafer Associates Inc., 1901 North Fort Myer Drive, Suite 800, Arlington, VA 22209, USA

Professor Herbert A. Pohl (Deceased)

ISBN-13:978-3-540-17502-5 e-ISBN-13:978-3-642-71867-0
DOI: 10.1007/978-3-642-71867-0

Library of Congress Cataloging-in-Publication Data. Energy transfer dynamics. Includes Index. 1. Physics. 2. Polarons. 3. Electron-phonon interactions. 4. Biophysics. 5. Fröhlich, H. (Herbert) 1905-. I. Fröhlich, H. (Herbert), 1905-. II. Barrett, T. W. (Terence William), 1939-. III. Pohl, Herbert A. (Herbert Ackland), 1916-1986.
QC71.E63 1987 539.7'54 87-4704

2153/3150-543210

Preface

On three occasions and at different locations, conferences were held to honor the eightieth birthday of Professor Herbert Fröhlich: on the 18th December, 1985, in Liverpool, England; on the 14th February, 1986, in Stuttgart, Germany; and on the 8th March, 1986, on the Palm Coast, Florida. This Festschrift is a compilation of the papers of those conferences.

Fröhlich's choice of problems, from the earliest days, was couched in the physics of intrinsically interacting systems of excitation. One example, in which he set the course of research which is still followed, concerned dielectric breakdown, developed from the 1930's over several decades. The interacting systems are the electrons (receiving energy from an electric field) and lattice atom motion (taking energy from the electrons via "electron-phonon" interaction, hence heat dissipation). There is a threshold field above which the latter cannot keep up with the former, and the combined system (electrons plus phonons) "runs away"; that is to say, collectively it switches to a new state.

While this general idea is straightforward, the actual theories of high field electronic conductivity, electron-phonon interaction, the role of defects, and effects of spatial heterogeneity posed major new research challenges in the then newly developing field of solid state physics. Fröhlich and his students played a leading role in solving these problems over the next three decades, establishing cornerstones and directions in the field. Clearly, high field transport in semiconductors, and a number of phenomena depending on electron-lattice coupling, utilize methods derived from their work.

But of all the studies during this period, perhaps the two most exciting—because of their simultaneous conceptual clarity, yet deep theoretical subtlety—were his seminal papers on the polaron (with Pelzer and Zienau, 1950) and on a collective electron-phonon model for superconductivity (1951). The polaron comprises an electron moving with a coherently "attached" lattice polarization, the whole being a collective state. Subsequently, it was realized by Feynman that this was a prototypical many-body problem, like the quantum electrodynamic interaction of an electron with radiation field oscillators. It may be said that the methods of quantum field theory flowered in treating the polaron problem.

More or less simultaneously, Fröhlich examined the coupling of two electrons to a phonon and showed that this could provide a coherent attractive interaction between them, and on that basis constructed a theory of the superconducting state. A major prediction of this theory, though it awaited key additions by Bardeen, Cooper and Schrieffer in 1957, was that the transition temperature in isotropic forms of the same superconductor would be inversely proportional to the square root of the isotropic mass, which was observed experimentally at just about that time!

In more recent years, Fröhlich turned to fundamental problems in biophysics. In 1968 he applied to the case of a biomolecular system a fundamental concept originally proposed by Penrose of energy transfer to the oscillatory modes of a system. Energy might be received by the molecule via a mechanism similar to that of Bose-Einstein condensation in superfluids or superconductors, except that the order arises when the energy or metabolic drive is made sufficiently large, rather than by lowering the temperature. Bose-Einstein condensation is present whenever the largest eigenvalue of the one-particle reduced density matrix is an extensive rather than an intensive quantity, or whenever—for a spatially uniform system with periodic boundary conditions—a finite fraction of the particles have identical momenta. Calculations of the long-wave elastic vibrations in such a molecular system led to the prediction of microwave effects and the studies discussed in this Festschrift. Other calculations based on the vibrational activity of molecular subgroups led to predictions concerning molecular energy transfer of higher energy in the tenths of electron volt range. The mechanism of energy transfer for the latter higher range has been recognized as being very close to the Davydov soliton theory of vibronic energy transfer and the Holstein small polaron concept, a developed version of which is discussed in a number of the following chapters, and which is fundamental to the field of bioenergetics.

Truly, to have the intuition and theoretical skill to perceive and explore ideas of such seminal importance is a rare quality, and Herbert Fröhlich's contributions to science over half a century are distinguished by this ability.

The editors would like to express their gratitude to the United States Naval Air Systems Command HQ for financial support in convening these conferences, to Professor Otis C. Dermer, Oklahoma State University, and to Mrs. D. Wilbanks of the Computerized Technical Composition Section, Naval Research Laboratory, Washington, D.C., without whose assistance this Festschrift would not have been possible.

James A. Krumhansl, New York
Terence W. Barrett, Washington, D.C.

Herbert A. Pohl

It is a painful duty to report that one of the editors of this book, Dr. Herbert Ackland Pohl, died very suddenly in his laboratory at the Massachusetts Institute of Technology in Cambridge, Massachusetts, on June 21st, 1986.

Dr. Pohl was born on February 17th, 1916, in Lisbon, Portugal, as an American citizen. He earned the Ph.D. in chemical physics at Duke University, and served as a faculty member at Johns Hopkins University until he joined the U.S. Navy to work at the Naval Research Laboratory in Washington, D.C. during World War II. Thereafter he was with the duPont Company in Wilmington for some years working on synthetic fibers, and later in the Atomic Energy Division. He was a faculty member successively at Princeton University, Brooklyn Polytechnic Institute and Oklahoma State University, where he served as professor of physics from 1964 to 1981. He was a visiting professor at the University of Uppsala (1963), Cambridge University (1971), and the Massachusetts Institute of Technology (1984-6), and a visiting research scientist at Woods Hole, Massachusetts, in 1976.

His interests in biophysics were wide-ranging, and led him to become editor of the *Journal of Biological Physics* in 1977. He was in constant collaboration with American and European colleagues. His best known book, *Dielectrophoresis*, is used in the United States, Europe and the Soviet Union, and cited in several chapters of this book. One such chapter, by Pohl and collaborators, reviews much of his recent ongoing research.

Dr. Pohl is survived by his wife Eleanor, of Stillwater, Oklahoma, three sons, two daughters, and 12 grandchildren. His many friends will remember his love of ideas, puns (preferably scientific), humanity, good company and Mark Twain humor. He had a remarkable determination to stay professionally active in research as an emeritus professor and will be missed by colleagues and friends on two continents.

Otis C. Dermer, Stillwater, Oklahoma
Terence W. Barrett, Washington, D.C.

Table of Contents

Part I Physics

Part II Biophysics

List of Contributors

Dr. T.W. Barrett
W.J. Schafer Associates, Inc.
1901 North Fort Myer Dr.
Suite 800
Arlington, VA 22209

Dr. I.J. Bigio
Center for Nonlinear Studies
Los Alamos National Laboratory
Los Alamos, NM, 87545 USA

Professor F.W. Cummings
Department of Physics
University of California
Riverside, CA 92512 USA

Professor E. Del Giudice
Dipartimento di Fisica dell'Universitá
Via Celoria 16-20133 Milano, Italia

Professor S. Doglia
Dipartimento di Fisica dell'Universitá
Via Celoria 16-20133 Milano, Italia

Professor D.G. Frood
Department of Physics
Lakehead University
Thunder Bay
Ontario P7B5E1, Canada

Professor T.J. Gallagher
Department of Electrical Engineering
University College
Dublin 2, Ireland

Professor T.B. Grimley
Donnan Laboratories
University of Liverpool
P.O. Box 147
Liverpool L69, 3BX
England, UK

Professor J.B. Hasted
Birbeck College
University of London
Malet Street
London WC1E 7HX
England, UK

Professor G.J. Hyland
Department of Physics
University of Warwick
Coventry
England, UK

Professor A.K. Jonscher
Dielectrics Group
King's College (KQC)
University of London
Strand, London WC2R 2LS
England, UK

Professor F. Kaiser
Institut für Angewandte Physik, Theorie
Technische Hochschule
6100 Darmstadt, FRG

Professor D.B. Kell
Department of Botany & Microbiology
University College of Wales
Aberystwyth, Dyfed SY23 3DA
Wales, UK

Professor J.A. Krumhansl
Laboratory of Atomic and Solid State Physics
Cornell University
Ithaca, NY 14853 USA

Professor P.T. Landsberg
Department of Electrical Engineering
University of Florida
Gainesville, FL 32611, USA

Dr. W.A. Lin
Center for Studies in Statistical Mechanics
The University of Texas at Austin
Austin, TX 78712, USA

Professor M. Milani
Dipartimento di Fisica dell'Universitá
Via Celoria 16-20133 Milano, Italia

Professor W. Nagl
Universität Kaiserslautern
Postfach 3049
6750 Kaiserslautern, FRG

Professor J.J. O'Dwyer
Physics Department
State University of New York College at
Oswego, NY 13126, USA

Professor R. Pethig
Institute of Molecular and
Biomolecular Electronics
University College of North Wales
Dean Street, Bangor, Gwynedd LL57 1UT
Wales, UK

Dr. W.T. Phillips
Francis Bitter National Magnet Laboratory
Massachusetts Institute of Technology
Cambridge, MA 02139, USA

Professor H.A. Pohl
Francis Bitter National Magnet Laboratory
Massachusetts Institute of Technology
Cambridge, MA 02139, USA

Professor M. Pollak
Department of Physics
University of California
Riverside, CA 92521, USA

Dr. J. Pollock
Francis Bitter National Magnet Laboratory
Massachusetts Institute of Technology,
Cambridge, MA 02139, USA

Professor F.A. Popp,
Universität Kaiserslautern,
Postfach 3049,
6750 Kaiserslautern, FRG

Professor E.W. Prohofsky,
Department of Physics
Purdue University
West Lafayette, IN 47907, USA

Professor L.E. Reichl
Center for Studies in Statistical Mechanics
The University of Texas at Austin
Austin, TX 78712, USA

Professor S. Rowlands
Faculty of Medicine
University of Calgary
Calgary, Alberta, T2N 4N1, Canada

Professor B.K.P. Scaife
Engineering School
Trinity College
Dublin 2, Ireland

Professor W.G. Scaife
Department of Mechanical & Manufacturing
Engineering
Trinity College
Dublin, Ireland

Professor H.P. Schwan
Department of Bioengineering
University of Pennsylvania
Philadelphia, PA 19104-6392, USA

Professor A.C. Scott
Department of Electrical and
Computer Engineering
The University of Arizona
Tucson, AZ 85721, USA

Professor G.L. Sewell
Department of Physics
Queen Mary College
University of London
Mile End Road
London E1 4NS
England, UK

Professor C.W. Smith
Department of Electronic and
Electronic Engineering
University of Salford
Salford, M5 4WT
England, UK

Professor H.J. Stapleton
Physics Department & Materials
Research Laboratory
University of Illinois at Urbana-Champaign
Urbana, IL 61801, USA

Professor L. Tisza
Department of Physics
Massachusetts Institute of Technology
Cambridge, MA 02139, USA

Professor G. Vitiello
Dipartimento di Fisica dell'Universitá
Salerno, Italia

Part I

Physics

1. Nonadiabatic Processes in Surface Physics

T. B. Grimley

ABSTRACT

Nonadiabatic processes are both frequent and important in surface physics, but unlike the electron-phonon interaction leading to superconductivity in the bulk, surface processes often depend on the interaction of the electrons with only a few moving nuclei. This makes it feasible to formulate the problems in terms of the adiabatic electron states, and so to treat also scattering problems where a few nuclei change their positions by arbitrarily large amounts. Three examples are discussed; two are fully quantum mechanical, and one is semi-classical.

INTRODUCTION

The most important manifestation of nonadiabaticity in solid state physics is superconductivity, whose quantitative and qualitative understanding begins with Fröhlich's paper in September 1950 (Fröhlich, 1950). From a formal point of view, the nonadiabatic process is described rather simply by the Born-Oppenheimer (1927) system of coupled equations, but since the adiabatic electron states depend on the vibrational coordinates (i.e., on the positions of all the nuclei), their determination is difficult. Because of this, it is usual to formulate the nonadiabatic problem using a fixed basis set of one-electron states, and this was in fact Fröhlich's approach.* But there are many nonadiabatic processes at metal surfaces that are due to the motions of a few nuclei, or even the motion of a single nucleus, and for these processes, an approach using the adiabatic electron states is practicable. Among processes in this category are:

(i) The electronic damping of adsorbate vibrations.

(ii) The sticking and inelastic scattering of light reactive atoms with energy loss to the electron system.

In ii the moving nucleus changes its position by arbitrarily large amounts so that it is not feasible to use a fixed basis set of electron states to treat this problem. Furthermore, ii throws up a fundamental question in quantum mechanics to which

*Frölich's idea that the dynamic electron-phonon interaction leading to an (indirect) electron-electron interaction is responsible for superconductivity was not readily accepted by some distinguished theoretical physicists as a critical reading of the discussion of his paper in Physica, **19**, 755 (1953) will show.

a definitive answer has not yet been given: what is the sticking coefficient on a cold solid in the limit of zero kinetic energy of the incoming gas atom? There are, however, many processes where the movements of the relevant nuclei can be treated classically to provide a time-dependent perturbation of the electron system. Among them are:

(iii) The ionization (neutralization) of hyperthermal atoms (ions) in collisions with metal surfaces.

In this paper I shall outline the theoretical approaches which we (G. P. Brivio, V. C. Jyothi Bhasu, K. L. Sebastian and I) have used to treat these problems, and to compute numerical results.

ELECTRONIC DAMPING OF ADSORBATE VIBRATIONS

This is a particularly simple example of the electron-vibration interaction which can be formulated by using a fixed basis set of electron states, and matrix elements of a coupling operator evaluated at the equilibrium geometry. The damping is due to the creation of low-energy electron-hole pairs (Brivio and Grimley, 1977) and is therefore important for vibrational modes with large amplitudes on atoms involved in the metal-adsorbate bonds.

Let $\vec{r}$ stand for the co-ordinates of all electrons in the system, and let Q be one of the normal co-ordinates of the adsorbate on the rigid substrate. The Hamiltonian is

$$H(\vec{r},\vec{Q}) = K_I + V_{II} + H_{EI}(\vec{r},Q) \tag{1}$$

where K_I is the kinetic energy of the absorbate ions, V_{II} the Coulomb interaction of all the ions, and H_{EI} the electron-ion Hamiltonian whose eigenstates $|n>$ with energies E_n are the adiabatic electron states depending parametrically on Q in general, but which are needed here only for the equilibrium configuration $Q = 0$. The Golden Rule formula for the lifetime τ_1 of the first excited state of the mode Q with frequency Ω_Q and reduced mass M is

$$1/\tau_1 = (\pi/M\Omega_Q) \sum_{m\neq 0} |<0| \frac{\partial H_{EI}(\vec{r},0)}{\partial Q} |m>|^2 \, \delta(E_m - \hbar\Omega_Q) \,. \tag{2}$$

Upon going over to second quantization in a basis set $\{|\mu>\}$ with creation and destruction operators c_μ^+ and c_μ, the matrix element in Eq. 2 is

$$<0| \frac{\partial H_{EI}(\vec{r},0)}{\partial Q} |m> = \sum_{\mu,\nu} <\mu \mid \frac{\partial V(\vec{r},0)}{\partial Q} \mid \nu>< 0 \mid c_\mu^+ c_\nu \mid m> \tag{3}$$

where $V(\vec{r},Q)$ is the unscreened Coulomb potential of an electron in the field of the adsorbate nuclei. Many-body effects leading to the screening of this potential are described by the Bethe-Salpeter amplitudes $<0|c_\mu^+ c_\nu|m>$. If the basis set $\{|\mu>\}$ are Hartree-Fock (HF) orbitals, and if excited electron states are formed by creating electron-hole pairs on the HF ground state, the Bethe-Salpeter amplitudes are either 0 or 1 (1 if $c_\mu^+ \, c_\nu|m>=|0>$, 0 otherwise), and the states $|m>$ that dissipate the vibrational energy have only one electron-hole pair present on the HF ground state. In this case, because $\hbar\Omega_Q$ is small on the scale of electronic excitation energies Eq. 2 reduces to

$$1/\tau_1 = (\pi\hbar/M)|\sum_{i,j} \tilde{n}_{ij}\ (\epsilon_F) < j\ |\frac{\partial V(\vec{r},0)}{\partial Q}|i> |^2 \tag{4}$$

where $\{|i>\}$ is a locally complete, not necessarily orthogonal, set of spin orbitals adequate for an embedded cluster description of the bonding of the adsorbate to the metal, and $\tilde{n}_{ij}\ (\epsilon)$ is the spectral density of the charge- and bond-order matrix defined on this set. Only its value at the Fermi level ϵ_F is needed.

An alternative formula to Eq. 3 is

$$<0|\frac{\partial H_{EI}(\vec{r},0)}{\partial Q}\ |m> = \lim_{t_1 \to 0} \int d\vec{r}_1\ \{f^*_\mu\ (\vec{r}_1) f_\nu(\vec{r}_1) \tag{5}$$

$$+ \int d4\, d4'\ f^*_\mu\ (4)\ \frac{\delta \sum(44')}{\delta U(1)}\ f_\nu\ (4')\} \frac{\partial V(\vec{r},0)}{\partial Q}$$

where f_μ is the Dyson orbital associated with the HF orbital ϕ_μ, $\delta\sum/\delta U$ is the functional derivative of the electron self-energy, and I have used the notation $1 \equiv (\vec{r}_1 t_1)$, and so on. If we assume that $\delta \sum/\delta U$ is the time-local in the sense that

$$\frac{\delta \sum(44')}{\delta U(1)} = \frac{\delta \sum(\vec{r}_4 \vec{r}_4')}{\delta U(\vec{r}_1)}\ \delta(t_4 - t_4')\ \delta\ (t_4 - t_1)$$

then Eq. 5 reduces to

$$<0|\frac{\partial H_{EI}(\vec{r},0)}{\partial Q}\ |m> = \int d\vec{r}_1 d\vec{r}_2 f^*_\mu\ (\vec{r}_1)\ \frac{\partial V_{eff}(\vec{r}_1,\vec{r}_2,0)}{\partial Q}\ f_\nu(\vec{r}_2) \tag{6}$$

where

$$V_{eff}(\vec{r}_1,\vec{r}_2,Q) = V_I(\vec{r}_1,Q)\ \delta\ (\vec{r}_1 - \vec{r}_2) + \sum(\vec{r}_1,\vec{r}_2,Q) \tag{7}$$

is a nonlocal potential in which the bare Coulomb potential V_I of all the nuclei in the system is screened by the electron self-energy. Equation 4 for the lifetime now has $V_{eff}(\vec{r}_1,\vec{r}_2,0)$ in place of $V(\vec{r},0)$.

We have made numerical calculations for a hydrogen atom on a simple metal using the HF self-energy in Eq. 7. We found $\tau_1 = 1.1 \times 10^{-11}\ s$, giving an infrared linewidth of 3 cm^{-1}, and others (Persson and Hellsing, 1982) have found a similar number by using Eq. 6 with V_{eff} replaced by the local-density version of the Kohn-Sham *local* potential, and the Dyson orbitals replaced by the Kohn-Sham orbitals. I note that there is an isotope effect, $\tau_1 \approx M$, yet to be observed.

STICKING AND INELASTIC SCATTERING WITH ENERGY LOSS TO THE ELECTRON SYSTEM

Both processes, sticking and inelastic scattering, proceed with the creation of low-energy electron-hole pairs. Our (Brivio and Grimley, 1979, 1983, 1985, and to be published) approach is via the theory of the gettering of a gas in a closed vessel, one wall of which is the metal, the others inert. By reducing the area of the metal wall, we can make the gettering rate as small as we please so that the gettering rate is obtained essentially exactly in the quasi-particle approximation. In this case the key quantity is the imaginary part of the gas atom self-energy evaluated at the initial-state energy. The gettering Hamiltonian has an exact representation in terms of the stuck states and elastic scattering states, and because of this, the gettering T-matrix that determines the gas atom's self-energy, and hence the rates of sticking and inelastic scattering, can be calculated exactly for nontrivial models.

Let $\vec{x}$ stand for all the coordinates needed to describe the system atom + metal. $\vec{x}$ contains the position $\vec{R}$ of the gas atom, which will be displayed when necessary. If H_0 is the Hamiltonian when the atom and metal are noninteracting, and $U(\vec{R})$ the elastic part, and $W(\vec{x})$ the inelastic part of the scattering potential, the full Hamiltonian is

$$H = H_0 + U + W . \tag{8}$$

Suppose that at time $t = 0$ the system is in an eigenstate $\Psi_i(\vec{x})$ of the Hamiltonian $H_0 + U$, where Ψ_i describes the gas atom making elastic collisions with the metal in its lowest electronic state. For $t > 0$, the inelastic part of the potential acts to cause the initial state Ψ_i to decay, and other eigenstates of $H_0 + U$ to become populated. These "other eigenstates" describe the gas atom either stuck to, or scattering off, the metal with the electron system excited. I assume that the elastic potential $U(\vec{R})$ does not change when the electron system is excited.

It is now a simple matter to show that, in the quasi-particle approximation, the probability $p_i(t)$ to observe the initial state at time t is

$$p_i(t) = \exp\{-2\Gamma_i(E_i)t/\hbar\} \tag{9}$$

where $-\Gamma_i(E_i)$ is the imaginary part of the gas atom's self-energy evaluated at the initial state energy E_i. Because Γ_i separates (see below) into the sum of sticking contributions and inelastic scattering contributions Γ_i^{stick} and Γ_i^{inel}, the sticking and inelastic scattering rates are

$$R_{\text{stick}} = 2\Gamma_i^{\text{stick}}(E_i)/\hbar, \quad R_{\text{inel}} = 2\Gamma_i^{\text{inel}}/\hbar . \tag{10}$$

The sticking coefficient is

$$s = R_{\text{stick}}/(R_{\text{stick}} + R_{\text{inel}} + \nu_{\text{coll}}) \tag{11}$$

where ν_{coll} is the number of collisions per unit time the gas atom makes with the metal surface.

The Hilbert space of the gettering Hamiltonian is spanned by the eigenstates $H_0 + U$. These are the initial states $|i\rangle$, and the other eigenstates $|f\rangle$ with energies E_i and E_f. These states and energies are of course $\vec{R}$-dependent. The full Hamiltonian is

$$\begin{aligned} H &= |i\rangle E_i \langle i| + \sum_f |f\rangle E_f \langle f| + W \\ W &= \sum_f (|i\rangle\langle i|W|f\rangle\langle f| + |f\rangle\langle f|W|i\rangle) \\ &\quad + \sum_{f,f'} |f\rangle\langle f|W|f'\rangle\langle f'| . \end{aligned} \tag{12}$$

The eigenstates of $H_0 + U$ are product states $|n;\vec{R}>|\gamma_n(\vec{R})>$ describing the gas atom in a state γ on an electronically adiabatic potential energy surface $\epsilon_n(\vec{R})$ derived from the electronic state $|n;\vec{R}>$. The simpler notation $|n,\gamma>$ will also be used.

From the Born-Oppenheimer system of coupled equations the matrix elements of the operator W are

$$<\beta,m|W|n,\gamma> = (\hbar^2/M)\int d\vec{R}\beta_m(\vec{R})W_{mn}(\vec{R})\gamma_n(\vec{R}) \tag{13}$$

$$W_{mn}(\vec{R}) = <m;\vec{R}|\nabla_R|n;\vec{R}>\nabla_R + \frac{1}{2}<m;\vec{R}|\nabla_R^2|n;\vec{R}>.$$

In terms of the gettering T-matrix (a column matrix) with elements defined by

$$<f|T(\epsilon)|i> = <f|W|i> + \sum_{f'}<f|W|f'>(\epsilon-E_{f'})^{-1}<f'|T(\epsilon)|i> \tag{14}$$

the formula for Γ_i is

$$\Gamma_i(E_i) = \pi\sum_f|<f|T(E_i)|i>|^2\,\delta\,(E_i - E_f). \tag{15}$$

Equations 10-15 are the basic equations of our approach but there are in practice some simplifications we can make because only low-energy electron-hole pairs are excited. An important one is that all electronically adiabatic potential energy surfaces have the same shape, being only shifted vertically from $\epsilon_0(\vec{R})$ by the electronic excitation energy. But of course we cannot escape having to calculate $W_{mn}(\vec{R})$ for all $\vec{R}$ in the range of the atom-metal force, and so far we have made computations only for some simple models of a metal-hydrogen atom system. Of particular interest is the sticking coefficient, which we find has a maximum value of about 0.3 at $E_i = 21$ meV, behaves like $\sqrt{E_i}$ as $E_i \rightarrow 0$, and goes to zero also as $E_i \rightarrow \infty$. The high-energy behavior is expected on general grounds from the matrix elements in Eq. 13. The low-energy behavior too can be established for any short-range potential, and our $\epsilon_0(\vec{R})$ *is* short-ranged. For a long-range potential, the low energy behavior of the sticking coefficient is controversial.

IONIZATION OF HYPERTHERMAL ATOMS IN COLLISIONS WITH METAL SURFACES

I have in mind the ionization of Na atoms with kinetic energies in the range 30-500 eV reflected from tungsten surfaces, which has been studied experimentally (Overbosch and Los, 1981a, 1981b). For such fast particles, we can in the first approximation assume that the nucleus moves on its classical trajectory, and therefore simply provides a known time-dependent perturbation of the electron system. The important effect is that, as the atom approaches the metal surface, the image interaction causes the ionization level of the atom to rise, and the affinity level to fall. A situation of most interest is when the ionization level is below the Fermi level ϵ_F when the atom is far from the surface, rises to go above ϵ_F at the atom's classical turning point, but has crossed ϵ_f beyond the range of the atom-metal electron hopping. Now, on the atom's inward journey, its ionization level is above ϵ_F

at distances where electron hopping to vacant metal levels is possible, and the ionization level depopulates ($Na \rightarrow Na^+$). On the ion's outward journey, the ionization level does not fall below ϵ_F again until the atom-metal distance is so great that electron hopping from occupied metal levels is impossible. Hence re-neutralization ($Na^+ \rightarrow Na$) cannot occur, and consequently the probability of positive ionization in the collision is essentially unity. On the other hand, if the ionization level crosses ϵ_F *within* the range of atom-metal electron hopping, then on the ion's inward journey we have $Na \rightarrow Na^+$ as before, but on its outward journey re-neutralization ($Na^+ \rightarrow Na$) *does* occur, and the probability of ionization in the collision is essentially zero.

This simple discussion is defective because it does not take account of the shifting and broadening of the ionization level through its hopping interaction with the metals levels, and neither does it tell us anything about the electronic state of the metal when the collision is over. But it is a useful picture nevertheless. To have a detailed theoretical model, consider (Grimley, Jyothi Bhasu, and Sebastian, 1983) a gas atom with orbitals $\{|A>\}$ interacting with a solid, whose electron Hamiltonian is H_s, *via* a set of orbitals $\{|\alpha>\}$ in the space spanned by the eigenfunctions of H_s. The Hamiltonian for the problem is written in the form

$$H = \sum_A E_A(t) n_A + \frac{1}{2} \sum_{A \neq B} J_{AB} n_A n_B + \sum_{A,\alpha} \{ f_{A\alpha}(t) C_A^+ C_\alpha$$
$$+ f_{\alpha A}(t) C_\alpha^+ C_A \} + H_s . \tag{16}$$

Here E_A is the orbital energy of $|A>$ when all Coulomb interactions between electrons in the set $\{|A>\}$ are neglected, C_A, C_α (C_A^+,C_α^+) are the usual Fermion destruction (creation) operations, $n_A = C_A^+ C_A$, J_{AB} is the Coulomb repulsion integral for electrons in $|A>$ and $|B>$, and $f_{A\alpha}(t)$ represents the hopping interaction between the atom orbital $|A>$ and the substrate orbital $|\alpha>$. The time-dependence of $f_{A\alpha}$ is determined from the motion of the atom's nucleus on its classical trajectory. E_A is also time-dependent in this trajectory approximation because of the effects of the image interaction on the ionization and affinity levels of the gas atom.

I work in the time-dependent Hartree-Fock approximation according to which an initial Slater determinant of one-electron states $|j_k>$ evolves in time as a single Slater determinant through the time evolution of the orbitals $|j_k>$ under the Hartree-Fock one-electron Hamiltonian. Thus the equation of motion of the evolution operator of the system is

$$i\hbar \partial U(t, t_0)/\partial t = H_{HF}(t) U(t, t_0), \quad t_0 \rightarrow -\infty \tag{17}$$

where H_{HF} is the Hartree-Fock version of Eq. 16,

$$H_{HF} = \sum_A E_A(t) n_A + \frac{1}{2} \sum_{A \neq B} J_{AB} \{ <n_A(t)> n_B + n_A <n_B(t)>$$
$$- <n_A(t)><n_B(t)> \} + \sum_{A,\alpha} \{ f_{A\alpha}(t) C_A^+ C_\alpha + f_{\alpha A}(t) C_\alpha^+ C_A \} + H_s \tag{18}$$

where for example $<n_A(t)>$ is the expectation value of n_A in the evolving Slater determinant. Since also in the time-dependent Hartree-Fock approximation

$$|j_k(t)> = U(t, t_0)|j_k>, \quad t_0 \to -\infty \tag{19}$$

we have, for the time-dependent occupancy of a gas atom orbital

$$<n_A(t)> = \sum_{B=occ.} |<A|U(t, t_0)|B>|^2 + \int_{-\infty}^{\epsilon_F} d\epsilon |<A|U(t, t_0)|\epsilon>|^2 \tag{20}$$

where as indicated, the sum on B goes over the occupied gas-atom orbitals, and the integral is over the occupied one-electron states $|\epsilon>$ of the semi-infinite metal.

We require the occupation numbers of the atomic orbitals after the collision when the gas atom is far from the metal, namely $<n_A(\infty)>$. According to Eq. 20 therefore we need the matrix elements $<A|U(\infty,t_0)|B>$ and $<A\ |U(\infty,t_0)|\epsilon>$. Our approach (Grimley, Jyothi Bhasu, and Sebastian, 1983) has been to integrate the equation of motion Eq. 17 numerically in the representation afforded by the initial states to find $U(t,t_0)$, $t > t_0$, and so to follow the changes in the occupation numbers with time through the collision. The calculation is more convenient in an interaction picture because the initial conditions are simpler to specify. Present-day supercomputers allow us to take account of the fact that, as the electronic state evolves in time, so does the Hellmann-Feynman force on the gas atom's nucleus, so that it is not necessary to assume that the classical trajectory is determined from the ground-state electronically adiabatic potential energy surface, but we have not made this refinement in our computations.

CONCLUSION

In this paper honoring Professor Fröhlich I have outlined theoretical methods that have been successful in treating some nonadiabatic processes at metal surfaces. None of these processes has the subtleties of the phenomenon of superconductivity, although they are not entirely without conceptual difficulties. On the other hand, they are primary processes in surface physics, easy to observe experimentally, and of the utmost importance in technology. At the present time, and as a result of extensive computations on theoretical models, their essential features are beginning to appear, and useful concepts are being formed.

REFERENCES

M. Born and J.R. Oppenheimer, Ann. Phys. **84**, 457 (1927).

G.P. Brivio and T.B. Grimley, J. Phys. C: Solid State Phys. **10**, 2351 (1977).

G.P. Brivio and T.B. Grimley, Surface Sci. **89**, 226 (1979).

G.P. Brivio and T.B. Grimley, Surface Sci. **131**, 475 (1983).

G.P. Brivio and T.B. Grimley, Surface Sci. **161**, L537 (1985).

G.P. Brivio and T.B. Grimley, Phys. Rev. to be published.

H. Fröhlich, Phys. Rev. **79**, 845 (1950).

T.B. Grimley, V.C. Jyothi Bhasu and K.L. Sebastian, Surface Sci. **124**, 305 (1983).

E.G. Overbosch and J. Los, Surface Sci. **108**, 99 (1981a).

E.G. Overbosch and J. Los, Surface Sci. **108**, 117 (1981b).

M. Persson and B. Hellsing, Phys. Rev. Lett. **49**, 662 (1982).

2. Dielectrics – The Fröhlich Connection

D. G. Frood, T. J. Gallagher, and B. K. P. Scaife

The contribution of Herbert Fröhlich to the science and technology of electrical insulation—or dielectrics—is as pervasive as it is significant. This has come directly through his own individual work and, indirectly, through his teaching and writing. Many workers in this subject are indebted to him for help and advice given over a long and productive career.

It is sobering to realize that his achievements in dielectrics make up only a portion of Fröhlich's scientific activity. As well as the ability to make original contributions he has that rare quality of being able to stimulate others to study his work and encourage them to develop and extend it. Truly, "When kings are at work, carpenters are kept busy" ("Wenn die Könige bauen, haben die Kärnen zu tun."—Schiller).

Broadly speaking Fröhlich has been concerned with two aspects of this subject, namely electrical conduction and breakdown at high fields, and dielectric polarization. It is impossible to separate completely one topic from another: often a development in one area initiates activity in another.

In his first foray into electrical breakdown (Fröhlich, 1937) he addressed the problem of breakdown in ionic crystals. His criterion for the occurrence of breakdown required that a rate, A, at which an electron gained energy from the external field should just balance the rate, B, at which energy was transferred from the electron to the lattice. He assumed that A would be a function of the electron energy, U_{elec}, and of the field strength, E, whereas B was taken to be a function of U_{elec} only. Thus breakdown was expected to occur when the following condition was satisfied

$$A\,(U_{elec}, E) = B\,(U_{elec})\,. \tag{1}$$

An important problem at the time was the calculation of B. A major advance was made in a joint paper on the mean free path of electrons in polar crystals (Fröhlich and Mott, 1939). This was a study, following on the work of Lyddane and Herzfeld (1938), of the manner in which a mobile electron in an ionic lattice interacts with the electric field set up by the normal modes of vibrations of the ions. A key result was that the electron interacts only with the longitudinal optical mode. This study was to have a considerable seminal influence in later years. In particular it led to a detailed treatment of the dielectric polarization of ionic crystals by Szigeti (1949) which, in its turn, was to provide a basis for investigations into electron-lattice interactions (Fröhlich, 1952, 1954) and ionic lattice vibrations (Fröhlich, 1949, 1958; Born and Huang, 1954). The topic of high-field breakdown was to be a subject on which Fröhlich worked for over twenty years (Fröhlich, 1939a, 1939b, 1939c, 1941, 1942a, 1943, 1947a, 1947b;

Fröhlich and Seitz, 1950; Fröhlich and Simpson, 1954; Fröhlich and Paranjape, 1956; 1961).

Under the aegis of the British Electrical and Allied Industries Research Association Fröhlich commenced a long-term investigation of the theory of dielectrics. One of the first publications resulting from this program (Fröhlich, 1942b) dealt with the theory of dielectric loss of a dilute solution of long-chain molecules, containing dipoles, in solid paraffin. Fröhlich was able to account for the dependence of the dielectric relaxation time on the molecular chain length that had been observed experimentally (Jackson, 1935; Sillars, 1938; Pelmore, 1939). This work was later extended (Fröhlich, 1946a) to order-disorder transitions in dipolar solids.

Fröhlich and Sack (1944) made one of the earliest attempts to develop a theory of dielectric absorption in dipolar materials that took account of dipole-dipole coupling in a satisfactory way.

Van Vleck and Weisskopf (1945) put forward a theory for the absorption of electromagnetic waves in dipolar gases due to thermally induced transitions between rotational levels. Fröhlich (1946b) provided a phenomenological theory of such (resonance) absorption that had an elegant simplicity and directness—a characteristic feature of his researches. Another good example of his approach to problems is provided in his study with Sack (1946) of the rheological properties of dispersions of elastic spheres in a viscous fluid. They applied a self-consistent procedure which stemmed directly from the method adopted by Onsager (1936) in his theory of the static, or zero frequency, dielectric constant of a dipolar fluid.

Kirkwood's (1939) development of Onsager's theory was generalized by Fröhlich (1948), who showed that the mean of the square of the dipole moment, $<M^2>$, of a macroscopic dielectric sphere, of volume V, is determined by the equation

$$<M^2> = 3kT\,(3V/4\pi)\,(\epsilon_s - 1)\,/\,(\epsilon_2 + 2)\,, \tag{2}$$

which is now taken as the starting point in any calculation of the static permittivity, ϵ_s, from a molecular point of view.

The appearance in 1949 of Fröhlich's *Theory of Dielectrics*, as well as filling a long-felt need for an up-to-date and authoritative account of dielectric theory, provided a considerable stimulus to further research both theoretical and experimental. It is no mere repository of already published work; apart from the many novel physical insights, it contains several original ideas. Two are worthy of notice: one is the concept of the polariton and the other is the notion that a vanishingly small frequency of optical mode lattice vibrations (mode softening) may be associated with transition to a ferroelectric, or permanently polarized state (Fröhlich, 1980).

The calculation of the mean free path of electrons in ionic crystals eventually led Fröhlich (1950) to the introduction of quantum field theory techniques into solid state physics, through the concept of what is now known as the large polaron. Further developments of these ideas (Fröhlich, 1952) resulted in the construction of a Hamiltonian function that was to provide the basis of modern theories of superconductivity.

A short but characteristically succinct paper with Pelzer (Fröhlich and Pelzer, 1955) showed that the frequencies of infinite-wavelength polarization waves (plasma oscillations) in dielectric bodies could be determined from the solutions of the equation

$$[\alpha(w)]^{-1} = 0, \tag{3}$$

in which α is the complex, frequency-dependent polarizability.

Fröhlich applied dielectric theory to the calculation of the energy loss of low-energy (1–10 eV) electrons moving in dielectric media such as water (Fröhlich and Platzman, 1953). Later he developed a theory for the energy loss of fast charged particles in solids (Fröhlich, 1959).

In recent years Fröhlich has concentrated much of his interest on biological problems, but this has not meant abandonment of dielectrics, as an examination of the proceedings of a recent symposium will show (Fröhlich and Kremer, 1983).

REFERENCES

M. Born, and K. Huang, *Dynamical Theory of Crystal Lattices*, Clarendon Press, Oxford, 1954.

H. Fröhlich, Proc. R. Soc. London **A160**, 230 (1937).

H. Fröhlich, Proc. R. Soc. London **A172**, 94 (1939a).

H. Fröhlich, Phys. Rev. **56**, 349 (1939b).

H. Fröhlich, Rep. Progr. Phys. **6**, 411 (1939c).

H. Fröhlich, Proc. R. Soc. London **A178**, 493 (1941).

H. Fröhlich, Phys. Rev. **61**, 200 (1942a).

H. Fröhlich, Proc. R. Soc. London **54**, 422 (1942b).

H. Fröhlich, Nature **151**, 339 (1943).

H. Fröhlich, Proc. R. Soc. London **A185**, 399 (1946a).

H. Fröhlich, Nature **157**, 478 (1946b).

H. Fröhlich, Proc. R. Soc. London **A188**, 521 (1947a).

H. Fröhlich, Proc. R. Soc. London **A188**, 532 (1947b).

H. Fröhlich, Trans. Faraday Soc. **44**, 238 (1948).

H. Fröhlich, *Theory of Dielectrics*, Clarendon Press: Oxford, 1st ed. 199; 2nd ed. 1958.

H. Fröhlich, Phil. Mag. **41**, 221 (1950).

H. Fröhlich, Proc. R. Soc. London **A215**, 291 (1952).

H. Fröhlich, Advan. Phys. **3**, 325 (1954).

H. Fröhlich, *In Max Planck Festschrift, 1958*, W. Frank, Ed., VEB Deutscher Verlag der Wissenschaften, Berlin, 1959, p. 227.

H. Fröhlich, Soviet Phys. Solid State **3**, 491 (1961).

H. Fröhlich, Proc. R. Soc. London **A371**, 102 (1980).

H. Fröhlich, and F. Kremer, Eds., *Coherent Excitations in Biological Systems*, Berlin-Heidelberg-New York-Tokyo, 1983.

H. Fröhlich, and N.F. Mott, Proc. R. Soc. London **A171**, 496 (1939).

H. Fröhlich, and B.V. Paranjape, Proc. Phys. Soc. London **B69**, 21 and 866 (1956).

H. Fröhlich, and H. Pelzer, Proc. Phys. Soc. London **A68**, 525 (1955).

H. Fröhlich, and R.L. Platzman, Phys. Rev. **92**, 1152 (1953).

H. Fröhlich, and Sack, R.A., Proc. R. Soc. London **A182**, 388 (1944).

H. Fröhlich, and R.A. Sack, Proc. Phys. Soc. London **185**, 415 (1946).

H. Fröhlich, and F. Seitz, Phys. Rev. **79**, 526 (1950).

H. Fröhlich, and J.H. Simpson, Adv. Electronics **2**, 187 (1950).

W. Jackson, Proc. R. Soc. London **A150**, 197 (1935).

J.G. Kirkwood, J. Chem. Phys. **7**, 811 (1939).

R.H. Lyddane, and K.F. Herzfeld, Phys. Rev. **54**, 846 (1938).

L. Onsager, J. Am. Chem. Soc. **58**, 1486 (1936).

D.R. Pelmore, Proc. R. Soc. London **A172**, 502 (1939).

R.W. Sillars, Proc. R. Soc. London **A169**, 66 (1938).

B. Szigeti, Trans. Faraday Soc. **45**, 155 (1949).

J.H. Van Vleck, and V. Weisskopf, Rev. Mod. Phys. **17**, 227 (1945).

3. Some Comments on the Role of Electrodes in Dielectric Breakdown of Solids

J. J. O'Dwyer

INTRODUCTION

Dielectric breakdown in solids may be broadly classified as "thermal" or "purely electrical". In thermal breakdown one considers the joule energy input from the power source as being partly used in raising the temperature of the dielectric and partly conducted away. Breakdown occurs when the temperature gradient cannot provide adequate thermal conduction to remove the joule heat; the temperature of the dielectric (and consequently the current density) then rises in an uncontrolled way. The applied voltage at this point is the critical voltage, and for reasonable material parameters, the time interval to the occurrence of the instability is of the order of seconds or longer. In purely electrical breakdown the time interval for voltage collapse is of the order of microseconds or less. In modelling this phenomenon, one therefore proposes processes that feature electronic instability.

The distinction given is not absolute since electronic processes may well be instrumental in bringing about a thermal instability, and there is no doubt that the final stages of purely electrical breakdown involve vaporization of the solid dielectric. However, in the present stage of dielectric breakdown theory the incomplete models of the process do fall into one or other of these two categories. It is our purpose in the following to discuss the effect of the electrodes on various models of breakdown. For both model types the geometrical configuration is important; for thermal breakdown the heat capacity and the thermal conductivity of the electrodes play a deciding role, while for purely electrical breakdown the electron- (or hole-) emitting characteristic is the important feature.

THERMAL BREAKDOWN

The basic differential equation for thermal breakdown is (see, e.g., O'Dwyer, 1973)

$$\sigma E^2 = C\frac{\partial T}{\partial t} - \nabla \cdot (K \nabla T) \tag{1}$$

where the electrical conductivity σ may be a function of temperature T and electric field E, and the thermal conductivity K may be a function of temperature. The specific heat per unit volume C is usually a constant. Time-dependent solutions to eq. 1 have also usually invoked the quasi-steady-state condition

$$\nabla \cdot (\sigma \vec{E}) = 0\,. \tag{2}$$

The role of the electrodes is then to provide both electrical and thermal boundary conditions for the solutions to eq. 1.

When a voltage is applied to a homogeneous isotropic dielectric by an electrode system of arbitrary shape, the electric field at the initial instant of time is determined by the electrode geometry and the dielectric constant. However, the situation is much more complicated thereafter; the conductivity of the dielectric results in joule heat generation, and the consequent temperature increase (which may be different in different parts of the dielectric) renders the dielectric no longer isotropic insofar as electrical and thermal conductivities are concerned. The basic differential equation (eq. 1) is a difficult nonlinear equation. Exact steady-state solutions have been given by Wintle (1981) for plane parallel electrodes and certain specially chosen functional forms for $\sigma(T)$ and $K(T)$. All of these solutions show qualitative similarity. All others including those for field-dependent electrical conductivity and for time-dependent current transients have required computing.

A well known set of solutions by Moon (1931) computed steady-state critical voltages for temperature-independent thermal conductivity K_o and electrical conductivity given by an Arrhenius function $\sigma = \sigma_o \exp(-\phi/kT)$. The thermal properties of the plane parallel electrodes were introduced by postulating a constant surface conductivity λ such that the rate of heat transfer per unit area from the dielectric to the electrode was $\lambda\ (T_1 - T_o)$ where T_1 is the dielectric temperature adjacent to the electrode and T_o is the electrode temperature (assumed constant). Computed solutions result in a universal result for dimensionless critical voltage $\nu = V\ (k\sigma_o/8\phi K_o)^{1/2}$ as a function of dimensionless dielectric thickness $c = \lambda d/2K_o$. The major features of the solutions are

$$\nu = \text{const if } c >> 1$$

$$\nu \propto c^{1/2} \text{ if } c << 1 \qquad (3)$$

corresponding respectively to "thick" and "thin" dielectrics. Clearly the thermal contact properties of the electrode determine just what is meant by thick or thin. If $\lambda >> 2K_o/d$ the dielectric is thick, or, to express the same thing in different words, there is good thermal contact between the dielectric and the heat sink. Since the critical voltage is constant in this case the precise value of the surface conductivity is unimportant for the case of good thermal contact. A thin dielectric is one for which $\lambda << 2K_o/d$, and in this case, there is poor thermal contact between the dielectric and the heat sink. In this instance the precise value of the surface conductivity does matter since $\lambda \rightarrow 0$ would correspond to complete thermal isolation and no stable solution for any voltage.

Another computed solution was given by Beers and O'Dwyer (1983) for the case of cylindrical electrodes. At the initial instant of voltage application, the electric field is determined by the cylindrical geometry alone, and has its minimum value at the outer electrode. When the steady state is reached the minimum electric field is approximately half way between the inner and outer electrodes; this remains true whether the inner electrode makes either good or poor thermal contact with the dielectric. However, there are other differences. For good thermal contact at the inner electrode, the electric field increases substantially between the initial instant and the steady state; this occurs because the cooling effect of the inner electrode keeps the electrical conductivity low in its immediate vicinity. The opposite is true for poor thermal contact at the inner electrode; in the steady state the electric field near the inner electrode may only be slightly in excess of its value at the outer electrode.

Another interesting effect for cylindrical electrode geometry showed a closely linear relation for the logarithm of current density versus the square root of voltage (Schottky plot) over more than a decade of current. This result arose entirely from the geometric and thermal properties of the electrode system since the dielectric conductivity was field independent.

PURELY ELECTRICAL BREAKDOWN

At room temperature or lower, the type of breakdown that occurs in most dielectrics has been called "purely electrical", and it develops so fast that an electronic instability is implied. The calculation of a critical field for ionic crystals by Fröhlich (1937) was a major advance in this area; the Fröhlich field was critical in the sense that it represented the onset of failure of detailed balance—collision ionization could no longer be balanced by collision recombination. Fröhlich's result was

$$E_c = E_o \, [1 + 2 \, \{\exp\, (h\omega/kT) - 1\}]^{1/2}$$

where ω is the angular frequency of longitudinal optical modes, and E_o is field strength characteristic of the particular dielectric and containing no disposable parameters. Experimental results for a series of alkali halides were closely fitted by the Fröhlich formula, and there is little doubt that this was one of the reasons that led to the establishment of the concept "intrinsic breakdown strength". From the experimental point of view the chief evidence to support this idea was the apparent independence of the breakdown field on extraneous physical parameters (excluding of course the temperature).

Experimental evidence on the effect of thickness on dielectric breakdown strength was slow in coming, mainly because very large voltages are required to break down thick dielectrics. However, the work of Watson et al. (1965) clearly showed a pronounced thickness effect amounting to a factor of about 5 on breakdown strength over two decades of thickness of NaCl. Extrapolation of the work of other investigators extended this to a factor of about 10 on the breakdown strength over three decades of thickness. There is probably no other solid dielectric for which breakdown strength measurements exist over such a wide range of thickness, but measurements on various dielectrics (see e.g. Budenstein et al. (1969)) over about one decade of thickness all support the existence of the thickness dependence of the breakdown strength.

The electrodes therefore play a role in determining dielectric breakdown strength. This role may be as small as simply providing boundary conditions for the field within the dielectric, or more significantly the charge-emitting properties of the electrode may be important. An example of the first type of electrode effect was provided by a calculation (O'Dwyer, 1973) of the variation of electric field strength between electrodes that could accompany constant current density in the presence of collision ionization. For a postulated maximum critical field, thin dielectrics are able to sustain a higher mean field than thick dielectrics; this conclusion stems from considering the electrodes only as providing boundary conditions for the electric field.

More realistically one should consider the role of the electrode as a charge injector. Even if an electrode is blocking at low fields it will undoubtedly emit, by either Schottky or Fowler-Nordheim mechanisms, at high fields. The effect of an emitting cathode has been quantitatively described in a model by O'Dwyer (1973) and Klein and Solomon (1976). The positive feedback that produces instability begins with Fowler-Nordheim emission of electrons from the cathode. For sufficiently high field within the dielectric, collision ionization commences to be significant at a field, given in order of magnitude, by the Fröhlich critical field. The positive holes formed by the collision ionization process are assumed to drift relatively slowly back towards the cathode. The space charge due to these holes distorts the field making it larger in front of the cathode; this in turn increases the electron emission from the cathode, and thereby completes the positive feedback loop. In this model the field is always greatest immediately in front of the electrode, and this must also be the site of maximum deposition of energy in the dielectric by collision ionization. The discharge that accompanies breakdown should therefore begin at the cathode on the basis of this model.

Since no metal electrode can have a perfectly flat plane surface, it is of interest to examine the effect of microscopic asperities. To this end the emission associated with spherical electrodes into a model dielectric has been investigated by O'Dwyer (1986). The smaller electrode and the spherically symmetric electric field around it should behave in a manner qualitatively similar to an asperity on a plane cathode surface. Let $j\ (E_c)$ represent the functional form for the current density emission from the cathode, and let σ be the intrinsic conductivity of the dielectric (by this we mean the conductivity in the absence of space charge). The condition that there be no space charge is $j\ (E_c) = \sigma E_c$; for voltages lower than the one that satisfies this condition there is heterocharge near the cathode, and for higher voltages there is homocharge. In the heterocharge condition the field is more intense near the electrode than would be expected from geometrical considerations alone. For homocharge the field at the electrode is *less* intense than would be dictated by geometry alone; a reduction by an order of magnitude is readily attainable for reasonable parameters. Moreover, in this latter case the maximum field strength occurs some distance in front of the electrode.

To summarize we can say that an electrode asperity protects itself to some extent from the high field that might be expected from geometry alone; it does so by emitting charge carriers to form homocharge, and the onset of this occurs at relatively lower voltages for dielectrics of lower intrinsic conductivity. Since the maximum field occurs some distance in front of the electrode, this will also be the region for maximum collision ionization. One would therefore anticipate that the beginning of a breakdown discharge would not occur precisely at the electrode surface, but some small distance out from it.

Finally, it should be noted that the least mobile charge species is responsible for space charge phenomena, while high-mobility electrons would cause collision ionization. The relatively immobile species could be either electrons or holes hopping between deep traps. The comments on the effect of electrode asperities therefore apply equally to either cathode or anode.

REFERENCES

B.L. Beers, and J.J. O'Dwyer, J. Appl. Phys. **54**, 4083 (1983).

P.P. Budenstein, P.J. Hayes, J.L. Smith, and W.B. Smith, J. Vac. Sci. Tech. **6**, 289 (1969).

H. Fröhlich, Proc. R. Soc. London **A160**, 230 (1937).

N. Klein, and P. Solomon, J. Appl. Phys. **47**, 4364 (1976).

P.H. Moon, Trans. Am. Inst. Elec. Eng. **50**, 1008 (1931).

J.J. O'Dwyer, *The Theory of Electrical Conduction and Breakdown in Solid Dielectrics,* Clarendon Press, Oxford, 1973.

J.J. O'Dwyer, Proc. 2nd Int. Conf. Conduction and Breakdown in Solid Dielectrics (1986), in press.

D.B. Watson, W. Heyes, K.C. Kao, and J.H. Calderwood, IEEE Trans. Elec. Insul. **E1-1**, 30 (1965).

H.J. Wintle, J. Appl. Phys. **52**, 4181 (1981).

4. Multipolar Fluctuations in a Dielectric Sphere

B. K. P. Scaife

ABSTRACT

Formulae are presented for the mean-square fluctuations in the multipole moments of (a) an isolated dielectric sphere, and (b) a sphere surrounded by a concentric spherical shell with the same dielectric properties.

In his *Theory of Dielectrics*, Fröhlich (1949) applied Einstein's (1910) theory of fluctuations in macroscopic parameters to the problem of the calculation of the mean-square dipole moment of a dielectric sphere.

Einstein showed that when the Helmholtz free energy, $F(u_1, u_2, \ldots)$, of a system depends upon macroscopic parameters $u_1, u_2, \ldots$, the probability that the macroscopic state of the system lies in the intervals $u_1, u_1 + du_1$; $u_2, u_2 + du_2$; ... is

$$\exp[-\beta F(u_1, u_2, \ldots)]\, du_1 du_2 \ldots / \int \exp \int [-\beta F(u_1, u_2, \ldots)]\, du_1 du_2 \ldots, \tag{1}$$

where $\beta^{-1} = kT$, k is Boltzmann's constant, and T is the absolute temperature.

If the free energy can be expressed in the form

$$F(u_1, u_2, \ldots) = \frac{1}{2} a_1 u_1^2 + \frac{1}{2} a_2 u_2^2 + \ldots + \frac{1}{2} a_j \,.\, u_j^2 + \ldots, \tag{2}$$

then the mean-square value of any one of the variables, say u_j, has the value

$$<u_j^2> = 1/\beta a_j = kT/a_j, \tag{3}$$

In Fröhlich's application of this result the macroscopic variables were the Cartesian components M_x, M_y, and M_z of the total dipole moment of a dielectric sphere, and

$$F(M_x, M_y, M_z) = (M_x^2 + M_y^2 + M_z^2)/2\alpha, \tag{4}$$

where α, the polarizability of the sphere, is given by

$$\alpha = 4\pi\epsilon_0 a^3 \;(\epsilon - 1)/(\epsilon + 2), \tag{5}$$

in which ϵ_0 is the absolute permittivity of free space, a is the radius of the sphere, and ϵ is the static, or zero-frequency, relative permittivity. From eqs. 3 and 4 we find that

$$< M_x^2 > = < M_y^2 > = < M_z^2 > = kT\alpha. \tag{6}$$

If the dielectric sphere consists of an assembly of particles with charges q_k at coordinates x_k, y_k, z_k, then

$$M_x = \sum_k q_k x_k, \ M_y = \sum_k q_k y_k, \ M_z = \sum_k q_k z_k. \tag{7}$$

The electrical state of the dielectric sphere can be more precisely described if we introduce multipole moments. We define the *polar* multipole moments m_{nm}^c and m_{nm}^s by the equations

$$m_{nm}^c = \sum_k q_k r_k^n P_n^m (\cos\theta_k) \cos m\phi_k \tag{8a}$$

$$m_{nm}^s = \sum_k q_k r_k^n P_n^m (\cos\theta_k) \sin m\phi_k, \tag{8b}$$

in which r_k, θ_k, ϕ_k are the spherical coordinates corresponding to x_k, y_k, z_k, with $z_k = r_k \cos\theta_k$, $y_k = r_k \sin\theta_k \sin\phi_k$, $x_k = r_k \sin\theta_k \cos\phi_k$, and P_n^m is an associated Legendre function $0 \leqslant m < n$). We shall assume that the total charge of the sphere vanishes so that

$$m_{00}^c = \sum_k q_k = 0. \tag{9}$$

Furthermore

$$M_x = m_{11}^c, \ M_y = m_{11}^s, \ M_z = m_{10}^c. \tag{10}$$

(Note that m_{n0}^s vanishes for all n.) When a system of charges, with multipole moments described by eqs. 8a and 8b, is placed in an electric potential

$$\sum_{n=1}^{\infty} \sum_{m=0}^{n} r^n P_n^m \left(\cos\theta\right) \left[A_{nm}^c \cos m\phi + A_{nm}^s \sin m\phi\right], \tag{11}$$

the interaction energy, U_i, takes the form:

$$U_i = \sum_{n=1}^{\infty} \sum_{m=0}^{n} \left[m_{nm}^c A_{nm}^c + m_{nm}^s A_{nm}^s\right]. \tag{12}$$

A dielectric sphere placed in a potential described by eq. 11 will have induced in it polar multipoles with moments

$$m_{nm}^c = -\alpha_{nm} A_{nm}^c \tag{13a}$$

$$m_{nm}^s = -\alpha_{nm} A_{nm}^s \tag{13b}$$

in which the generalized polarizability α_{nm} has the value

$$\alpha_{nm} = 4\pi\epsilon_0 a^{2n+1} \zeta_m \frac{n(\epsilon - 1)}{(n\epsilon + n + 1)} \frac{(n+m)!}{(n-m)!}, \quad \zeta_m = \begin{cases} 1, & m = 0, \text{ or} \\ \frac{1}{2}, & m \neq 0 \end{cases} \tag{14}$$

This equation may be derived by standard techniques of electrostatic theory. The derivation of this and other formulae in this note will be published in detail elsewhere.

The expression eq. 12 for the interaction energy can therefore be written

$$U_i = -\sum_{n=1}^{\infty}\sum_{m=0}^{n} \alpha_{nm} \left[\left(A_{nm}^c\right)^2 + \left(A_{nm}^s\right)^2 \right] \tag{15a}$$

$$= -\sum_{n=1}^{\infty}\sum_{m=0}^{n} \alpha_{nm}^{-1} \left[\left(m_{nm}^c\right)^2 + \left(m_{nm}^s\right)^2 \right] . \tag{15b}$$

The free energy F of the dielectric sphere with a given set of multipoles (Fröhlich's self energy) is obtained from

$$F = \frac{1}{2} \sum_{n=1}^{\infty}\sum_{m=0}^{n} \alpha_{nm}^{-1} \left[\left(m_{nm}^c\right)^2 + \left(m_{nm}^s\right)^2 \right] , \tag{16}$$

a result which may be verified by direct calculation of the energy in the field outside and within the sphere.

From eqs. 2, 3, and 16 we conclude that

$$< \left(m_{nm}^c\right)^2 >_{vac} = kT\alpha_{nm}, \ 0 \leqslant m \leqslant n , \tag{17a}$$

$$< \left(m_{nm}^s\right)^2 >_{vac} = kT\alpha_{nm}, \ 1 \leqslant m \leqslant n. \tag{17b}$$

These formulae can be extended to include the case in which the dielectric sphere is surrounded by a concentric spherical shell with the same dielectric properties and an outer radius b equal to $1/\eta$ times the inner radius a. The results are

$$< \left(m_{nm}^c\right)^2 >_{med} = kT\alpha_{nm}\left[1-\eta^{2n+1}f_n(\epsilon)\right] / \left[1-f_n(\epsilon)\right] , \tag{18a}$$

$$< \left(m_{nm}^s\right)^2 >_{med} = < \left(m_{nm}^c\right)^2 >_{med}, \ 1 \leqslant m \leqslant n. \tag{18b}$$

where

$$f_n(\epsilon) = n(n+1)(\epsilon-1)^2 / \left[(\epsilon n + \epsilon + n)(\epsilon n + n + 1)\right]. \tag{19}$$

It is straightforward to deal with the case when the sphere and the shell have different dielectric properties.

For a dielectric consisting of molecules each with the same dipole moment μ we note that

$$m_{20}^c = \sum_j (3\mu_{jz}R_{jz} - \mu_j \cdot R_j) , \tag{20}$$

in which $\vec{R}_j = \vec{i}X_j + \vec{j}Y_j + \vec{k}Z_j$ denotes the position of dipole $\vec{\mu}_j$.

Equations eqs. 17 and 18 may be of use in computer calculations of the static relative permittivity.

REFERENCES

A. Einstein, Ann. Phys. **33**, 1275 (1910).

H. Fröhlich, *Theory of Dielectrics,* Clarendon Press, Oxford, 1949, p. 40.

5. Nonlinear Physics of Electronic and Optical Materials for Submicron Device Applications

T. W. Barrett

With 12 Figures

ABSTRACT

The density matrix approach to the constitutive relation between electric fields and the current density in organic polymers is compatible with a nonlinear transport equation. The derivation of such an equation is fundamental to any understanding of conduction in submicron devices, in organic polymers and in multiple quantum well structures—which may be considered the ultimate submicron devices—and also to understanding the nonlinear optical properties of those same polymers and materials. As a theoretical platform for such a derivation, physical models are introduced based on the long Josephson junction and the nonlinear line of pendula. A nonlinear transport equation is presented based on a density matrix formalism and a nonlinear Hamiltonian for charge transport in organic conductors. The equation can describe both local and displacive transitions and gives a physically intuitive picture of the concept of off-diagonal long-range order and Bose-Einstein condensation.

INTRODUCTION

It is well known that the interplay of currents and voltages in Josephson circuitry is similar in form to that encountered in certain mechanical systems, and, in particular, to the simple pendulum line. The ability to understand and visualize the behavior of Josephson junctions in terms of mechanical devices has greatly facilitated analysis of their dynamic function (Fulton, 1975, 1977). The same mechanical devices can illuminate conduction in organic polymers due to (a) the fluxon of the long Josephson junction being a nonlinear (soliton) propagation phenomenon; and (b) both (i) the neutral soliton (in polymers that have not undergone redox reactions), and (ii) the charge domain wall soliton (in polymers that have undergone redox reactions), have been implicated in spin waves (in the case of the former) and charge conduction (in the case of the latter). The analytical mechanics involved in these processes suggests that the methods used to study the Josephson junction fluxon are uniquely applicable to the study of the conduction in submicron devices and polymers where the conduction is inherently nonlinear (Trullinger et al., 1978; Barrett, 1983a).

The models constructed for the study of the Josephson effect include those of Scott (1969), Waldram et al. (1970), Sullivan and Zimmerman (1971), Yamashita and Ondera (1979), Yamashita et al. (1979), and Fulton (1977). These models have been very useful in developing physical intuition concerning the Josephson effect and agree in their choice of physical and mathematical descriptions.

The advantages of numerical analysis based on mechanical models are many. Most notably, the performance of the model can be readily altered by a change of variables. Furthermore, the fluxon (soliton) in the long Josephson junction is firmly established, but the equation of transport in very small devices and in polymer and multiple quantum well systems is presently under theoretical discussion. As the soliton is a mathematical entity, the abstract conditions and interplay of variables applying to the fluxon also apply to the nonlinear transport equation in polymers and multiple quantum well structures. Examination of nonlinear (soliton) propagation in the simple pendulum line is useful in that its behavior is described in very familiar physical variables and is fundamental to both the long Josephson junction and conduction in organic polymers. Thus, the intention here is to underpin the conditions for transport equations in polymer systems, multiple quantum well structures, and very small devices with the established conditions for nonlinear propagation in the long Josephson junction and the pendulum line.

QUESTIONS REQUIRING ANSWERS

There are a number of presently unanswered questions concerning the new electrical materials developed in the last ten years. The following are some examples:

1. Can superlattices act as the gain region of devices?
2. Can the ballistic regime of device materials in which the transit distance is within the mean free path of the electron/hole/exciton/polaron/bipolaron be adequately described?
3. Can the carrier effective mass theory be reformulated with respect to temporal limitations?
4. Can the needed quantum mechanical corrections be made to Drude conductivity of nonclassical particles so that conduction in polymeric materials, nonlinear optical materials, and multiple quantum well structures can be described?
5. Can the Einstein and Nyquist relations be reformulated?

In the following sections a first attempt will be made to provide a theoretical platform for answering some of these questions.

PHYSICAL MODEL

The phenomenological equations describing the macroscopic behavior of superconducting barriers are (Josephson, 1965):

$$\partial\phi/\partial x = (2ed/\hbar c)H_y;\ \partial\phi/\partial t = (2e/\hbar)V;\ J_z = j_1\sin\phi + \sigma V, \tag{1}$$

where ϕ is the phase difference between two superconductors, H is the magnetic field in the barrier, V is the voltage across the barrier, $d = 2\lambda + l$, λ is the penetration depth, and l is the barrier thickness. If the barrier is regarded as having a capacitance C per unit area, and using Maxwell's equations and eq. 1, above, we obtain:

$$[\partial^2/\partial x^2 - (1/\bar{c}^2)\,(\partial^2/t^2) - (\beta/\bar{c}^2)\,(\partial/\partial t)\,]\phi = [\,1/(\lambda_0^2)\,]\sin\phi, \tag{2}$$

where $\bar{c}^2 = c^2/4\pi dC$ is the phase velocity in the barrier, $\lambda_0^2 = \hbar c^2/8\pi edj_1$ is the penetration depth, and $\beta = 4\pi d\bar{c}^2\sigma/c^2 = \sigma/C$ is a damping constant.

If the barrier is assumed to be of unit length in the y-direction, then the total free energy of the barrier is (Lebwohl and Stephen, 1967):

$$F = (\hbar j_1/2e)\int dx\,[\,(1-\cos\phi) + 1/2\,\lambda_0^2\,(\partial\phi/\partial x)^2 + 1/2\,(\lambda_0/\bar{c})^2\,(\partial\phi/\partial t)^2\,], \tag{3}$$

where the first term is the coupling energy for two superconductors and the others the electromagnetic energy in the barrier. The rate of dissipation of energy is:

$$dF/dt = -\sigma\int V^2\,dx. \tag{4}$$

We may note, here, in anticipation of conclusions to follow, that removal of any terms in the left-hand side of eq. 3 will reduce energy dissipation.

Turning now to a more fundamental model, we find that recognition of an analogy to the long Josephson junction occurred when Anderson (1964) pointed out that solutions of eq. 2 representing vortex lines in the barrier are obtained as solutions of

$$\partial^2\phi/\partial x^2 = (1/\lambda_0^2)\sin\phi, \tag{5}$$

which, except for sign, is the equation of a pendulum. Figure 1 shows a sketch of a simple model with moment arm, I_c, moment of inertia, M, provided by a fly wheel, damping coefficient, D, provided by a paddle wheel arrangement, and phase angle ϕ measured from vertical downward. The rod is subject to an external torque I.

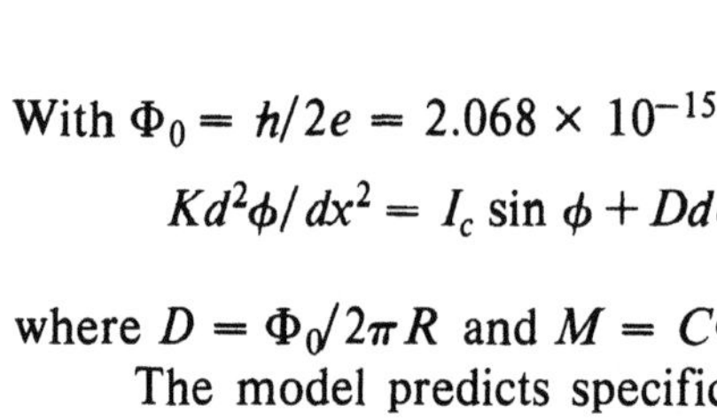

Fig. 1 — After Fulton (1977).

With $\Phi_0 = h/2e = 2.068 \times 10^{-15}\ Wb$, we have:

$$Kd^2\phi/dx^2 = I_c\sin\phi + Dd\phi/dt + Md^2\phi/dt^2 + I, \tag{6}$$

where $D = \Phi_0/2\pi R$ and $M = C\Phi_0/2\pi$.

The model predicts specific current-voltage relations. In a steady state there is an average current balance given by:

$$<I> = <I_c\sin\phi> + <V>/R. \tag{7}$$

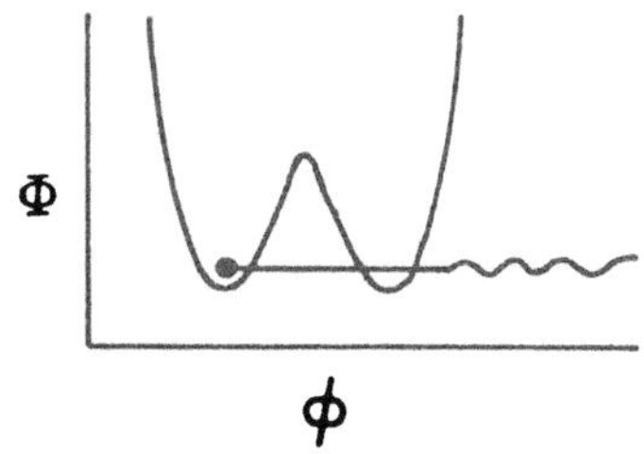

Figure 2.

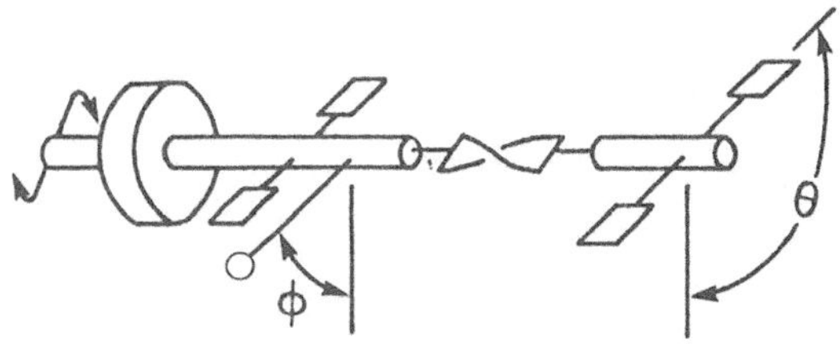

Fig. 3 — After Fulton (1977).

If I and V are decomposed into steady and alternating components: $I = <I> + <I_{ac}>$ and $V = <V> + <V_{ac}>$, then with I fixed, $<I_c \sin \phi> = <V_{ac}^2>/R$, so a departure from linear ohmic behavior indicates that the voltage has an alternating component. On the other hand, with a radio frequency, f, applied near the junction frequency $<V>/\Phi_0$, the ac supercurrent locks into phase with applied current, so that $<V> = \Phi_0 f$, giving a linear relation.

The characteristic particle in a double potential (Φ) well with harmonic spring connection to neighbors (Fig. 2) possesses an inductively connected element with phases: $\phi = 2\pi \int (V/\Phi_0)dt$ and $\Theta = 2\pi \int (V_z/\Phi_0)dt$, where V and V_z are the voltages across the junction and an external passive element z, i.e., the nearest neighbor (Fulton, 1977) (Fig. 3):

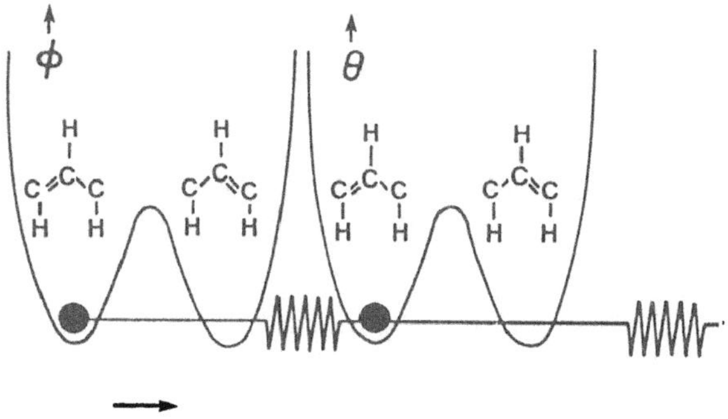

Fig. 4 — Phase diagram for two monomers of the linear polymer, *trans*-polyacetylene, which possesses two degenerate conformational ground states.

In the polymer analogy, the phases ϕ and Θ are associated with the conformational energies of the monomer under consideration and its nearest neighbor respectively. Similarly, V and V_z are the conformational potentials of the monomer under consideration and its nearest neighbor (Fig. 4). Figure 4 depicts the two isomerized states of *trans*-polyacetylene in illustration of the two conformations available in the monomer. The two-potential well in the case of polymers with two conformations of unequal energies is formed of unequal wells (Fig. 5).

The physical representation of a polymer, such as *trans*-polyacetylene, undergoing a local monomer order-disorder-order transition, or isomerized-state transition, is shown in Fig. 6:

In terms of the pendulum line analogy describing this dynamic change in the monomer of a polymer: an inductance, L, from the nearest neighbor gives a pendulum bar twist, LI_c/Φ_0, balancing the pendulum in a horizontal position. For

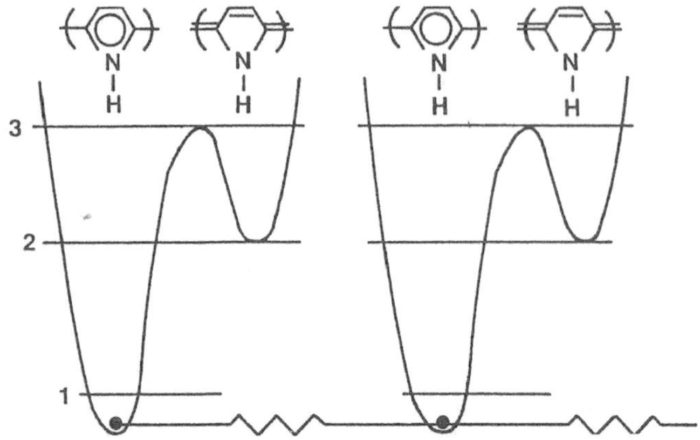

Fig. 5 — Two-potential or two-phase well for two monomers of a nondegenerate three-state polymer system (polypyrrole).

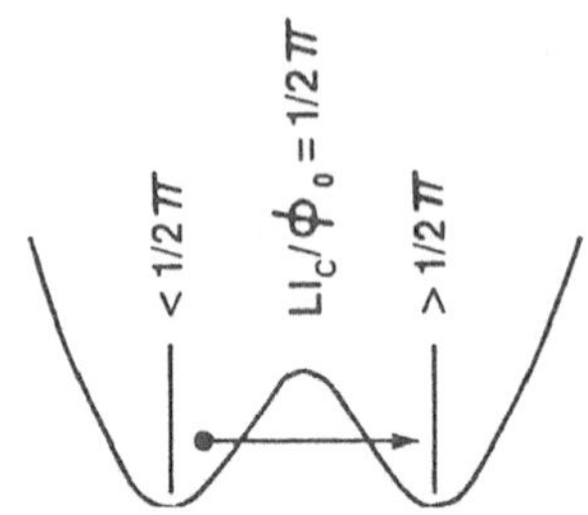

Figure 6.

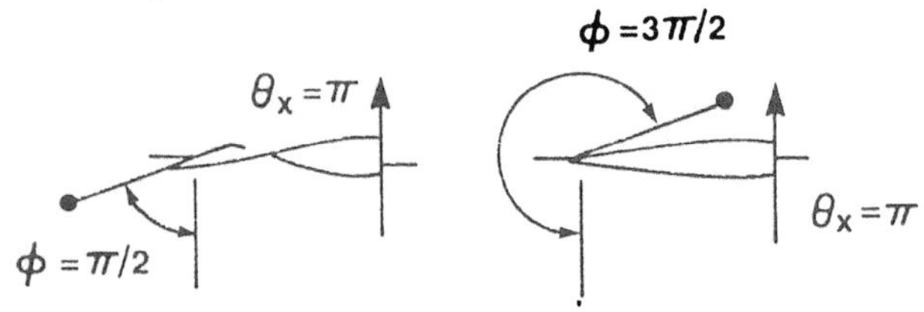

Fig. 7 — After Fulton (1977).

$LI_c/\Phi_0 << 1/2\pi$, Φ follows Θ, but at $LI_c/\Phi_0 = 1/2\pi$ the pendulum is no longer stable, and for $LI_c/\Phi_0 > 1/2\pi$ a flip to the other state occurs completing the transition (Fulton, 1977) (Fig. 7). This corresponds to a local monomer transition in a polymer.

Transport along the line can also be described. It is well known that in a uniform pendulum array of infinite length, a full twist or vortex can be moved from place to place with no cost of energy (Scott, 1969; Waldram et al., 1970; Lebwohl and Stephen, 1967). Application of a weak torque in the vicinity of a vortex causes a pendulum to rotate in the direction of the applied torque and the twist moves along the array away from the torque. This is the propagating soliton of the displacive mode of the system (Fig. 8). If two solitons are propagated along a pendulum array, then in an overdamped system the two will be annihilated. In an underdamped system the two pass right through each other and continue on with their relative energies intact.

In the polymeric system, the mechanism involves (1) a local monomer order-disorder-order transition, and (2) a moving displacive transition. The mechanism for passing from 1 to 2 is exhibited in the demonstration that for a system described by the equation (Pitaevskii, 1959)

$$(\partial^2\psi/\partial X^2) + (\partial^2\psi/\partial y^2) - (\partial\psi/\partial t)\,(|\psi|^2 - 1)\,\psi = 0, \tag{8}$$

the change of the complex function, ψ, with time t, is:

$$i\hbar\partial\psi/dt = \hat{L}\,\psi, \tag{9}$$

where $\hat{L}$ is a nonlinear operator and a function of a second variable, which is time

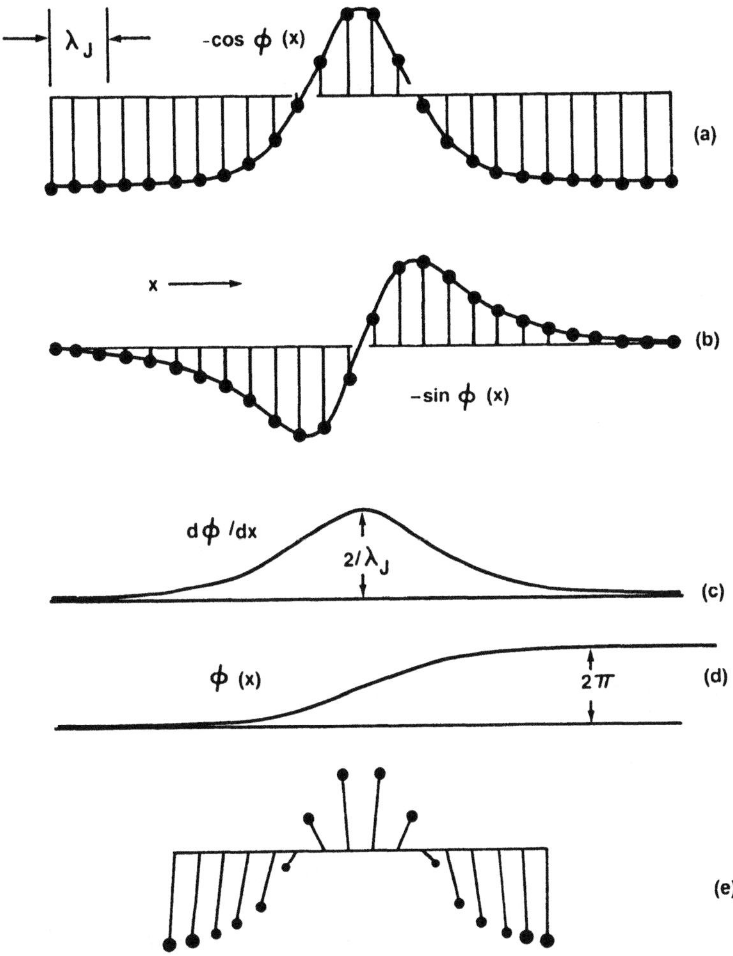

Fig. 8 — Various views (a) to (d) of a segment of an infinitely long pendulum array containing a vortex. (a): Front view (cos ϕ (x)), (b): Top view (− sin ϕ (x)), (c): $d\phi/dx$, (d): $\phi(x)$. (e): A vortex trapped in an array of finite length by the excessive weight of the end pendulums. (From Fulton, 1977.)

in the example considered here. As this second variable increases, the antiHermitian part of $\hat{L}$ becomes dominant (as compared with the Hermitian). In this case the equation that describes the polymer transition goes from a Schrödinger type to a diffusive or displacive type (Pitaevskii, 1959, 1961; Usui, 1969). In effect, the antiHermitian part of $\hat{L}$ is larger, the faster the system relaxes. $\hat{L}$ cannot be Hermitian in the case of a nonlinear propagating wave (soliton), because density would be conserved at the local site, when what is occurring is a change in density from one local site to another. If $\hat{L}$ is Hermitian, then the wave is static, i.e., pinned or trapped, and only local transitions are possible.

The antiHermitian part of $\hat{L}$ describes the process whereby thermodynamic equilibrium is approached, i.e., the process whereby the density of states at monomer 1 in a polymer chain are converted into the density of states at the nearest neighbor, monomer 2, etc. The nonlinear operator, $\hat{L}$, acting at a local site is:

$$\partial E/\partial t + \nabla Q = 0, \tag{10}$$

where Q is an energy loss vector. In the case of "superconductivity" or decrementless conduction between monomers 1 and 2, etc.

$$\nabla Q_1 + \nabla Q_2 = Q, \tag{11}$$
$$\partial E_1/\partial t + \partial E_2/\partial t = 0.$$

In a more realistic situation, a submicrometer device will have loss to the environment and therefore

$$\nabla Q_1 + \nabla Q_2 + \nabla Q_e = 0, \tag{12}$$
$$\partial E_1/\partial t + \partial E_2/\partial t + \partial E_e/\partial t = 0,$$

where the subscript, e, signifies the environment.

In the case of the anharmonic oscillator model for polymeric transport, we have:

$$\partial\Phi/\partial t = \hat{L}\Phi, \text{ or} \tag{13}$$

$$\Phi(x-x_z) = (K/2)\underset{(ll')}{n.n.}\left\{\sum_l \underbrace{[Bx^3+Ax]}_{\substack{\text{CHANGE IN } x\\ \text{DESCRIBED BY}\\ \text{HERMITIAN}\\ \text{COMPONENT OF } \hat{L}}} - \underbrace{C[x(l)-x(l')]^2}_{\substack{\text{CHANGE IN } x\\ \text{DESCRIBED BY}\\ \text{ANTIHERMITIAN}\\ \text{COMPONENT OF } \hat{L}}}\right\}, \tag{14}$$

where x is associated with the bonding arrangements in the monomer, Φ is the configurational potential, and l is the length of the polymer. Equation 14 describes *both* local and displacive transitions and will be used in the following consideration of the form a transport equation should take. If $A < 0$, $B > 0$ and $C > 0$, $B/C >> 1$, then a *local transition* occurs (from one degenerate state to another, or from a ground state to an ionized state), If $A > 0$, $B > 0$ and $C > 0$, $B/C << 1$, then a local *and a displacive transition* occur (Schneider and Stoll, 1973, 1975, 1976, 1978, 1981). *Thus, the excited state remains trapped until or unless a displacive transition occurs.*

In matrix notation, the Hermitian component of the associated quantum mechanical Hamiltonian is:

$$H_{ll} = (\hbar^2/m\delta^2)\, c_{ll}\, \delta^{-1/2} + \text{kinetic energy component}, \tag{15}$$

and the antiHermitian component is:

$$H_{ll'} = (-\hbar^2/2m\delta^2)\, c_{ll'}\, \delta^{-1/2} + \text{kinetic energy component}, \tag{16}$$

where the c are the vector components of the state of the polymer system; δ is a length of the polymer that is the monomer length in the limit; and the total Hamiltonian is:

$$H_T = H_{ll} + H_{ll'} + \text{kinetic energy components}. \tag{17}$$

The model presented above for conduction in organic polymers can be related to one consisting of noninteracting, free (Bloch) electrons scattered by "random" rigid impurity centers (Kohn and Luttinger, 1957; Luttinger and Kohn, 1958). A collection of ν electrons moving under the influence of an external field can be represented by a density matrix (Kohn and Luttinger, 1957; Barker, 1982):

$$\rho = (1/\nu) \sum_{i=1}^{\nu} a_m^i(t) a_n^{i*}(t) = Av\{ |\psi><\psi| \}, \tag{18}$$

obtained from the time-independent wave functions of the electrons defined:

$$\psi_i(x,t) = \sum_n a_n^i(t) \psi_n(x). \tag{19}$$

A Hamiltonian can be defined by using eq. 17:

$$H = H_0 + H_T = H_0 + \Phi(x - x_z) = H_0 + \lambda.V, \tag{20}$$

where $\lambda = LI_c/\Phi_0$ in the pendulum line analogy.

The temporal variation (Liouville equation) of ρ is given by:

$$i\hbar(\partial\rho/\partial t) = [H,\rho]. \tag{21}$$

With $\lambda = 0$, $\Phi(x - x_Z) = 0$ (eq. 14), $H_T = 0$, $H = H_0$ and (Kohn and Luttinger, 1957):

$$\rho = Ke^{-\beta H}; \beta = 1/kT; K^{-1} = Tr(e^{-\beta H}). \tag{22}$$

and the linear Boltzmann transport equation is recovered. The density matrix can be developed in ascending powers of the strength of the scattering potential, or, as in eq. 22, in terms of the familiar Boltzmann equation valid in the limiting case of weak or very dilute scatterers. All the results of Kohn and Luttinger's quantum theory of electrical transport phenomena are also recovered, but with the important difference that H_T describes nonlinear displacive transport.

It follows, therefore, that while the possible existence of two states in a polymer (giving the Hermitian component to the wave equation) is a necessary condition for charge or electron transport, sufficient conditions require the existence of vibrational energies to produce the displacive transition (antiHermitian component of the wave equation). *Only if the antiHermitian component of the equation of state exceeds the Hermitian component will a perturbation become dispersive or displacive, i.e., describable by a transport equation* (Barrett, 1983b).

The question of nonlinear transport was addressed by Fröhlich (1973), Mahan (1981), and Bilz et al. (1981). A mean polarization or high dipole moment was treated as a metastable state. Perturbation by an external field results in (i) the adoption of a ground state, and (ii) excitation of incoherent vibrations in the metastable state leading to coherent vibrations in the ground state. Excitation of coherent vibrations results in energy transfer in a precise direction, i.e., long-range "attraction" to a distant site occurs. A supply of energy is then required to return (lift) the material into the metastable state. This picture can also be interpreted in terms of local and displacive transitions.

The same picture applies in the case of transport in giant macromolecules and involves the concept of off-diagonal long-range order (ODLRO). The concept of ODLRO was originally proposed by Penrose (1951) and Penrose and Onsager (1956) to generalize the mathematical description of Bose-Einstein condensation so as to be applicable to a system of interacting particles. Bose-Einstein condensation is present whenever the largest eigenvalue of the one-particle reduced-density matrix is an extensive rather than an intensive quantity, or whenever for a spatially uniform system with periodic boundary conditions a finite fraction of the particles have identical momenta.

ODLRO is present whenever one of the reduced-density matrices exhibits some degree of factorization which may be due to the existence of a macroscopic eigenvalue. Yang (1962) demonstrated that the Bardeen, Cooper and Schrieffer (BCS) superconducting ground state is unique in that it is the only ground state which gives rise to the maximum possible eigenvalue in the reduced-density matrix for the system, and the onset of such an order leads to a new thermodynamic phase of the system. Bose-Einstein condensation is the simplest form of an ODLRO.

Cummings and Johnston (1966) demonstrated that the occurrence of ODLRO is completely independent of the existence or nonexistence of a condensation into a single-momentum mode; hence it is of interest to biophysicists, since the existence of ODLRO is not dependent upon freezing. These authors equated the existence of ODLRO with the existence of coherence and proposed that coherent radiation, superconductivity, and superfluidity are analogous phenomena. Writing two decades later, these authors would surely have included charge conduction and vibronic energy conduction in organic polymers as also analogous. The pure "superfluid" state of each is associated with the complete factorization of an approximately defined reduced-density matrix. The coherent states of each have the form $\prod_k e^{\alpha k} \Theta_k | 0 \rangle$ where Θ_k is an operator characteristic of the "basic group" and $| 0 \rangle$ is the ground state of the normal fluid. Thus, ODLRO first occurs in a ground state on which a generator for the coherent radiation states operates. The description is therefore identical to that provided by eq. 14 but without the physical intuition added by a physical model (Barrett, 1980).

These same concepts are inherent in the theory of dielectrics. Fröhlich (1968) suggested that biomacromolecules with recurrent bonds and biological dipolar membranes have a branch of longitudinal electric modes (long-range Coulomb interaction) in a frequency region between 10^{11} and $10^{12}\,s^{-1}$. If energy is supplied above a mean rate to such branches, then a steady state is reached in which a single mode of a branch is very strongly excited and the supplied energy is not thermalized but stored. The stored energy is highly ordered so that long-range phase correlations occur. The analogy can then be made again between the low-temperature condensation of a Bose gas and this steady state.

In the excited state of such a mode, the mean value of the square of the electric polarization may be large and hence lead to deformations and stresses in macromolecules and membranes, and also to a relative stabilization of the excitation, so that a relatively small supply of energy is required to maintain it.

NONLINEAR OPTICAL EFFECTS

Three- and four-wave mixing are optical effects that have important applications (Pepper, 1982, 1986; Shkunov and Zel'dovich, 1985; Fisher, 1983) and that are dependent upon the nonlinear interaction of incident field and material (Figs. 9 and 10).

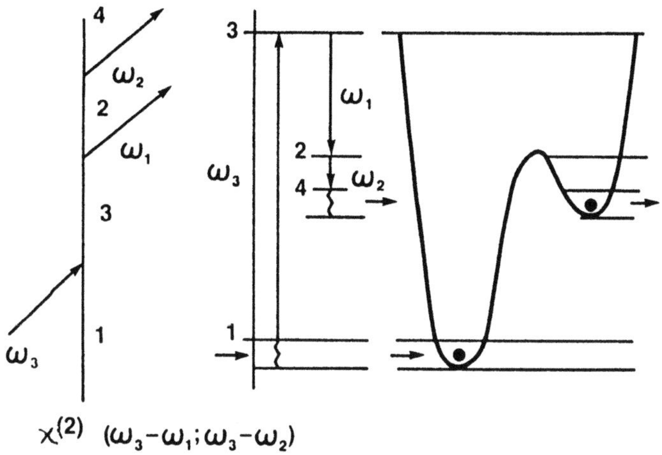

Fig. 9 — An example of parametric mixing using a nondegenerate, three-level polymeric system for generation of the second-order nonlinear susceptibility.

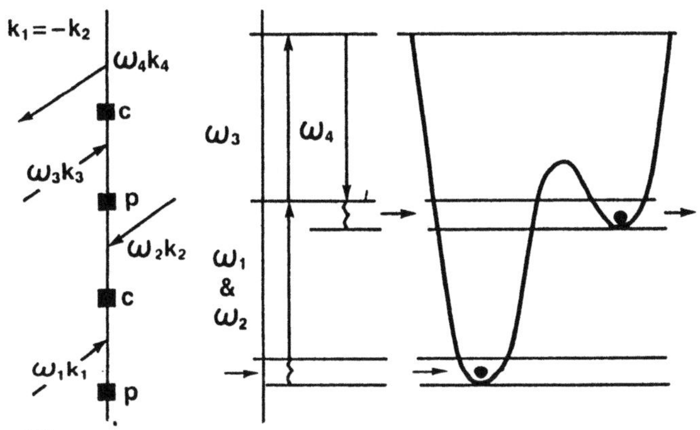

$\chi^{(3)}(\omega;\omega,\omega,-\omega)$ p:population difference; c:coherent superposition.

Fig. 10 — An example of backward-going phase conjugation by degenerate four-wave mixing using a nondegenerate, three-level polymeric system for generation of the third nonlinear susceptibility.

The polarization of the medium is (Lam, 1982; Lam and Abrams, 1982):

$$\mathbf{P}(x,t) = \int_{-\infty}^{+\infty} d^3 V \rho \mu + c.c., \tag{23}$$

where μ is the electric dipole moment, and V is the velocity of response, or

$$<\mathbf{P}> = Av\{<\psi|\mathbf{P}|\psi>\} = Tr\ (\rho\mathbf{P}). \tag{24}$$

Nonlinear polarizations are defined by an expansion in the density matrix operator and the polarization:

$$\rho = \rho^{(0)} + \rho^{(1)} + \rho^{(2)} + \ldots \tag{25}$$

$$<\mathbf{P}> = <\mathbf{P}^{(0)}> + <\mathbf{P}^{(1)}> + <\mathbf{P}^{(2)} + \ldots$$

QUANTUM WELL EFFECTS

The first quantum size effects have been produced in a photoelectrochemical system using a superlattice photoelectrode (Nozik et al., 1985). The electrode contained 40 alternating layers of undoped GaAs and $GaAs_{0.5}P_{0.5}$ each 250 Å thick. The barrier heights for electrons in the conduction band and for holes in theVlence band were 0.28 eV and 0.30 eV. Electrodes were fabricated by placing the superlattice structure in contact with a GaAs substrate. The electrodes were then immersed in an electrolyte cell. The photocurrent as a function of wavelength shows maxima corresponding to electron transfer into the electrolyte from the discrete energy levels in the quantum well. High quantum yields are expected from use of multiple quantum well structures because the higher energy of the electrons created in the higher-energy quantum levels are not degraded into heat before the electrons are used in *this conversion process.*

MULTIPLE QUANTUM WELL STRUCTURES

Multiple quantum well structures are novel structures with many important applications (cf. Döhler, 1983). The mechanism of conduction in multiple quantum well structures is like that in neither a metal, nor an inorganic or organic semiconductor, owing to the presence of miniconduction bands with low interband scattering permitting Bloch oscillations or a periodic crystal potential. Only a first approximation to a pendulum line analogy can presently be made (Fig. 11).

On account of the presence of minibands in these materials, carrier movement can be mediated by tunneling mechanisms. The carriers can be in a state of high kinetic energy with the velocity distribution narrowly peaked. The carrier momentum and energy can be nonstationary and the electron energy can be a dependent variable. When the electrons have high enough velocity, they escape coulombic scattering. In such cases the transport might be quasiballistic rather than

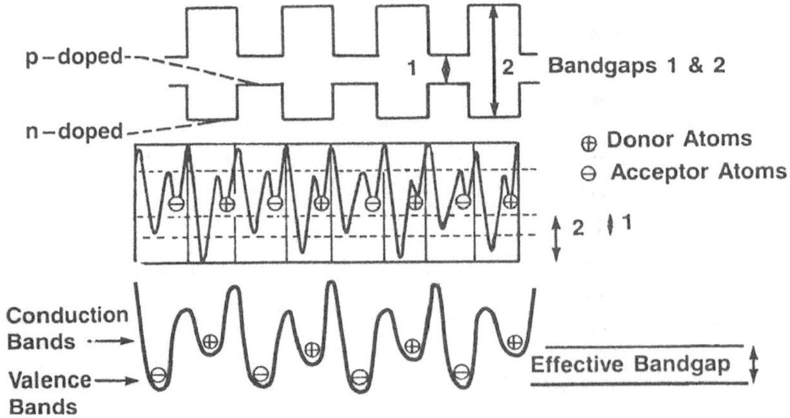

Fig. 11 – *n*- and *p*-doped superlattice in the line of pendula format indicating donor (+) and acceptor (–) atoms.

diffusive (Barker, 1985; Luryi and Kastalsky, 1985) and such structures demonstrate hot electron effects (Shur and Eastman, 1979, 1980).

The existence of ballistic transport, in which the interaction between carriers and phonons can be ignored, is currently questioned. Monte Carlo calculations of carrier transport alone may not be entirely appropriate, as they do not describe the interaction between the collective and the individual aspects of gases required (Grondin et al., 1983; Lugli and Ferry, 1985). (The model described by eq. 14 above, however, does provide this description.) Although ballistic electron transport has been detected in GaAs test structures when cooled to 140 K (Robinson, 1986), there is, at present, no ballistic transport of use in a practical transistor. If used, it is likely that the effect will be limited by energy relaxation being slower than momentum relaxation, i.e., the displacive transfer of energy time scale will be slower than the local momentum change time scale. This will cause nonlinearity and field effects. At the present time, therefore, as far as future devices are concerned, it is more appropriate to talk of "quasiballistic" transport.

The existence of hot electron effects in barrier devices is supported by the results of Monte Carlo calculations (Barker, 1985). Monte Carlo algorithms have been used for modeling small semiconductor devices in one or two dimensions when nonstationary transport and hot electrons are present (Lebwohl and Price, 1971; Zimmerman, 1981). The inclusion of quantum effects in Monte Carlo simulations permitted the calculation of scattering rates for high fields as the imaginary part of the complex-valued self-energy by the use of the optical theorem (Ferry and Barker, 1979). A modified Monte Carlo method, the trajectory integral method (Socha and Krumhansl, 1985), permits calculation of electron-electron scattering rates dependent upon the shape of the distribution function. The theoretical problem in hot electron behavior is to define the manner in which the distribution function of the electrons is modified in the presence of a field (Ferry, 1982). The carrier distribution is Maxwellian before satiation effects. After satiation effects appear and the carrier energy exchange is not sufficiently rapid, a displaced Maxwellian distribution with momentum balance equation treatment is possible (Fröhlich and Paranjape, 1956; Bosch and Thim, 1974). This crossover takes place in the physical model when the conditions change from local to displacive transitions (cf. eq. 14 above).

Other electron behaviors besides hot electron effects are implicated in multiple quantum well heterostructures and submicron devices, such as drift velocity overshoot and quantum size effect (Reggiani, 1985; Hamaguchi, 1985). Monte Carlo simulation has demonstrated that electron drift is very anisotropic in the overshoot region. Kash et al. (1985) have also demonstrated in GaAs-AlGaAs quantum wells the generation and decay of a large nonequilibrium optical phonon population. Several mechanisms of energy loss are being proposed (Fischetti, 1984; Porod and Ferry, 1985a,b).

The preliminary analysis given here offers an alternative approach to the continuum approximation for the calculation of the normal vibration modes. Maradudin (1984) has pointed out the need for a molecular, rather than a continuum, point of view in the treatment of superlattice phonons; and Krumhansl and Alexander (1983) have noted with respect to related matters, that it is important to distinguish between initial and final states on the one hand, and transitory intermediate configurations on the other. They note that chemistry, thermodynamics, and structural analysis provide information about the former, but shed little light on the latter, in which dynamics and tunneling concepts dominate. It is also because of the dynamics that the final states are achieved. The accent, therefore, on Monte Carlo simulation of electron transport perhaps results in a blindness toward the importance of phonon energies in the treatment of transport in periodic structures. It is for this reason that Fig. 11 provides an antidote to a continuum approximation. However, although the picture shown in Fig. 11 offers a comparison and contrast with other nonlinear conduction mechanism descriptions, a more definitive treatment must await the results of future data-gathering experiments.

NONLINEAR CONDUCTION IN INORGANIC SUBSTANCES (NONMETALLIC AND NONPOLYMERIC)

The origin of nonlinear effects in inorganic substances can be light-induced rearrangement of nuclear coordinates, which is inherently slow—on the millisecond scale—or to satiation effects in a two-level system (Miller et al., 1981). The latter mechanism is the one underlying the erformance of most inorganic materials used in degenerate four-wave mixing and will be considered here.

The nonlinear dispersion in inorganic semiconductors has been attributed to (1) direct saturation of independent damping-broadened band states (in GaAs), or (2) a dynamic Burstein-Moss shift of the absorption edge (in InSb) (Wherrett and Higgins, 1982). Optical bistability has been observed in two inorganic materials: InSb (Miller et al., 1979) and GaAs (Gibbs et al., 1979b). In InSb optical signal amplification ("transphasor action") has been achieved.

Some of the features of the nonlinearities studied are: the nonlinearity increases as the pumping frequency approaches the material bandgap, the linear absorption is low at the frequencies used; and the nonlinear coefficient is negative.

The physical origin of the nonlinearity is different for the two materials mentioned (Wherrett and Higgins, 1982). In GaAs the refraction is attributed to

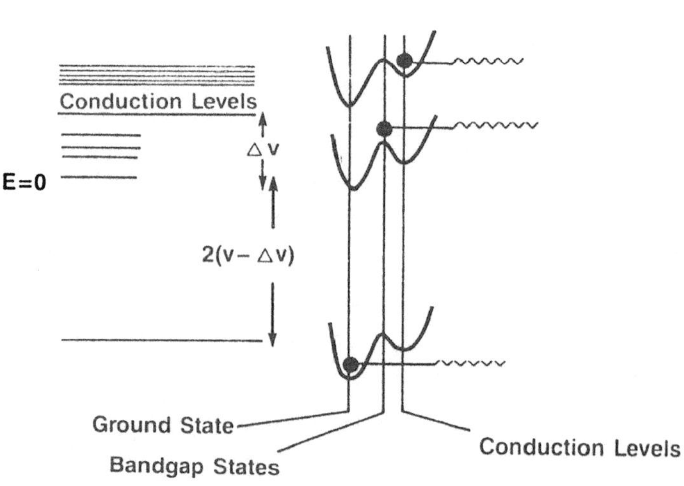

Figure 12.

saturation of a discrete excitonic absorption line (Gibbs et al., 1979a). In InSb, band effects provide the dominant mechanism.

The first model (direct saturation), which is applied to GaAs, describes a set of uncoupled two-level systems or interband transitions each of which can be saturated. This saturation of absorption provides the mechanism for the nonlinear contributions to the refractive index. To obtain absorption below the bandgap, broadening of the two-level system is required. This broadening is due to a damping or dephasing time of the material.

In InSb there is no indication of excitonic absorption. However, owing to the smaller bandgap (relative to GaAs), band effects are more important. The origin of the absorption is not treated by the blocking model, but the carriers are assumed to thermalize rapidly within each band. This results in an intensity-dependent shift in the effective bandgap called the dynamic Burstein-Moss shift (Burstein, 1954; Moss, 1954). This shift is associated with the intensity-dependent contribution to the refractive index.

In the case of Burstein-Moss refraction, the excited carriers are assumed to scatter rapidly, providing a thermal distribution and an intensity-dependent shift in the band edge.

Wherrett (1983) more recently considered five theoretical models of the origins of nonlinear susceptibility: (1) the saturation model, in which the effect of the excited carriers prevents further interband transitions; (2) the interband blocking model, in which there is a mechanism similar to that in the previous model but with independent thermal distributions of the excited carriers occurring within the valence and conduction bands; (3) the intraband model, based on free carrier linear dispersion theory; (4) the nonparabolic model, in which the group velocity of the free carriers is considered nonlinear in the momentum; and (5) the density matrix model of Kohn and Luttinger (1957), Luttinger and Kohn (1958), Armstrong et al. (1962), and Butcher and McLean (1963). Wherrett concludes that the same physical processes, namely, interband transitions, are represented in the calculations of $\chi^{(3)}(\omega_1, -\omega_2, \omega_3)$.

SUBMICROMETER DEVICES

The trend towards the development of submicron devices requires the replacement of the classical Boltzmann theory by concepts involving long-range order (Barker and Ferry, 1980a,b; Ferry and Barker, 1980). When particle collision duration becomes comparable with the scattering time (the mean free path), the isolated scatterer model required by Boltzmann theory fails. The carrier plasma frequency, ω_p, and the optic-phonon frequency are material characteristics that largely determine the material response. Just as saturation in absorption implies an intensity-dependent nonlinearity, so does miniaturization, and by similar mechanisms.

The Boltzmann transport equation assumes a distribution function unmodified by the field. Not only is this untrue for hot electrons, it is also untrue in the case of submicron devices as well as superlattices. In all these instances, there is a field dependence of the mobility and of the dielectric dispersion. Barker and Ferry (1980b) have argued that for large devices

$$\tau_c << \tau << \tau_L , \tag{26}$$

where τ_c is the collision duration, τ is the mean free time, and τ_L is the transit time. It is also necessary for many collisions to occur within a driving cycle, i.e., $\omega\tau << 1$.

Grondin et al. (1983) split τ_L into a τ_R, which is a relaxation time, and a τ_h that is the time to reach global equilibrium, i.e., the time for the nonequilibrium transport to arrive at the end of the nonlinearly joined line. τ_R is also divided into τ_m, a momentum relaxation time of the local transitions in the monomer units of the nonlinear line, and τ_δ, an energy relaxation time of the nonlinear coupling in the present model.

In terms of the transition probability:

$$P(i \rightarrow f) = < \psi_i \,|\, \mu_{\text{if}} \,|\, \psi_f > / (\omega_{\text{if}} - \omega - i\Gamma), \tag{27}$$

a nonlinearity can be introduced by making Γ a function of τ_m and τ_ξ, i.e., $\Gamma = \Gamma(\tau_m, \tau_\xi) = \Gamma(\tau_R)$. When $\tau \nless \tau_R$ (due to a limiting τ_ξ), the functions $P(i \rightarrow f)$ and $\mathbf{P}(\tau_h) = \sum_i P_i\,(i \rightarrow f)$ are nonlinear with respect to τ. As τ is a function of perturbation energy and together with τ_R defines the steady state, then the nonlinearity is an intensity-dependent nonlinearity. By reason of energy conservation a $\tau_E = \tau_\xi + \tau_\epsilon, \tau < \tau_E$, can be defined, where τ_ϵ defines the dynamic response of the environmental contribution to the device's nonlinearity. The A, B, and C constants of eq. 14 above would then have a τ_R dependence.

An early approach to addressing this situation is the quantum transport model of Kohn and Luttinger (1957) and others (Kubo, 1957, 1966; Dresden, 1961; Chester, 1963; Luttinger, 1968; Barker, 1979a), which consists of free (or Bloch) electrons scattered by "random" or rigid impurity centers. A density matrix is developed in ascending powers of the strength of the scattering potential. The familiar Boltzmann equation represents an approximation valid in the limiting cases of weak or very dilute scatterers. Luttinger and Kohn (1958) observed that it is possible to arrange the solution of the density matrix in such a way that it involves first the effects of one scatterer, then of pairs, then of triplets, etc., in a manner similar to a virial expansion of equilibrium properties. This is called a density expansion utilizing "scattering operators". Such a representation is, however, entirely linear, but can be adapted to a nonlinear process (see above).

There are other differences associated with submicron devices. For example: (a) the well-known Einstein relation between the diffusion coefficient, the temperature, and the ohmic mobility (Einstein, 1905, 1906), (b) the Nyquist relation between noise spectral density, resistance, and temperature (Nyquist, 1928), and (c) the more general fluctuation-dissipation theorem (Callen and Welton, 1951; Kubo, 1957; Van Vliet and Van der Ziel, 1977), are all valid only in thermal equilibrium and fail in the case of nonlinear systems (Nougier, 1979). There is also speculation that submicrometer devices have feature sizes that could support phase transitions and other nonequilibrium cooperative processes (Barker, 1979b).

POLYMERIC CONDUCTORS

Since the discovery in 1973 that the inorganic polymer poly(sulfur nitride), $(SN)_x$, conducts at room temperature like a metal ($\approx 1 \times 10^3\ (\Omega\,\mathrm{cm})^{-1}$), and at 0.3 K becomes a superconductor (although highly unstable), and the discovery in 1977 that polyacetylene with a conductivity of $< 10^{-5}\ (\Omega\,\mathrm{cm})^{-1}$ can be made highly conducting ($\sim 10^3\ (\Omega\,\mathrm{cm})^{-1}$) by exposure to oxidizing or reducing agents, there has been much effort to use polymers in electrical devices. There is now no question that the mechanism of conduction in these polymers is different from that used by metals and in nonpolymeric semiconductors. Polyacetylene, poly-*p*-phenylene and polypyrrole conduct by spinless charge carriers, rather than with unpaired electrons (Brédas and Street, 1985).

In many cases studied, ionization creates an excited state, different from the ground state. The ionized state can be considered to provide the lowest unoccupied levels of a conduction band, and the ground state to provide the highest occupied levels of a valence band. Localized electronic states are created in the bandgap between the states when it is energetically favorable to localize the charge that appears on the chain.

If, on oxidation, an electron is removed from the polymer chain, a local distortion or relaxation of the lattice occurs due to strong electron-phonon coupling and a polaron or radical ion is created. If a second electron is removed from the chain, and it is energetically more favorable that the electron is taken from the originally created polaron, then a bipolaron is created. On the other hand, if it is unfavorable to take the electron from the originally created polaron, but from somewhere else on the chain, then two polarons are created where one existed.

The bipolaron is a di-cation or pair of like charges and associated with a strong lattice distortion or relaxation. It has been compared (Brédas and Street, 1985) with the Cooper pair in the BCS theory of superconductivity. Its formation indicates that the energy gained from the lattice is larger than the Coulomb repulsion between the electrons (Anderson, 1975).

There is yet another mechanism for conduction, by which the conduction and valence bands are associated with either one of the two degenerate states. This situation occurs in polyacetylene, and if the chain contains an odd number of electrons, then a neutral soliton or radical π electron is created leading to the appearance of a localized electronic level in the bandgap. On oxidation or reduction, the spin density in the neutral soliton becomes a charge density in a charged soliton.

Apart from the aspects of interchain interactions and the amorphous nature of these materials, which are important, in all cases considered here, namely, materials supporting the formation of polarons, bipolarons, or solitons, the current carriers are mobile spinless charge carriers in collective motion, not electrons. All bands are also totally filled or empty. Thus the mechanism of conduction is quite different, as already noted, from that of metals and inorganic semiconductors. A lowest common denominator for organic electrical conduction appears to be (1) the availability of two states (either ionized or degenerate), and (2) strong electron-phonon coupling. There is no weak interaction or scattering of the charge carriers. Therefore, the description of charge transport in these systems by the Boltzmann equation appears unlikely and the concepts described above in connection with the pendulum line apply.

The nonlinear optical effects in organic polymers are due to the polarization of the π-electron cloud (Kerr effect) giving a very fast response time, as opposed to the slower rearrangement of nuclear coordinates, or absorption satiation in a two-level system. The nonlinearity arises from the intensity-dependent transition from local to displacive changes described above.

Experimental evidence has confirmed the prediction by Su and Schrieffer (1980) that in *trans*-polyacetylene, $(CH)_x$, an electron-hole pair should evolve into a pair of solitons within an optical phonon period of about 10^{-13} s. (Shank et al., 1982; Vardeny et al., 1982). The same changes seen on photoexcitation also occur in redox reactions ("doping"), which implies photogeneration of soliton-antisoliton pairs or the creation of states at the midgap. There are two relevant time scales for photoexcitation in *trans*-$(CH)_x$: the picosecond (10^{-12} s.) regime aforementioned during the formation and decay of photoexcited states, and a nanosecond (10^{-9} s.) regime of residual excitation decay. Photoconductivity on the picosecond time scale implies rapid nonradiative recombination of oppositely charged solitons. Furthermore, electron-hole pairs in encompassing neighboring polymer chains are also expected to be photoexcited (interchain excitation). This excitation is slower.

The applications of such organic polymers as nonlinear optical materials are many and diverse (cf. Garito and Singer, 1982; Williams, 1983).

DENSITY MATRIX FORMULATION OF TRANSPORT

Butcher and McLean (1963) extended the constitutive relation between the current density and the electric field in a solid to include terms in all powers of the field. It was shown that the density matrix of the charged particles in a solid can be used to calculate the nth-order conductivity tensor. The second- and third-order tensors describe nonlinear effects for the radiation fields. As a generalization: for low frequencies and odd values of n, the nth-order conductivity tensor may be expressed in terms of the $(n+1)$th derivative of the one-electron energy with respect to the wave vector. A précis of the argument is as follows.

The current density, $\mathbf{J}(t)$, may contain contributions from higher powers of the electric field than the first:

$$\mathbf{J}(t) = \sum_{n=1}^{\infty} \mathbf{J}^{(n)}(t), \tag{28}$$

and, more generally:

$$\mathbf{J}_{\mu}^{(n)} = \int d\omega_1 \ldots \int \sigma^{(n)}_{\mu_{\alpha_1}\cdots\alpha_n}(\omega_1,\ldots\omega_n)\, E_{\alpha_1}(\omega_1)\ldots E_{\alpha_n}(\omega_n) \exp\left[-it \sum_{m=1}^{n} \omega_m\right] d\omega_n \tag{29}$$

where the tensor $\sigma^{(n)}_{\mu_{\alpha_1}\cdots\alpha_n}$ is the nth order conductivity tensor.

The generation of the second harmonic of some fundamental frequency ω in a nonlinear medium depends on the tensor $\sigma^{(2)}_{\mu_{\alpha\beta}}(\omega, \omega)$ which has symmetry properties. On the other hand, the generation of radiation with frequency $\omega_1 + \omega_2$ from two fields of frequencies ω_1 and ω_2 depends on $\sigma^{(2)}_{\mu_{\alpha\beta}}(\omega_1,\omega_2)$, which has fewer symmetry properties.

An associated vector potential is defined:

$$\mathbf{A}(t) = -c \int_{-\infty}^{t} E(\tau)\,d\tau, \tag{30}$$

and the Hamiltonian of the charged particle system is:

$$H = H_0 - (1/c)\,\mathbf{A}(t)\boldsymbol{\pi} + (\lambda/2c^2)\,\mathbf{A}^2(t), \tag{31}$$

where H_0 is the unperturbed Hamiltonian, $\boldsymbol{\pi} = \sum_{j=1}^{N} (q_j/m_j)\mathbf{p}_j, \lambda = \sum_{j=1}^{N} (q_j^2/m_j)$, and m_j, q_j, and $\mathbf{p}_j$ are the mass, charge, and momentum of the jth particle.

The equation of motion under the influence of an electric field is:

$$(d\rho(t)/dt) = (1/i\hbar)\,[H,\rho(t)], \tag{32}$$

where $\rho(t)$ is the density matrix defined as:

$$\rho(t) = \rho_0 + \sum_{n=1}^{\infty} (i\hbar c)^{-n} \int_{-\infty}^{0} dt_1 \int_{-\infty}^{t_1} dt_2 \ldots \tag{33}$$

$$\ldots \int_{-\infty}^{t_{n-1}} [\pi_{\alpha_1}(t_1), [\pi_{\alpha_n}(t_n), \ldots [\pi_{\alpha_n}(t_n), r_0] \ldots]\, A_\alpha(t+t_n)$$

$$\ldots A_{\alpha_n}(t+t_n)\, dt_n$$

where $\boldsymbol{\pi}(t) = \exp(itH_0/\hbar)\,\boldsymbol{\pi}\exp(-itH_0/\hbar)$. In the case of device miniaturization, it should be noted that Barker and Ferry (1980a, b) have pointed out that the density matrix cannot satisfy a Liouville equation of this form unless the device is totally decoupled from the environment. If it is not decoupled, then an additional Hamiltonian and density matrix for the device-environment interactions must be added.

The higher-order multipole interactions give rise to additional contributions to the current density that depend on the spatial variation of the electric field. The conductivity tensors then depend on the wavelength of the radiation in the solid, as well as its frequency. There is thus spatial dispersion as well as temporal dispersion.

When the constitutive equation is substituted into Maxwell's equations, the nonlinear terms couple together the Fourier components of the electric field at different frequencies. The picture is thus, again, a phenomenology which can be reduced to the pendulum line picture and Feynman diagram form.

DIELECTRIC BREAKDOWN

Models for the theory of dielectric breakdown demonstrate underlying principles in common with other nonlinear phenomena. In Fröhlich's (1947) treatment, current-field interactions are described by:

$$J = \sigma E = e^2 \tau_1 N_1 E/m, \tag{34}$$

where τ_1 is the average time of relaxation, E is the electric field, and energy is transferred to a material's electrons from the field at a rate:

$$A = JE \text{ per unit volume}\ . \tag{35}$$

A fundamental tenet is that to obtain a stationary state the electrons must be able to transfer energy to the lattice vibrations at the same rate. The total number, N_1, of electrons in the conduction levels is:

$$N_1 = C_1(T)e^{-V/kT}\ . \tag{36}$$

The number of electrons in excited isolated levels is:

$$N_2 = C_2(T)e^{-V/kT + \Delta V/kT} \tag{37}$$

and the ratio C_1/C_2 is approximately equal to the ratio of the number of energy levels in the interval kT of the conduction levels to the corresponding number in the isolated levels. Only the N_1 electrons in the conduction levels contribute to the energy transfer, A, from the field to the lattice.

A balance equation can be derived:

$$(e^2 \tau_1 \tau_2/mh\nu)\ (C_1/C_2)\ (1/n(T_0))\ E^2\ e^{-\Delta V/kT} = e^{h\nu/kT_0 - h\nu/kT} - 1, \tag{38}$$

where $n(T_0) = 1/(e^{h\nu/kT_0} - 1)$ and $1/\tau_2$ is the average value of $P(E)$, which is the transition probability from the level of E to the level of $E + h\nu$.

For $E > E^*$ no solution exists because the rate of energy transfer A from the field to the electrons is always larger than the rate of energy transfer B from the electrons to the lattice. Hence E^* is the breakdown field.

In the context of the present discussion, if the material has a periodic structure, then the breakdown field, E^*, is that field at which either the material exhibits displacive transitions, i.e., becomes conductive in the way of an organic polymer described above, or becomes an inorganic semiconductor.

The propagation of a wavefront or charge density wave in organic materials is a mechanism whereby the lattice or chain cools off a local heat spot (Miller, 1986). If the wavefront is pinned over some time interval, the energy input from the source is not disseminated but must be shared by the monomer units or particles behind the wavefront. These are "heated up" and disordered motion results.

CONCLUSION

A completely general approach to nonlinear transport (Eq. 14) has been presented based on the physical intuition of the joined pendulum line. This picture finds application in understanding the unique conduction mechanisms of electrical conduction in organic polymers and also their nonlinear optical properties. The picture or model can also be applied to multiple quantum well structures and submicron devices (heterostructures), but, due to the present lack, or difficulty, of experimental investigation of the relevant physical parameters of those structures, relating the model to physical parameters must remain at this time unfinished.

Negative differential resistance in the two-energy-level system of GaAs (Gunn effect) implies a change in the sign of the effective mass of the electrons. The Gunn effect can also be treated by the picture given above. The negative resistance in multiple quantum well structures also implies this change. However, even without detailed experimental study of the multiple quantum well structures, it is safe to predict that there will be a close analogy.

The concepts presented are completely general and offer a platform based on the physical intuition provided by the model, of such subjects—analogous in a mechanical sense—of Bose-Einstein condensation and off-diagonal long-range order. All of these topics are part of the larger subject of energy transfer with energy conservation.

REFERENCES

P.W. Anderson, *Lectures in the Many-Body Problem*, E.R. Caianello (ed.), Vol. 2, Academic, New York, 1964.

P.W. Anderson, Phys. Rev. Lett. **34**, 953 (1975).

J.A. Armstrong, N. Bloembergen, J. Ducuing, and P.S. Pershan, Phys. Rev. **127**, 1918 (1962).

J.R. Baker, in *Physics of Nonlinear Transport in Semiconductors*, D.K. Ferry, J.R. Barker and C. Jacoboni, Eds., NATO Advanced Study Institutes Series, Vol. 52, Plenum, New York, 1979a, pp. 127-151.

J.R. Barker, in *Physics of Nonlinear Transport in Semiconductors*, D.K. Ferry, J.R. Barker and C. Jacoboni, Eds., NATO Advanced Study Institutes Series, Vol. 52, Plenum, New York, 1979b, pp. 589-606.

J.R. Barker, Physica **B134**, 22 (1985).

J.R. Barker, *Handbook on Semiconductors, Volume 1, Band Theory and Transport Properties*, W. Paul, Ed., North Holland Publishing Co., Amsterdam, 1982, pp. 617-659.

J.R. Barker and D.K. Ferry, Solid-State Electronics **23**, 519 (1980a).

J.R. Barker and D.K. Ferry, Solid-State Electronics **23**, 531 (1980b).

T.W. Barrett, Adv. Biol. Med. Phys. **17**, 1 (1980).

T.W. Barrett, Am. Phys. Soc., January (1983a).

T.W. Barrett, *Soliton Propagation in One-Dimensional Organic Conductors*, U.S. Naval Air Development Center Memorandum Report No. NADC-83074-60, 1983b (Sept.).

H. Bliz, H. Büttner, and H. Fröhlich, Z. Naturforsch. **36b**, 208 (1981).

R. Bosch and H. Thim, IEEE Trans. Electr. Dev. **ED-21**, 16 (1974).

J.L. Brédas and G.B. Street, Acc. Chem. Res. **18**, 309 (1985).

E. Burstein, Phys. Rev. **93**, 632 (1954).

P.N. Butcher and T.P. McLean, Proc. Phys. Soc. London **81**, 219 (1963).

H.B. Callen and T.A. Welton, Phys. Rev. **83**, 34 (1951).

G.V. Chester, Rep. Progr. Phys. **26**, 411 (1963).

F.W. Cummings and J.R. Johnstone, Phys. Rev. **151**, 105 (1966).

G.H. Döhler, Sci. Am. **249**, 144 (Nov. 1983).

M. Dresden, Rev. Mod. Phys. **33**, 265 (1961).

I. Egri, R. Matz, H. Lüth, and A. Stahl, Surface Sci. **128**, 51 (1983).

A. Einstein, Ann. Phys. **17**, 549 (1905).

A. Einstein, Ann. Phys. **19**, 289, 371 (1906).

D.K. Ferry, *Handbook on Semiconductors, Volume 1, Band Theory and Transport Properties*, W. Paul, Ed., North Holland Publishing Co., Amsterdam, 1982, pp. 563-597.

D.K. Ferry and J.R. Barker, Solid State Commun. **30**, 361 (1979).

D.K. Ferry and J.R. Barker, Solid-State Electronics, **23**, 545 (1980).

M.V. Fischetti, Phys. Rev. Letter. **53**, 1755 (1984).

R.A. Fisher, (Ed.,) *Optical Phase Conjugation*, Academic, New York, 1983.

D. Frank and A. Stahl, Solid State Commun. **52**, 861 (1984).

H. Fröhlich, Proc. R. Soc. London **A188**, 521 (1947).

H. Fröhlich, Int. J. Quantum Chem. **11**, 641 (1968).

H. Fröhlich, Coll. Phenom. **1**, 101 (1973).

H. Fröhlich and B.V. Paranjape, Proc. Phys. Soc. London **B69**, 21 (1956).

T.A. Fulton, IEEE Trans. Magn. **MAG-11**, 749 (1975).

T.A. Fulton, in *Superconductor Applications: SQUIDS and Machines*, B.B. Schwartz and S. Foner, Eds., Plenum, New York, 1977, pp. 125-187.

A.F. Garito and K.D. Singer, Laser Focus, 59 (February, 1982).

H.M. Gibbs, S.L. McCall, T.N.C. Venkatesan, A.C. Gossard, A. Passner, and W. Wiegmann, Appl. Phys. Lett. **35**, 451 (1979b).

H.M. Gibbs, H.C. Gossard, S.L. McCall, A. Passner, W. Wiegmann, and T.N.C. Venkatesan, Solid State Commun. **30**, 271 (1979a).

L. Gotthard, A. Stahl, and G. Czajkowskt, J. Phys. C: Solid State Phys. **17**, 4865 (1984).

R.O. Grondlin, P. Lugli, and D.K. Ferry, Proc. SPIE Int. Soc. Opt. Eng. **439**, 18 (1983).

O. Hamaguchi, Physica **134B**, 87 (1985).

W. Huhn and A. Stahl, Phys. Stat. Sol. (b) **124**, 167 (1984).

K. Kash, J. Shah, D. Block, A.C. Gossard, and W. Wiegmann, Physica **B134**, 189 (1985).

W. Kohn and J.M. Luttinger, Phys. Rev. **108**, 590 (1957).

J.A. Krumhansl and D.M. Alexander, in *Structure and Dynamics: Nucleic Acids and Proteins*, E. Clementi and R.H. Sarma, Eds., Adenine, New York, 1983, pp. 61-80.

R. Kubo, J. Phys. Soc. Jpn. **12**, 570 (1957).

R. Kubo, Rep. Prog. Phys. **29**, 225 (1966).

J.F. Lam, Opt. Eng. **21**, 219 (1982).

J.F. Lam and R.L. Abrams, Phys. Rev. A: Gen. Phys. **26**, 1539 (1982).

P. Lebwohl and J. Price, Appl. Phys. Lett. **19**, 530 (1971).

P. Lebwohl and M.J. Stephen, Phys. Rev. **163**, 376 (1967).

P. Lugli and D.K. Ferry, IEEE Electron Device Lett. **EDL-6**, 25 (1985).

S. Luryi and A. Kastalsky, Physica **B134**, 453 (1985).

J.M. Luttinger, *Mathematical Models in Solid-State and Superfluid Physics*, Oliver and Boyd, New York, 1968.

J.M. Luttinger and W. Kohn, Phys. Rev. **109**, 1892 (1958).

G.D. Mahan, *Many-Particle Physics*, Plenum, New York, 1981.

A.A. Maradudin, in *The Physics of Submicron Structures*, H.L. Grubin, K. Hess, G.J. Iafrate and D.K. Ferry, Eds., Plenum, New York, 1984, pp. 159-169.

A. Miller, D.A.B. Miller, and S.D. Smith, Adv. Phys. **30**, 697 (1981).

D.A.B. Miller and S.D. Smith, Opt. Commun. **31**, 101 (1979).

D.A.B. Miller, S.D. Smith, and A. Johnston, Appl. Phys. Lett. **35**, 658 (1979).

D.A.B. Miller, C.T. Seaton, M.E. Prise, and S.D. Smith, Phys. Rev. Lett. **47**, 197 (1981).

M.D. Miller, Phys. Rev. **B33**, 1641 (1986).

T.S. Moss, Proc. Phys. Soc. London **B67**, 755 (1954).

J.P. Nougier, in *Physics of Nonlinear Transport in Semiconductors*, D.K. Ferry, J.R. Barker, and C. Jacoboni, Eds., NATO Advanced Study Institutes Series, Vol. 52, Plenum, New York, 1979, pp. 415-465.

A.J. Nozik, B.R. Thacker, and J.M. Olson, Nature **316**, 51 (1985).

H. Nyquist, Phys. Rev. **32**, 110 (1928).

O. Penrose, Phil. Mag. **42**, 1373 (1951).

O. Penrose and L. Onsager, Phys. Rev. **104**, 576 (1956).

D.M. Pepper, Opt. Eng. **21**, 156 (1982).

D.M. Pepper, Sci. Am. **254**, No. 1, 74 (Jan. 1986).

L.P. Pitaevskii, Sov. Phys. JETP **8**, 282 (1959).

L.P. Pitaevskii, Sov. Phys. JETP **13**, 451 (1961).

W. Porod and D.K. Ferry, Phys. Rev. Lett. **54**, 1189 (1985a).

W. Porod and D.K. Ferry, Physica **B134**, 137 (1985b).

L. Reggiani, Physica, **B134**, 123 (1985).

A.L. Robinson, Science **231**, 22 (1986).

T. Schneider and E. Stoll, Phys. Rev. Lett. **31**, 1254 (1973).

T. Schneider and E. Stoll, Phys. Rev. Lett. **35**, 296 (1975).

T. Schneider and E. Stoll, Phys. Rev. B: Solid State **13**, 1216 (1976).

T. Schneider and E. Stoll, Phys. Rev. B: Solid State **17**, 1302 (1978).

T. Schneider and E. Stoll, in *Physics in One-Dimension*, J. Bernasconi and T. Schneider, Springer, Heidelberg, 1981, pp. 75-86.

A.C. Scott, Am. J. Phys. **37**, 52 (1969).

C.V. Shank, R. Yen, R.L. Fork, J. Orenstein, and G.L. Baker, Phys. Rev. Lett. **49**, 1660 (1982).

V.V. Shkunov and B.Y. Zel'dovich, Sci. Am. **253**, No. 6, 54 (Dec. 1985).

M.S. Shur and L.F. Eastman, IEEE Trans. Electri. Dev. **ED-26**, 1677 (1979).

M.S. Shur and L.F. Eastman, Electron. Lett. **16**, 522 (1980).

J.B. Socha and J.A. Krumhansl, Physica **B134**, 142 (1985).

A. Stahl, Solid State Commun. **49**, 91 (1984).

W.P. Su and J.R. Schrieffer, Proc. Nat. Acad. Sci. U.S.A. **77**, 5626 (1980).

D.B. Sullivan and P. Zimmerman, Am. J. Phys. **39**, 1504 (1971).

S. Takeno, Progr. Theor. Phys. **71**, 395 (1984).

S.E. Trullinger, M.D. Miller, R.A. Guyer, A.R. Bishop, F. Palmer, and J.A. Krumhansl, Phys. Rev. Lett. **40**, 206 (1978).

T. Usui, Progr. Theor. Phys. **41**, 1603 (1969).

K.M. Van Vliet and A. Van der Ziel, Solid-State Electron. **20**, 231 (1977).

Z. Vardeny, J. Strait, D. Moses, T.C. Chang, and A.J. Heeger, Phys. Rev. Lett. **49**, 1657 (1982).

J.R. Waldram, A.B. Pippard, and J. Clarke, Phil. Trans. R. Soc. **A268**, 265 (1970).

L.C. West and S.J. Eglash, Appl. Phys. Lett. **46**, 1156 (1985).

B.S. Wherrett, Proc. R. Soc. London **A390**, 373 (1983).

B.S. Wherrett and N.A. Higgins, Proc. R. Soc. London **A379**, 67 (1982).

D.J. Williams, (ed.), *Nonlinear Optical Properties of Organic and Polymeric Materials*, American Chemical Society, Washington, 1983.

T. Yamashita and K. Ondera, J. Appl. Phys. **50**, 4503 (1979).

T. Yamashita, Y. Ogawa, and Y. Ondera, J. Appl. Phys. **50**, 3547 (1979).

C.N. Yang, Rev. Mod. Phys. **34**, 694 (1962).

E. Zimmerman, J. Appl. Phys., 825 (1981).

6. Relativistic Characteristics of Electrons in Simple Nonparabolic Energy Bands

P. T. Landsberg*

ABSTRACT

An analogy is traced between a gas of electrons moving with relativistic velocities and a gas of nonrelativistically moving electrons in simple nonparabolic energy bands.

The works of Herbert Fröhlich have a width and depth that have been an inspiration to younger colleagues like myself. It is appropriate, therefore, for an occasion that honors him to pull out of the bank of models in theoretical physics two items from completely disparate areas, and show that they are basically identical in mathematical structure. The fields are special relativity theory and solid state theory. The analogy to be discussed is an accident without profound and deep implications. However, the facts to be shown up are surprising and they can be suggestive. They enable experiments, that have heretofore been considered appropriate to a relativistic plasma to be transferred to a solid state setting, or vice versa, and the same applies to the mathematical techniques involved (Zawadski et al. 1985; Paverman and Papuashvili, 1985; Landsberg, 1986).

We take from the shelf an isotropic but nonparabolic band defined by the dispersion relation between kinetic energy E and wave vector k by

$$\frac{\hbar^2 k^2}{2m_0^*} = E\left(1+\frac{E}{2E_0}\right). \tag{1}$$

Here m_0^* is the effective mass at the bottom of the band and $2E_0$ is a constant of the order of the energy gap between conduction and valence band. It has been a pretty successful model and has been widely used since it was put forward by Evan Kane in 1957. It has been used in connection with many III-V compounds, to study hot electron distribution, thermoelectric power, piezoresistance, and other properties that are much loved by the solid state physicists and, even more relevant perhaps, of great interest to engineers who design and manufacture devices. It is an area in which one would hardly expect to find any very close analogy with the special theory of relativity, which sits after all on a very different shelf in our model bank. There is, however, one fairly obvious point: particle-antiparticle pair creation requires a minimum energy of $2\epsilon_0 = 2m_0c^2$ where m_0 is the rest mass of a particle. Electron-hole pair creation requires an energy of E_G which, in our notation, is just $2E_0$. Is the analogy

$$E_0 \left(\equiv \frac{1}{2} E_G\right) \Longleftrightarrow \epsilon_0 \left(\equiv m_0 c^2\right) \tag{2}$$

*Permanent address: Faculty of mathematical studies, University, Southampton 509 SNK, England.

all there is to it?

To see that there is more to it than that, we work out from eq. 1 the velocity of an electron, $w = \hbar^{-1} \nabla_k E(k)$ and also its momentum-mass based on $p = \hbar k$:

$$\tilde{m} \equiv \frac{p}{w} = [2m_0^* E(1+E/2E_0)]^{1/2}/w .$$

For small kinetic energies E one finds

$$w = (2E/m_0^*)^{1/2}, \qquad \tilde{m} = m_0^* .$$

The mass is independent of energy, as expected from the parabolic approximation. In the general case we find, on the other hand,

$$w = \frac{(1+2E_0/E)^{1/2}}{1+E_0/E} v$$

$$\tilde{m} = \frac{E+E_0}{E_0} m_0 .$$

Thus the particle velocity v saturates at

$$v \equiv (E_0/m_0^*)^{1/2}, \tag{3}$$

as E is increased indefinitely, and the momentum-mass $\tilde{m}$ increases without limit. To the analogy eq. 2 one is thus tempted to add

$$v \left(\equiv \left(\frac{E_0}{m_0^*} \right)^{1/2} \right) <\!\!-\!\!-\!\!> c . \tag{4}$$

This is a compelling interpretation since the detailed algebra shows (Landsberg, 1986) that

$$\tilde{m} = m_0^* \left(1 - \frac{w^2}{v^2} \right)^{-1/2} \tag{5}$$

in full agreement with a usual result of relativity theory. The acceleration mass m^* has a different factor (Landsberg, 1986)

$$m^* = m_0^* (1 - w^2/v^2)^{-3/2},$$

which also occurs in relativity theory.

The gas of particles in the nonparabolic band considered here has a density of state

$$N(E) = \frac{gV}{2\pi^2} k^2 \frac{dk}{dE}$$

where V is the volume, g is the spin degeneracy per particle ($g = 2$). This yields

$$N(E) = \frac{gV}{4\pi} \left(\frac{2m_0^*}{h^2} \right)^{3/2} \left(1 + \frac{E}{E_0} \right) \left(E + \frac{E^2}{2E_0} \right)^{1/2} . \tag{6}$$

The number of particles is accordingly

$$n = \int N(E) f(E) dE \equiv N_c I, \qquad N_c \equiv \left(\frac{2\pi m_0^* k_B T}{h^2} \right)^{3/2} g , \tag{7}$$

which defines I and were $f(E)$ is the Fermi factor

$$f = \left[1+\exp\left(\eta - \frac{\mu - E_c}{k_B T}\right)\right]^{-1}, \quad \eta \equiv \frac{E - E_c}{k_B T}, \tag{8}$$

μ is the electrochemical potential and E_c is the energy at the bottom of the conduction band. Thus eq. 7 is apparently a standard formula, familiar from parabolic bands. But the difference arises from the integral ($\eta_0 \equiv E_0/k_B T$)

$$I = \left(\frac{2\eta_0}{\pi}\right)^{1/2} \eta_0 \int_0^\infty \left(1+\frac{\eta}{\eta_0}\right) [2\frac{\eta}{\eta_0} + \left(\frac{\eta}{\eta_0}\right)^2]^{1/2} fd\left(\frac{\eta}{\eta_0}\right)$$

which becomes the conventional Fermi integral only in the limit—the "nonrelativistic" limit or parabolic band limit—where $E_0 \to \infty$. The mean energy is obtained similarly. The "extreme relativistic limit" $E_0 \to 0$ corresponds to a vanishing energy gap and is probably less useful in the solid state context. In the relativistic case it gives us the statistical thermodynamics of gases whose components have zero rest mass (e.g., photons).

If now one makes the substitution

$$E = (\cosh\theta - 1)\, E_0, \quad \text{(definition of } \theta\text{)},$$

one comes to

$$n = \frac{4\pi g (m_0^* v)^3 V}{h^3} \int_0^\infty \frac{\sin h^2\theta \cosh\theta\, d\theta}{1+\exp\left\{\frac{1}{k_B} T\, [E_0(\cosh\theta - 1) - (\mu - E_0)]\right\}}. \tag{9}$$

This integral can be reduced to a modified Besssel function K_2 of order 2 and argument $E_0/k_B T$ in the case of nondegeneracy. This was first shown by Jüttner about 70 years ago (Chandrasekhar, 1939). He, however, was originally interested in the relativistic Boltzmann gas and not in nonparabolic bands in semiconductors, concepts of much more recent origin!

This early mathematical work was not discovered during the thirty years of nonparabolic band theories until now. It might have been found after all independently, or from the literature, by workers in solid state theory (Harman and Strauss, 1961; Bebb and Ratliff, 1971; Anderson, 1980). In the appendix some integrals will therefore be discussed in the solid state setting although they are identical with those relevant to the relativistic gas. One sees that, although one may be looking on widely different shelves in our model bank, still one may find models which have the most striking beauty of symmetry: that of identity.

Appendix

Some integrals for electrons in simple nonparabolic bands

In the nondegenerate limit eq. 9 can be covered by a partial integration to

$$n = \frac{4\pi g(m_0^* v)^3 V}{3h^3} \frac{E_0}{k_B T} \exp \frac{E_0 + \mu - E_c}{k_B T} \int_0^\infty \sinh^4\theta \exp\left(-\frac{E_0}{k_B T}\cosh\theta\right) d\theta \tag{A1}$$

$$= \frac{4\pi g(m_0^* v)^3 V}{h^3} \frac{k_B T}{E_0} K_2\left(\frac{E_0}{k_B T}\right) \exp \frac{E_0 + \mu - E_c}{k_B T}.$$

We have here used

$$\frac{\pi^{1/2}\left(\frac{1}{2}z\right)^\nu}{\Gamma\left(\nu+\frac{1}{2}\right)} \int_0^\infty \sinh^{2\nu}\theta \exp(-z\cosh\theta) = K_\nu(z)$$

where the K's are modified Bessel functions.

On the basis of the grand canonical ensemble, with grand partition function Ξ, one also has for a pressure P and volume V (Landsberg and Dunning-Davis, 1965)

$$PV = k_B T \ln\Xi = k_B T \sum_{j_1,j_2,j_3} \ln\left[1 + t(j_1,j_2,j_3)\right]$$

where

$$t(j_1,j_2,j_3) \equiv \exp\left[\frac{\mu - E_c - E(j_1,j_2,j_3)}{k_B T}\right].$$

This becomes by the replacement $\sum_{j_1,j_2,j_3} \rightarrow \frac{4\pi g V}{h^3}\int \dots p^2 dp$

$$PV = \frac{4\pi g V k_B T}{h^3}\int_0^\infty p^2 \ln\left[1 + t(p)\right] dp.$$

Hence

$$PV = \frac{4\pi g V}{3h^3}\int_0^\infty p(E)^3 f(E)\, dE$$

$$= \frac{4\pi g V}{3h^3} m_0^{*4} v^5 \int_0^\infty (x^2 + 2x)^{3/2} f\, dx$$

where $x \equiv E/E_0$. Using $x = \cosh\theta - 1$

$$PV = \frac{4\pi g V}{3h^3}(m_0^*)^4 v^5 \int_0^\infty \frac{\sinh^4\theta\, d\theta}{1 + \exp\left\{\frac{1}{k_B T}\left[E_0(\cosh\theta - 1) - (\mu - E_c)\right]\right\}}. \tag{A2}$$

In the nondegenerate limit one sees from eqs. A1 and A2 that the gas law

$$PV = nk_B T$$

holds for nondegenerate electrons in a simple nonparabolic band.

REFERENCES

W.W. Anderson, Infrared Phys. **20**, 363 (1980).

H.B. Bebb and C.R. Ratliff, J. Appl. Phys. **42**, 3189 (1971).

S. Chandrasekhar, *An Introduction to Stellar Structure*, University of Chicago Press, Chicago, 1939, p. 385.

T.C. Harman and A.J. Strauss, J. Appl. Phys. **32**, 2265 (1961).

P.T. Landsberg, Phys. Rev. B: Solid State, to appear (1986).

P.T. Landsberg and J. Dunning-Davis, in *Statistical Mechanics of Equilibrium and Non-Equilibrium*, J. Meixner, Ed., North Holland, Amsterdam, 1965, p. 36.

V.S. Paverman and N.A. Papuashvili, Sov. Phys. Semiconductors **19**, 852 (1985).

W. Zawadzki, S. Klahn and U. Merkt, Phys. Rev. Lett. **55**, 983 (1985).

7. Breakdown of Invariants in Nonlinear Systems

L. E. Reichl and W. A. Lin

With 4 Figures

The mechanisms by which intramolecular energy flow is affected by the detailed nonlinear structure of molecules are of considerable importance in a number of fields of science. Two questions of growing interest concern the role of nonlinearity in the functioning of biological molecules, and the feasibility of developing a mode-selective chemistry. For example, one possible means of coherent energy propagation is the soliton. There are, in fact, two distinct types of soliton, topological and nontopological. Topological solitons have been proposed as a mechanism for transport in polyacetylene (Su and Schrieffer, 1980). Nontopological solitons were first proposed by Davydov as a mechanism for energy propagation in the α-helix protein (Davydov, 1979; Hyman et al., 1981) and are the subject of several papers in this volume. Nontopological solitons, strictly speaking, only exist in integrable systems, but might be expected to maintain their integrity for long times in a quasi-integrable system. Quasi-integrability is a fairly recent concept and was first made mathematically rigorous in classical nonlinear systems by KAM (Kolmogorov, 1967; Arnold, 1963; and Moser, 1962). KAM showed that conservative nonlinear integrable systems, when subjected to a weak perturbation that renders them nonintegrable, will continue to exhibit integrable-like (quasi-integrable) behavior (provided conditions for the KAM theorem are satisfied) over a large measure of their phase space. Indeed, quasi-integrable behavior has been found to be quite common at low energy (in most cases) in classical anharmonic nonlinear systems such as the Henon-Heiles (Henon and Heiles, 1964) system. At low energy the anharmonicities have small effect and the ideas of KAM are well verified. Regions of quasi-integrability can also be observed in conservative nonlinear integrable systems perturbed by a dynamic external field. In all cases, the breakdown of quasi-integrability is due to the growth and overlap of resonance zones as some parameter for the system (energy when anharmonicities destroy integrability or external field amplitude in driven systems) is varied. In regions where quasi-integrability has been destroyed, the system begins to exhibit chaotic behavior.

To our knowledge, no KAM theorem has been proven for nonlinear quantum systems. And yet, the existence of quasi-integrable regimes in the phase space of molecules is essential if nontopological soliton propagation is to be a viable means of coherent energy flow in molecules and if mode-selective chemistry is to be developed. As we will show, there is evidence that a KAM-type mechanism for breakdown of invariants does exist in nonlinear quantum systems.

The system we will treat consists of a single particle of mass m trapped in an infinitely deep square well potential, of width $2a$, and driven by a monochromatic external force with amplitude λ and frequency ω. The classical version of this system exhibits a transition due to resonance overlap from quasi-integrable to chaotic behavior, but does so in a very simple manner. Because of this simplicity the transition point in the classical system can be accurately obtained using renormalization group analysis (Lin and Reichl, 1986). However, this system has the further advantage that the quantum version also has a relatively simple structure and as we shall show, a number of analytic results can be obtained for it.

Let us first consider the classical system. The Hamiltonian can be written

$$H = \frac{p^2}{2m} + \lim_{V_0 \to \infty} V_0 \left[\eta\,(x - a) + \eta\,(-x - a) \right] + \lambda x \cos \omega t \tag{1}$$

where p and x are the momentum and position respectively of the particle, and $\eta\,(x) = 1$ for $x > 0$ and $\eta\,(x) = 0$ for $x < 0$. This system is most easily studied in terms of action-angle variables $(J\,,\,\theta)$. If we introduce the canonical transformation

$$x = -a + \frac{2\theta a}{\pi} \operatorname{sign}[\sin\theta]\,, \tag{2}$$

$$p = \frac{\pi J}{2a} \operatorname{sign}[\sin\theta]\,, \tag{3}$$

(sign means sign of the argument) the Hamiltonian takes the form

$$H = \frac{\pi^2 J^2}{8a^2 m} - \frac{4\lambda a}{\pi^2} \sum_{\substack{n=-\infty \\ \text{odd}}}^{\infty} \frac{1}{n^2} \cos(n\,\theta - \omega t)\,. \tag{4}$$

From eq. 4 we see that the effect of the external field is to introduce an infinite number of traveling cosine potential wells into the action-angle phase space of the system. A given cosine potential well can trap those particles traveling at the same speed and thus change the topological structure of the phase space in the region of influence of the potential well. This trapping phenomenon is called a nonlinear resonance zone. Each cosine potential gives rise to a nonlinear resonance centered at a different value of the action variable J. To see this, we note that the nth cosine potential, $\cos(n\theta - \omega t)$, has an associated phase velocity $\dot{\theta} = \frac{\omega}{n}$. The action variable J_n^c for a particle traveling at this speed is determined from Hamilton's equation

$$\dot{\theta} = \left(\frac{\partial H}{\partial J} \right)_n = \frac{\pi^2 J_n^c}{4a^2 m} = \frac{\omega}{n}\,. \tag{5}$$

Thus the cosine potential, $\cos(n\theta - \omega t)$, creates a resonance zone centered at $J_n^c = \dfrac{4a^2 m\omega}{\pi^2 n}$ in action space. The width of the nth resonance zone is determined as follows. The Hamiltonian that describes the behavior of the phase space in the neighborhood of the nth resonance zone (as long as it has not overlapped any of its neighbors) is given by the single-resonance Hamiltonian

$$H = \frac{\pi^2 J^2}{8a^2 m} - \frac{4\lambda a}{\pi^2 n^2} \cos(n\theta - \omega t)\,. \tag{6}$$

In the rest frame of the nth cosine potential, eq. 6 takes the form of the pendulum Hamiltonian. The resonance region corresponds to the region of libration of the pendulum and from eq. 6 is easily shown to have a half-width $\Delta J_n = 8\sqrt{\lambda m a^3}/\pi^2 n$. Thus the particles trapped by the nth cosine potential lie in a region of width $2\Delta J_n$ centered at J_n^c. These results tell us that the resonance zone $n = 1$ is largest and lies highest in action. The remaining zones lie at lower energy as n increases and decrease in size as n increases.

As long as the resonance zones do not overlap, they are separated by so-called KAM surfaces. These are the quasi-integrable regions of the phase space. However, when overlap of two resonance zones occurs all KAM surfaces between them are destroyed and the system is free to wander chaotically throughout much of the region of influence of the overlapping resonance zones. The behavior we have been describing can be seen clearly in Fig. 1, where we display a strobe plot of 23 different trajectories for the case $a = 1$, $m = 1/2$, $\omega = 5$, and $\lambda = 0.05$. Figure 1

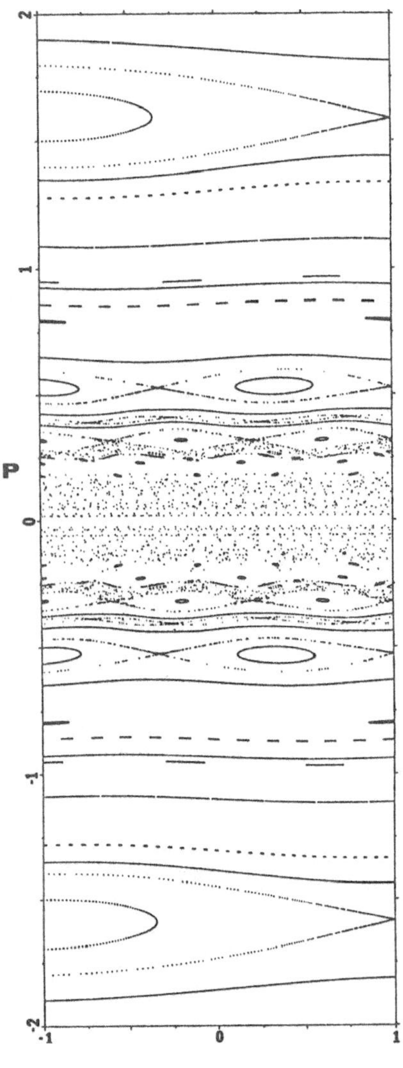

Fig. 1 — Strobe plot of phase space trajectories for the Hamiltonian in eq. 1 with $\lambda = 0.05$, $\omega = 5$, $m = 1/2$, $a = 1$ (Henon and Heiles, 1964).

shows such a plot of (p, x) space, but because of the simple relation between (p, x) and (J, θ) space (cf. eqs. 2 and 3), it is qualitatively similar to what we would see in a strobe plot of (J, θ) space. (To obtain a strobe plot we numerically solve the equations of motion obtained from eq. 1 and plot p and x after each period of the external field.) The resonance zones $n = 1$, 3, and 5 are clearly evident in Fig. 1 and also the chaos resulting from the overlap of the higher-order zones can be seen for small p. The solid lines separating resonance zones are KAM surfaces, i.e. the remnants of integrability of the unperturbed square well system. It is important to note that the picture we have just given of chaos due to resonance overlap does not tell the whole story. There are, in fact, more resonance zones present than just the primary zones that appear explicitly in the Hamiltonian of eq. 2. The 'interaction' of primary zones gives rise to a self-similar network of secondary zones, some of which can be seen in Fig. 1. This self-similar network is the basis upon which renormalization group calculations (Escande, 1982) of resonance overlap are obtained.

Let us now turn to the quantum version of this system (Reichl and Lin, 1986). Does a KAM-type mechanism for breakdown of invariants occur in the quantum system, and if so, how can we understand it? The Hamiltonian governing the behavior of a quantum particle trapped in an infinitely deep square well and driven by a dynamic external field may be written

$$\hat{H} = \hat{H}_0 + \lambda\,(\hat{x} - a) \cos \omega t \;, \qquad (7)$$

where $\hat{H}_0$ is the Hamiltonian for the particle in the square well when no external force is present. We will now assume that the square well lies in the interval $0 \leqslant x \leqslant 2a$. We denote eigenstates of $\hat{H}_0$ by $\{|n>\}$ with $n = 1, 2, 3, \ldots$, and we denote eigenvalues of $\hat{H}_0$ by $E_n^0 = \hbar\Omega\, n^2$, where $\Omega = \hbar\pi^2/8ma^2$ and $\hbar$ is Planck's constant. In the position representation $<x\,|\,n> = a^{-1/2}\sin(n\pi x/2a)$. If we let $|\psi(t)>$ denote the exact state of the system, and let $\psi_n(t) = <n\,|\,\psi(t)>$ denote the probability amplitude to find the system in the state $|\,n>$ at time t, then the Schrödinger equation in the energy picture takes the form

$$i\hbar\,\frac{\partial\psi_n(t)}{\partial t} = \hbar\Omega\, n^2\psi_n(t) - \frac{4a\lambda}{\pi^2}\cos\omega t \sum_{\substack{N=1\\(\text{odd})}}^{\infty} \frac{1}{N^2}\left(\psi_{n-N}(t) + \psi_{n+N}(t)\right) \qquad (8)$$

where we have used the condition $\psi_{-n}(t) = -\psi_n(t)$. If we introduce the generating function

$$F(\theta, t) = \sum_{n=-\infty}^{\infty} e^{in\theta}\,\psi_n(t) \;, \qquad (9)$$

and make use of eq. 8, we obtain

$$ih\,\frac{\partial F}{\partial t} = -\hbar\Omega\,\frac{\partial^2 F}{\partial\theta^2} - \frac{4a\lambda}{\pi^2}\sum_{\substack{N=1\\(\text{odd})}}^{\infty}\frac{1}{N^2}\left(\cos(N\theta - \omega t) + \cos(N\theta + \omega t)\right)F \qquad (10)$$

Both eqs. 9 and 10 are exact. Equation 10 describes the behavior of the system in the angle picture. The similarity with the classical system is clear. The effect of the external field is to induce an infinite number of traveling cosine potential wells in the phase space of the quantum system. Each cosine potential can act as a trap for energy levels, $|n>$. In order to locate the energy levels that are most strongly bound by the Nth cosine potential, we can write the Schrödinger equation (Reichl and Lin, 1986) but retain only one term in the sum. If we then transform to the rest frame of the cosine potential we obtain the Mathieu equation. The most strongly bound state of this Mathieu equation occurs at energy level $\tilde{n}_N = \omega/2\Omega N$. This locates the energy level at which the Nth nonlinear resonance zone in the quantum system is centered. A simple estimate for the half-width Δn_N of the Nth resonance zone can be found by requiring that the maximum energy of a trapped level in the rest frame of the Nth cosine potential, $\hbar\Omega\,(\Delta n_N)^2$, be equal to the depth of the potential well, $8a\lambda/\pi^2N^2$. This yields $\Delta n_N = (64\lambda\, ma^3/\hbar^2 N^2\pi^4)^{1/2}$. The total number of energy levels trapped in the Nth resonance zone is $2\Delta n_n$. Finally, we can obtain a rough estimate of the external field amplitude, λ_c, at which the resonance zones N and $N+2$ will overlap. The overlap condition is $\tilde{n}_N - \tilde{n}_{N+2} = \Delta n_N + \Delta n_{N+2}$. This yields $\lambda_c = ma\omega^2/4\,(N+1)^2$. We thus have obtained the following picture of the effect of the dynamic field on the particle in the square well. When $\lambda = 0$ the quantum numbers, n, are "good" quantum numbers. If the particle is initially in state $|n>$ it will remain there. However, when we turn on the external field, its effect is to create nonlinear resonance zones that can trap those levels $|n>$ within the region of influence of the resonance zone. Probability that was originally in one of the trapped levels in a given zone is free to spread among other levels in that zone since the state of the system oscillates in the cosine well. However, if a theorem analogous to the KAM theorem holds in the quantum picture, probability will be blocked by KAM invariants from flowing between two different resonance zones as long as overlap has not occurred. It is interesting to test these ideas. We have solved, numerically, the Schrödinger equation in which we have retained resonance zones $N = 1$ and $N = 3$. In the energy picture, this can be written

$$i\hbar \frac{\partial\psi_n(t)}{\alpha t} = \hbar\Omega\, n^2\psi_n(t) - \frac{2a\lambda}{\pi^2}\left[e^{-i\omega t}\psi_{n-1}(t) + e^{+i\omega t}\psi_{n+1}(t)\right] - \frac{2a\lambda}{9\pi^2}\left[e^{-i\omega t}\psi_{n-3}(t) + e^{+i\omega t}\psi_{n+3}(t)\right]. \quad (11)$$

The parameters in eq. 11 were chosen to have values $a = 10a_B$, $\omega = (2\pi)^2/1.692\tau_B$, $\lambda = \Lambda e^2/a_B^2$ (with $\Lambda = 4$ and 9), and $m = m_e$, where a_B is the Bohr radius, τ_B is the period of the first Bohr orbit, e is the electron charge, and m_e is the mass of the electron. Our theory predicts that the resonance $N = 1$ is located at $\tilde{n}_1 \approx 150$ and has half width $\Delta n_1 \approx 51$ when $\Lambda = 4$ while resonance $N = 3$ is located at $\tilde{n}_3 \approx 50$ with half width $\Delta n_3 \approx 17$. In Figs. 2 and 3, we show solutions to eq. 11 for $\Lambda = 4$.

In Fig. 2, our initial condition was $\psi_{130}(0) = 1$ and $\psi_n(0) = 0$ otherwise, while in Fig. 3 our initial condition was $\psi_{55}(0) = 1$ and $\psi_n(0) = 0$ otherwise. In these figures we see the spread of probability after about 49 external field periods. The probability has reached its maximum spread. We see that the resonance zones are well separated. Our theory predicts that overlap can occur at $\Lambda = 8.6$. In Fig. 4

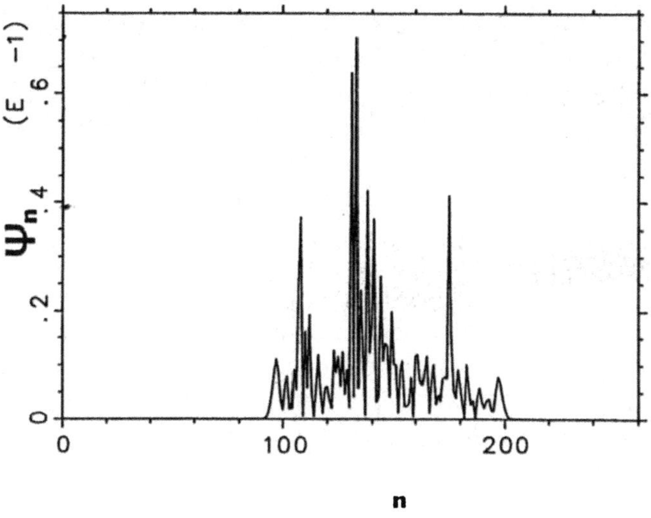

Fig. 2 — Solution to eq. 11 after time $t \approx 49$ $(2\pi/\omega)$ for $\Lambda = 4$, $\psi_{130}(0) = 1$ and $\psi_n = 0$ otherwise.

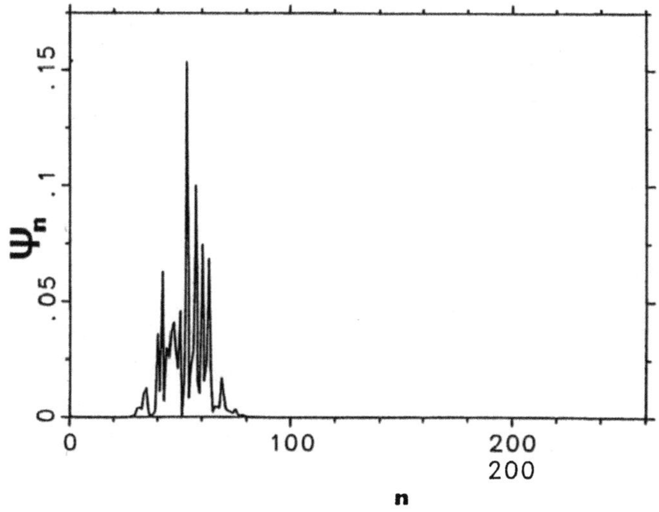

Fig. 3 — Solution to eq. 11 after time $t \approx 49$ $(2\pi/\omega)$ for $\Lambda = 4$, $\psi_{55}(0) = 1$ and $\psi_n(0) = 0$ otherwise.

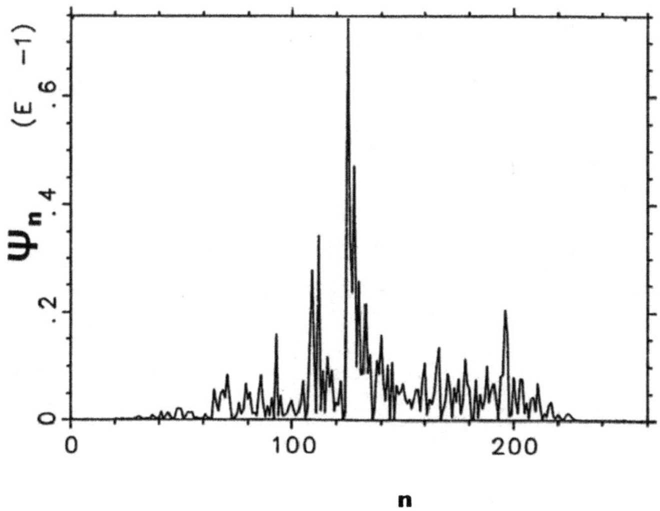

Fig. 4 — Solution to eq. 11 after time $t \approx 16$ $(2\pi/\omega)$ for $\Lambda = 9$, $\psi_{130}(0) = 1$ and $\psi_n(0) = 0$ otherwise.

we show the solution to eq. 5 for $\Lambda = 9$ and initial conditions $\psi_{130}(0) = 1$ and $\psi_n(0) = 0$ otherwise, but after about 16 periods of the external field. In all cases 250 amplitudes ψ_n were used in the integration and error in normalization remained below 6×10^{-5}. In this case overlap has definitely occurred and the probability is free to spread throughout the region of influence of the two resonance zones. (This type of behavior has also been observed by Berman, Zaslavsky, and Kolovsky (1981, 1982).)

The results we have obtained show that for nonlinear quantum systems a KAM-type mechanism for the breakdown of invariants ("good" quantum numbers) appears to exist. As we mentioned earlier, this same nonlinear resonance overlap mechanism is responsible for the onset of chaos in nonlinear conservative classical systems. We thus might expect that in some molecules, and especially highly symmetric molecules, there are regions of energy where the quantum system exhibits quasi-integrable behavior and soliton propagation can be maintained. However, this is a subject which still remains to be explored.

ACKNOWLEDGEMENT

The authors thank the U.S. Naval Air Systems Command HQ (Contract MDA903-85-C-0029) for support of this work.

REFERENCES

V.I. Arnold, Russ. Math. Surv. **18**, 985 (1963).

G.P. Berman, G.M. Zaslovsky, and A.R. Kolovsky, Sov. Phys. JETP **54**, 272 (1981).

G.P. Berman, G.M. Zaslovsky, A.R. Kolovsky, Phys. Lett. **87A**, 152 (1982).

A.S. Davydov, Phys. Scripta **20**, 387 (1979).

D.F. Escande, Phys. Scripta **T2**, No. 1, 126 (1982).

M. Henon and C. Heiles, Astron. J. **69**, 73 (1964).

J. Hyman, D. McLaughlin, and A. Schott, Physica **3D**, 23 (1981).

A.N. Kolmogorov, in *Foundations of Mechanics*, R. Abraham, Ed., W.A. Benjamin, New York, 1967, Appendix D.

W.A. Lin and L.E. Reichl, Physica **19D**, 145 (1986).

J. Moser, Nachr. Akad. Wiss. Göttingen II, Math. Phys., Kl. **1962**, 1 (1962).

L.E. Reichl and W.A. Lin, Phys. Rev. A: Solid State **33**, 3598 (1986).

W. Su, J. Schrieffer, and A. Heiger, Phys. Rev. B: Solid State **22**, 2099 (1980).

8. Dynamics of Interacting Electrons in Aperiodic Solids

M. Pollak

With 2 Figures

ABSTRACT

The paper addresses the dynamics of electrons in solids that lack the translational symmetry of crystals. It is shown that under certain conditions, small changes in temperature can change the response time of the electronic system by many orders of magnitude. The focus of the paper is on conditions under which the electronic eigenstates are localized, when Coulomb interactions between electrons in different states are important, and when the electronic system is near thermodynamic equilibrium. The Coulomb interactions can enhance greatly the times required for relaxation, owing to the need for collective transitions at low temperature. It is found useful to define a logarithmic time, partly because it allows one to assign reasonably sharply defined values to typical experimental times, or to typical biological lifetimes. Similarities and differences with spin glasses, systems that are frequently reported on in the literature, are discussed. In a speculative way, the paper explores whether the effects considered here may be useful for biological or technological memory or energy storage devices, particularly for small systems, and those far from equilibrium.

INTRODUCTION

In recent years there has been very considerable interest in so-called disordered solids. The stimulation of interest came partly for technological reasons, e.g. the prospects of solar energy utilization using amorphous silicon. However, much of the interest came from a desire to see what new properties to expect when the traditional confinement of solid state physics to translational symmetry is abandoned.

Perhaps the most important development in this area came with the works of Anderson (1958) and Mott (1967), who showed that the electronic states in noncrystalline solids can be either localized (i.e. confined to small volumes), or delocalized (i.e. extended more or less uniformly throughout the material). This contrasts with periodic solids, where electronic states are compelled by symmetry to be Bloch states, and thus be delocalized.

This paper is concerned with the dynamic properties of electrons in solids lacking translational symmetry. A striking feature here is the existence of an extremely broad spectrum of computed response times, which can be much longer than any experimental times. For example, Tauc's group at Brown University, and Kastners' group at MIT, measure similar processes in similar materials in the picosecond regime and hour regime, respectively. The very long response times and their possible application are the main topic of this paper.

It has, of course, been known for a long time that solids such as glasses are frozen in nonequilibrium configurations, unable to relax to the crystalline equilibrium state during any experimentally realizable time. It is worth mentioning also that Schrödinger (1944) had postulated that such a phenomenon can have far-reaching consequences in biology and can constitute the mechanism behind genetic memory and its stability. The discovery of DNA and its function later confirmed this picture. The term aperiodic solid, which will be used here to emphasize the lack of translational symmetry, was coined by Schrödinger.

While the existence of extremely long relaxation times of *atoms* in glasses has been a familiar phenomenon, *electrons* in solids have generally been considered to be rapidly responding, on the scale of usual experimental times. However, the response of electrons in aperiodic solids can also be slow, particularly (but not exclusively) when Coulomb interactions are important. The basic physical reason for the slow response is the lack of translational symmetry, or, more directly, localization. The lack of translational symmetry can be either intrinsic, e.g. in a glass, or due to symmetry breaking, e.g. in the case of a polaron (Fröhlich, 1954; Holstein, 1959). This paper focuses on the first case.

SOME BASIC CONSIDERATIONS

To compare transition rates between localized states in aperiodic solids to transition rates between band states in crystals, we first state some simple conditions. Localized states are assumed to be singly occupied owing to a large Hubbard energy. Transitions are assumed to be effected by phonons, via a deformation potential, and the dominant processes are assumed to be single-phonon absorption, or emission, so that the "golden rule" can be used for evaluating transition rates. The band functions have the form $\phi_k(r) = f(r)\exp(ik.r)$, and the localized states are described by $\phi_i(r) = F(r)\exp[-|(r_i - r)|/a]$, where $f(r)$ is a periodic function varying rapidly on an atomic scale; $F(r)$ is also a rapidly varying function, k is the usual k-vector, r_i is the position of the center of the localized state ϕ_i, and a is the localization radius. The perturbation Hamiltonian to be used in the golden rule is $p.g\exp(ik'.r)$, where p is the deformation potential tensor, g the amplitude of the strain tensor due to the phonon with wave vector k', and the dot denotes a scalar product. The rapidly varying functions f and F do not contribute to the matrix elements $<\phi_k|V|\phi_{k''}>$, $<\phi_i|V|\phi_j>$. The important part of the matrix element for the localized states then is $<\exp[-|(r_i - r)|/a] V\exp[-|(r_j - r)|/a]>$. The square of this matrix element, needed for the golden rule, is proportional to $\exp[-2r_{ij}/a]$, where $r_{ij} = |r_i - r_j|$. The analogous matrix element squared for the band case is unity, because conservation of crystal momentum requires $k + k' + k'' = 0$. The conclusion is that the transition rate w_{ij} between localized states i, j, as compared with the transition rate w_{kk}, between two band states k, k'', is reduced by a factor $\exp[-2r_{ij}/a]$, which can, of course, be extremely small. Another conclusion is that the spread of w_{ij}'s is enormous, because of the exponential dependence on r_{ij}, which is nonuniform in an aperiodic solid, and typically several times as large as a, in an Anderson (1958) localized system.

The above discussion omitted the consideration of the density of final states. This is proportional to the number of such phonons present that can effect the transition between i and j (or between k and k''). This consideration introduces a factor $\approx \exp(-\Delta_{ij}/kT)$ for transitions upward in energy (for $\Delta_{ij} > kT$), and a factor of unity for transitions downward in energy. Here Δ_{ij} is the energy difference between the states i and j, i.e., $\Delta_{ij} = E_i - E_j$. It should be noted that while there is obviously no activation energy for downward transitions, these can still be very slow because of the factor $\exp(-2r_{ij}/a)$.

While we have discussed only a specific type of transition, similar conclusions apply to other types. Common examples are multiphonon transitions where the electron can jump over a barrier, in a semiclassical fashion, or transitions between localized polaronic states. In the first case, the tunneling factor $\exp(-2r_{ij}/a)$ is absent, but there is an activation factor even for hops downward in energy, due to the barrier, so downward transitions can again have very low rates, and a very large spread if the barrier energy is a random variable (Pollak and Pike, 1972). The polaron transition rates are more complicated (Schnakenberg, 1968; Emin, 1974; Robertson and Friedman, 1976); at high temperatures there is an electronic tunneling factor and an activation factor for the atomic motion over a barrier along the configuration coordinate, and at low temperatures the atoms also tunnel across the barrier in configuration space. Downward transitions then have no activation factor, but there are two tunneling factors. The conclusion that downward transition rates, important in relaxation to equilibrium, can be very low, and have an extremely wide distribution, thus seems to be a general feature of localized electrons in aperiodic solids. In the next section, a more detailed discussion of the dynamics of noninteracting localized electrons in aperiodic solids is presented.

DYNAMICS OF NONINTERACTING ELECTRONS

The broad spectrum of relaxation is observable in various experiments, notably ac conductivity, and so-called dispersive transport. The conductivity is a fractional power of the frequency down to frequencies well below 1 Hz (for compilation of experimental data see Jonscher (1977)). The nonanalytic behavior of the conductivity at these low frequencies is a clear demonstration that relaxation rates comparable with such frequencies are still important. Dispersive transport experiments are of several kinds; basically they are a time-domain analog of the ac conductivity, but with the important difference that they relate to processes far from equilibrium. The various quantities measured in these experiments, e.g. photo-induced currents, exhibit a power law dependence on time (Pfister and Scher, 1978; Tauc, 1982; Street, 1982; Marshall, 1983; Morigaki, 1985), up to very long times (sometimes hours), indicating again that transition rates comparable to such inverse long times are still important.

Recently it has been suggested (Pollak and Ortuno, 1982; Davies et al., 1982; Grunewald et al., 1982) that under certain conditions the electrons can never reach equilibrium. (This condition has been referred to as describing the electron glass). In the following, a method to study such systems is developed. It will be confined to near-equilibrium conditions. It will also be assumed that the process at hand is that for which the transition rates are (Miller and Abrahams, 1960)

$$w_{ij} = \exp(-2r_{ij}/a)\exp(-\Delta_{ij}/kT) \quad (1)$$

for upward transitions, and

$$w_{ij} = w_0 \exp(-2r_{ij}/a) \quad (1a)$$

for downward transitions. Such an assumption does not restrict the generality of the method. The important element is that w_{ij} be an exponential function of random variables, in the above case r_{ij} and/or Δ_{ij}.

To investigate approach to equilibrium, it can be imagined that the temperature of a thermal bath is lowered very gradually, and the ability of the system to follow the reduction in temperature studied. Such a process simulates common experiments on real glasses. In the procedure followed here, the lowering of the temperature will occur in small steps ΔT, such that $\Delta T/T << 1$. This assures that conditions "near equilibrium" prevail. Imagine then that at time $t = 0$, the temperature is lowered from T_0 to $T_0 - \Delta T$, and that the system at $t = 0$ was in equilibrium, i.e. that the occupation of states i obeyed

$$f_i^0 = \left[\exp(E_i/kT_0) + 1\right]^{-1} \quad (2)$$

where E_i is the energy of state i. The approach to new equilibrium is governed by the rate equations

$$df_i/dt = \sum_j \left[f_j(1-f_i)w_{ji} - f_i(1-f_j)w_{ij}\right] = \sum_j (u_{ji} - u_{ij}) \quad (3)$$

with eq. 2 for initial conditions for the time-dependent occupation functions f. The "near equilibrium" conditions assure that $|f_k(t) - f_k^0| = \delta f_k(t) << f_k^0$. The quantities u_{ij} represent the actual transition rates, since they are weighted by the probabilities that i is occupied, and j empty, as is required for transitions from i to j to take place. In equilibrium, detailed balancing requires $u_{ij}^0 = u_{ji}^0$. Near equilibrium, $|u_{ij} - u_{ji}| << u_{ij}^0$. Since both the f's and the w's are exponential functions of random variables, one can define a random variable x_{ij} such that $u_{ij} = w_0 \exp(-x_{ij})$. For the process described by eqs. 1 and 1a, one has

$$x_{ij} = ln(w_0/u_{ij}) = 2r_{ij}/a + E_{ij}/kT, \quad (4)$$

where (Ambegoakar et al., 1971)

$$E_{ij} = \left[|E_i| + |E_j| + |E_i - E_j|\right]/2 \quad (5)$$

with the energies E_i, E_j measured from the thermodynamic potential.

It is also useful to define a logarithmic time

$$y = ln(w_0 t) \quad (6)$$

that has several desirable properties:

(1) y is compatible with the quantities x_{ij} (see eq. 4).

(2) the probability $w_0 \exp(-u_{ij}t)$ that a transition happened at $t < 1/u_{ij}$, plotted versus y, resembles a step function, with the value 1 for $t > (1/u_{ij})$, and 0 for $t < (1/u_{ij})$. Thus, on the scale of y it is a good approximation to assume that none of the transitions with $x_{ij} > y$, and all the transitions with $x_{ij} < y$, happened at y.

(3) On the scale of y, the concept of "experimental time," or of "biological time" y_e, corresponds to reasonably sharply defined values of y. For example, with $w_0 = 10^{13}\ s^{-1}$, y for 1 min is 31, and for 10 years is 46, which puts y_e in a rather small range.

(4) The quantity $\beta = (1/kT)$ appears directly in x_{ij}, and is therefore more compatible with y than with t. There must exist some set of transitions needed to bring the systems from a nonequilibrium state specified by the set of occupations $\{f_i\}$ to the equilibrium state specified by $\{f_i^0\}$. This set is not necessarily unique. Of interest is the set for which the largest $x_{ij} = x_m$ is as large as possible. If $x_m < y_e$, then equilibrium conditions are observable. An important point is that in contrast to the limited range of y_e, β , and thus x_{ij}, can be varied through a very large range. Thus there may exist a realizable β_g such that $x_m < y_e$ for $\beta < \beta_g$, and $x_m > y_e$ for $\beta > \beta_g$. One can then define the glass temperature $T_g = (1/k_{\beta_g})$. It is noted that since y_e is not sharply defined, neither is T_g. This is in keeping with the definition of T_g in most work on real glasses. (However, the above differs from the definition of the glass temperature usually used for spin glasses (Edwards and Anderson, 1975)).

The time evolution of the system is given by eq. 3 and by the initial conditions, eq. 2. However, condition (2) above allows (Pollak, 1984) us to obtain the time evolution in y without explicitly solving eq. 3. It occurs in such a way that transitions with $x_{ij} < y$ happen at or before y. This leads to a time-dependent bond percolation problem, as illustrated in Fig. 1. One can imagine "bonds" being formed as y progresses between all pairs of states i, j such that $x_{ij} < y$. With increasing y, more and more bonds are formed which interconnect sites into clus-

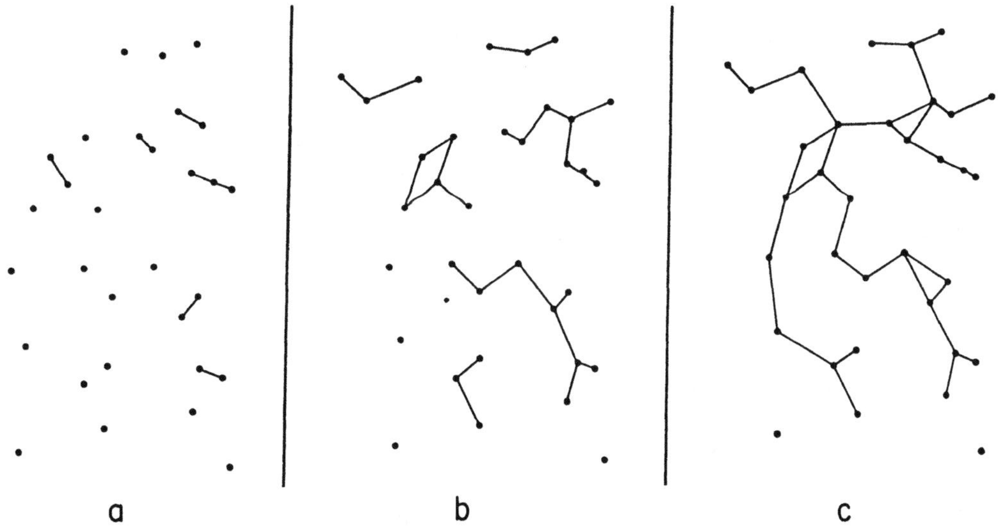

Fig. 1 — Schematic illustration of the time-dependent percolation process that describes relaxation to equilibrium for localized noninteracting electrons in aperiodic solids. The dots represent local electronic states, and the lines connect states between which transitions can occur at or before the logarithmic time y, explained in the text. The parts a, b, c, represent three different times; (a) $y << y_c$, (b) $y < y_c$, (c) $y > y_c$, where y_c is the time at which a macroscopic size cluster first becomes interconnected.

ters. States belonging to the same cluster are in local equilibrium with each other. There exists a value x_c of x_{ij} at which a macroscopic cluster suddenly occurs; x_c is calculable from the critical percolation probability, and from the distribution of x_{ij}. The occurrence of the macroscopic cluster happens at time $y_c = x_c$, and this is realizable if $y_c < y_e$. The realizability depends on T, since $x_c = x_c(T)$.

The point $y = y_c$ is physically of particular interest; at $y > y_c$ a finite fraction of the system, spanning the entire system, is in equilibrium (Pollak and Hunt, 1985). Although total equilibrium is not reached until every state is connected to the macroscopic cluster, equilibrium properties should become observable at $y = y_c$. For example, dc conduction can occur at this point, as can diffusion over macroscopic lengths, and the specific heat reaches its equilibrium value. The point is also interesting because various quantities connected with the percolation problem, such as cluster size, correlation length, or order parameter, which represent physical quantities, exhibit a critical behavior at x_c. For example, the polarizability is related to the correlation length, which, at critical percolation, is known (Stauffer, 1979) to have the critical exponent ≈ -0.85. One thus expects to observe (Pollak and Hunt, 1985) nonanalytic behavior of certain properties at $y = y_c$. It must be realized, however, that actual singularities are an artifact of the approximation stated in condition 2 above. On the scale of y the approximation is a very good one, but the discontinuities or divergences at y_c are slightly rounded off. Nevertheless, sharp features in the vicinity of y_c, such as critical exponents, should be observable. On the other hand, the scale of t expands the neighborhood of y_c so much that no sharp features are, under normal circumstances, expected on this scale. But the β scale is compatible with the y scale. As shown below, the same critical exponents should be expected at T_g as at y_c. Consider measuring some property for a fixed duration y_e, at different temperatures. The quantity of interest near the critical point is $(y_e - y_c)$. Since $x_{ij}(T)$ is generally a decreasing function of T, $dx_{ij}/dT < 0$. All bonds at y_e are characterized by $x_{ij} < y_e$. Denoting the largest of these by x^*, one has

$$y_e - y_c = x^* - x_c = x^* - (dx_{ij}/dT)_{x^*}(T_g - T) - x^* = -A(T - T_g), \qquad (7)$$

where $-A$ is the value of the derivative in eq. 7, and x_c has been expanded to first order in $(T - T_g)$ around x^*. Thus,

$$|T - T_g|^S \propto |y_e - y_c|^S, \qquad (8)$$

as stated.

DYNAMICS OF INTERACTING ELECTRONS

The above one-electron description becomes inapplicable when Coulomb interactions between electrons are important. To proceed, one must first have a description of the many-electron states. A possible starting point are the solutions O_i of the Hamiltonian without interactions. If a_1^+, a_k are the creation and annihilation operators of ϕ_1, ϕ_k, respectively, then the many-electron states can be described by

$$\Psi_I = A_{IK} S_K, \qquad (9)$$

$$S_K = a_i^+ a_j^+ \cdot\cdot\cdot\cdot\cdot a_m a_n S_0, \quad i \neq j \neq \cdot\cdot\cdot\cdot\cdot \neq m \neq n$$

where the capital letters denote many-electron configurations. The S's are Slater determinants formed of the ϕ's, S_0 is some reference configuration, usually chosen to be a lowest-lying S, and K specifies $\{i,j,\dots.m,n\}$. The number of a^{+}'s and a's in each term must be the same. Over a large regime of conditions there exists a semiclassical regime (Pollak and Ortuno, 1986), where for each I only one A_{IK}, say A_I, is significant, i.e., $A_I^* A_I \simeq 1$, while for all other K, $A_{IK}^* A_{IK} << 1$. Hence

$$\Psi_I \simeq A_I S_I \tag{10}$$

which is the same description as for a many-electron state without interactions. Although the small A_{IK} are (in the semiclassical regime) not important for the description of the many-electron states, they are crucial for the calculation of the (many-electOron) transition rates (Pollak,1981; Gosar, 1983a, 1983b; Pollak and Ortuno, 1985a,b).

While the states are very similar to the states in the absence of interactions, their energy levels are quite different. More importantly for the present purpose, the relative ordering of the energy levels is altered by the interactions, and therein lies the major cause for the large effect that interactions have on the electron dynamics.

Of main interest for relaxation, at usual temperatures and near equilibrium, are the low-lying states. In the absence of interactions these tend to be single-electron excitations, because without interactions the energy of a multi-electron excitation is the sum of the individual excitation energies of the excited electrons. (We mention in passing that the matrix element for multi-electron transitions in absence of interactions vanishes in any case, because the electron-phonon Hamiltonian causing transitions is a one-electron operator.) Furthermore, the low-lying states tend to be long range (i.e., large r_{ij}) excitations, since the configurational phase space favors them (Mott, 1968).

In the presence of interactions, long-range one-electron excitations tend to have *large energies* because such excitations cross the Coulomb gap (Pollak, 1970; Srinivasan, 1971; Efros and Shklooskii, 1975). The physical reason for this is similar to that of the the formation and motion of polarons. The interaction between electrons generates a polarization atmosphere around an electron by repelling other electrons. A long-range, *single-electron* transition does not permit the polarization atmosphere to move with the electron. The analogous motion of a polaron corresponds to a transfer of the electron with no change in the configurational coordinate (Fröhlich, 1954; Holstein, 1959). In contrast, the motion of a polaron requiring the lowest possible energy involves tunneling along the configuration coordinate. The analog here is a multi-electron transition shown schematically in Fig. 2. To comprehend the dynamics of interacting electrons in aperiodic solids, an important difference between this system and polarons must be understood. While in the case of polarons the "dressing" involves particles distinguishable from the electron, here the dressing is also by electrons. The result is that exchange is very important (as shown in Fig. 2), and affects greatly the transition rates (Pollak, 1981; Gosar, 1983a, 1983b; Pollak and Ortuno, 1985a,b).

To evaluate the electron dynamics one must know the many-electron spectrum, and the many-electron transition rates. An accurate theory exists for the latter, but the former is a formidable problem, and the spectrum is currently not

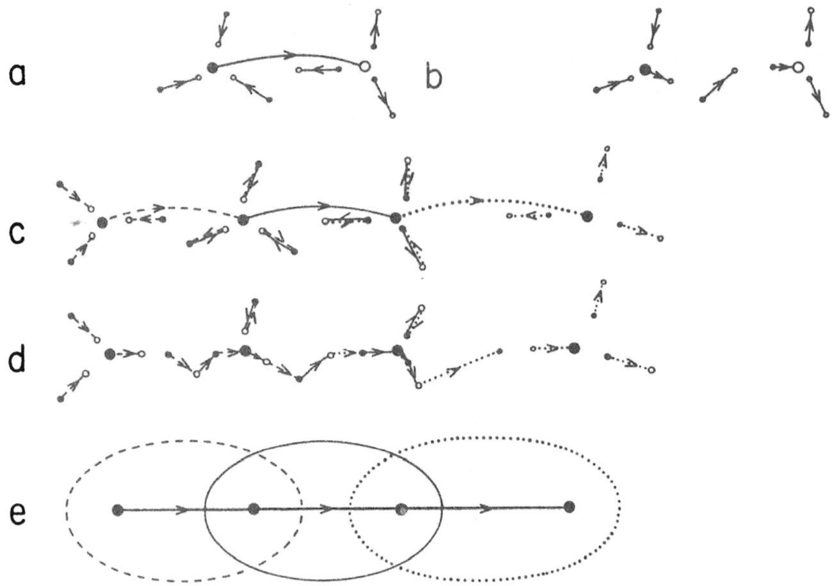

Fig. 2 – Schematic illustration of how equilibration of interacting electrons in the semiclassical regime may occur by many-electron transitions. (a) A polaron-like transition, where an electron moves between two sites, carrying with it a polarization atmosphere (or part of it) formed by other electrons. (b) A more favorable exchange configuration of the identical transition, in which the sum of the transition distances is minimized (the cascade configuration). (c) A sequence of three polaron-like transitions. (A sequence of macroscopic dimension is needed for equilibration. The parts belonging to the three transitions are distinguished by different types of lines.) (d) The same set of three transitions, in their cascade-type exchange configuration. (e) Three overlapping volumes that contain the three many-electron transitions. This part is intended to illustrate the basis for the approximate percolation theory mentioned in the text.

known. Various calculations in the literature use model spectra (Pollak and Hunt, 1985; Hunt, to be published). Probably the simplest model assumes that n-electron excitations are uniformly distributed over an n-dependent energy regime. While such an assumption has no real basis, it is convenient for calculations. Computations to obtain realistic spectra are being attempted.

The theory of the n-electron transition rates w_{IJ} is quite complex, so only the result is presented here (Pollak, 1981; Gosar, 1983a, 1983b; Pollak and Ortuno, 1985a,b):

$$w_{IJ} = w_0\, g^{n-1} \exp(-2r_{ij}/a) \exp(-E_{IJ}/kT)\,. \tag{11}$$

In eq. 11 the sum is over the n distances between the positions of the electrons in the states I and J. There are actually $n!$ such sums, corresponding to the different exchange configurations of the transition; the correct one to use in eq. 11 is the sum with the smallest value. Usually the individual distances in such a sum will be comparable, so it is convenient to write $r_{ij} \approx nr'$, where r' is some characteristic transfer distance. E_{IJ} is the energy difference between the states I and J. The factor $\exp(-E_{IJ}/kT)$ is absent in a downward transition. The quantity g involves interaction energies, random energies, and resonance energies, and is a measure of the relative importance of the Coulomb energy. It vanishes when the interaction vanishes. Note that eq. 11 then correctly predicts that only one-electron transitions are allowed. The prefactor, w_o', is a weak function of several arguments in comparison with the exponential dependences. For $n=1$, $w_0' = w_0$, and eq. 11 agrees with eq. 1. Equation 11 shows that the transition rate for many-electron transitions decreases very rapidly with the number of electrons participating in the transition.

The equation analogous to eq. 3 that describes the time evolution is the Pauli Master Equation,

$$df_I/dt = \sum_J (f_J w_{JI} - f_I w_{JI}) = \sum_J (u_{JI} - u_{IJ}). \tag{12}$$

An exclusion term analogous to $1-f_j$ (say) in eq. 3 is clearly not needed here, since the system can be in only one many-electron state at a time. The initial condition analogous to eq. 2 is here

$$f_i^0 = \exp(-E_I/kT_0) / \sum_I \exp(-E_I/kT_0) \tag{13}$$

Specification of "near equilibrium" is also similar to the noninteracting case, $|u_{JI} - u_{IJ}| << u_{JI}$, and $|f_I^0 - f_I| << f_I^0$.

It is again possible to develop a time-dependent percolation theory (Pollak and Ortuno, 1985b) for multi-electron transitions to describe the time evolution of the system, though this theory is at present not as well founded as the one-particle percolation theory. Figures 2c, d, e are helpful in visualizing the basis for such a theory. The n-electron excitations are confined to n-dependent volumes. To obtain percolation, a continuous path of such volumes, partially overlapping with each other (see Fig. 2e), must exist. Approximate theories using this picture were developed (Pollak and Hunt, 1985; Hunt, to be published), and show again that a temperature T_g, as defined above, exists. For the specific case examined (Hunt, to be published), T_g for the interacting system was about two orders of magnitude larger than T_g for a similar system without interactions. It is worth stressing that the results apply only to a model many-electron excitation spectrum, which may not be realistic. In addition, a certain feature of the dynamics of interacting electrons is neglected, namely the fact that interaction may correlate successive many-electron transitions (Pollak and Ortuno, 1986). Such correlations are hopefully not very important for large n, i.e. at very low T. However, if such correlations are important, they should raise T_g even higher.

SOME COMMENTS AND SPECULATIONS

The discussions in the previous sections brings out two important points: the time of relaxation to equilibrium of electrons in aperiodic solids can be extremely long, and it can be dramatically shortened by a small change in T near T_g. From the practical viewpoint this implies a possible mechanism for producing memory that can be reset by a slight change in temperature, or a possible energy storage mechanism in which the energy can be released by a small change in T. While these properties were discussed for "near equilibrium" conditions, this restriction is not necessary, and was used mainly for convenience. When initial conditions are far from equilibrium, the relaxation to equilibrium must proceed via "near equilibrium" conditions, so the relaxation cannot be shorter.

It is of interest to list experimental evidence for the various effects that were discussed above. Probably the most direct evidence for very slow relaxation comes from various dispersive transport experiments. These include (Pfister and Scher, 1978; Tauc, 1982, Street, 1982; Morigaki, 1985) time-resolved photoconductivity, photoluminescence, photoinduced absorption, and time-of-flight experiments. Usually, the initial conditions are far from equilibrium. The experiments generally begin with a rapid light-induced electron-hole generation, so $|f_i(t=0) - f_i^0| << f_i^0$ is not satisfied for all i. Such experiments demonstrate extremely well both the very broad spectrum of relaxation times, and the existence of very long relaxations. Processes from picoseconds to hours are observed in similar materials, as already mentioned. Another group of studies that clearly demonstrates the enormous spectrum of relaxation rates, including very slow ones, are ac conductivity experiments (Jonscher, 1977). These experiments are usually performed in the linear response regime, i.e. near equilibrium, and in systems where interactions are not important. One system where there seems to be overwhelming (though not unanimous) agreement (Shklovskii and Efros, 1979; Böttger and Bryskin, 1985; Efros and Pollak, 1985) about the importance of interactions is impurity conduction. Evidence for the importance of very slow processes in these systems exists from ac conductivity experiments (Pollak and Geballe, 1961; Paalanen et al., 1986).

The predicted glass transition per se has not yet been observed, though experiments to find it are under way in several laboratories. However, glass transitions have been observed in the related systems of spin glasses (van Hemmen and Morgenstern, 1983), and, as will be argued below, the conditions for a glass transition are enhanced in the systems considered here, as compared to spin glasses. These consists of randomly positioned spins that interact with each other, usually via exchange. The Hamiltonian for the spin system is

$$H = \sum_{i<j} j_{ij}\, s_i\, s_j + mB \sum_i s_i \tag{14}$$

where j_{ij} are coupling constants with random magnitudes and, possibly signs, between nearest neighbor spins s_i, s_j. B is an external magnetic field, and m is the magnetic moment. The Hamiltonian for the system under discussion can be formally written in the same way, in the semiclassical regime, if one makes the proper choices for j, s, B, and m. The choices depend somewhat on the particular model. Most directly, eq. 14 can be used for a model in which the states i are randomly located, all $E_i = E$ are identical, there are half as many electrons as states, and there is a (positive) charge of $-e/2$ at the site of each state to satisfy overall neutrality. Then $j_{ij} = e^2/\kappa r_{ij}$, $s_i = -1/2$ when i is occupied by an electron, $1/2$ when unoccupied, and $E = mB/2$; κ is the dielectric constant. Models more representative of real systems have E_i randomly fluctuating about a mean value E, and the ratio K of electrons to states is $0 < K < 1$, not necessarily $K = 1/2$. The fluctuating E_i adds an extra term $m \sum_i B_i\, s_i$ to eq. 14, where $B_i = 2(E_i - E)$ can be considered to be a randomly fluctuating field. If $K \neq 1/2$, charge neutrality requires in this model a charge $-e[1/2 + (K - 1/2)] = -e(1/2 + b)$ at each site. So long as $s_i = \pm 1/2$, the extra terms $(e^2 b/\kappa) \sum (1/r_{ij})\, s_i + (b^2 e^2/\kappa) \sum (1/r_{ij})$ appear in eq.

14. The second term is a constant, does not contribute to physical effects, and can be left out. The first term can be written $mB'\sum s_i + m\sum B'_i s_i$, where $B' = (e^2 b/m\kappa)\sum < 1/r_{ij} >$, $B'_i = (e^2 b/m\kappa)\sum (1/r_{ij} - < 1/r_{ij} >)$. B′ and B'_i play roles of an applied field, and another random field, respectively.

Other models exist in the literature where the compensating positive charge is distributed differently. E.g., in a model appropriate for p-type impurity conduction, charges $-e$ are at random locations outside the positions of the localized electronic states. In a model for inversion layers, the compensating charge is continuously distributed. A frequently used model has the states and the compensating charges on a lattice, the aperiodic nature of the system arising from the random E_i. The essential features of all these models are similar, and they should have substantially the same behavior.

One can now summarize the formal differences between the spin glass and the interacting localized electrons.

(1) the interaction j_{ij} in the spin glass is taken to be effective only between nearest neighbors, while the present analog, $e^2/\kappa r_{ij}$, has a long-range effect.

(2) There exists a randomly fluctuating field $B + B_i'$ at every site, which is normally absent in spin glasses. However, certain models for spin systems, the so called random-field models (see review by Young and Bhatt, 1986), use such a potential.

(3) In considering the dynamics of the two systems, it is very important to realize that in the spin systems there are no constraints on the flipping of individual spins, other than energy considerations. In contrast, the transition of an individual electron mandates the flip of two "spins", so that $\sum_i s_i$ is conserved, as required by particle conservation.

Although no exact evaluation of these effects on a glass transition exists, strong arguments can be made that they all enhance the likelihood of a glassy state. The first effect does so by increasing the effectiveness of interactions. The second effect has been studied in its own right (the so called random field model), and one knows that it can produce a glass transition in its own right (Young and Bhatt, 1986). The third effect eliminates the potentially fastest transitions (single spin flips), and cannot but slow down relaxation. In fact, as discussed in a previous Section, and indicated by dispersive transport experiments, it also can alone produce a glassy state. It is probably the most important of the three effects.

It is interesting to speculate whether the effects inherent in the electron glass transition might be utilized for memory, or energy storage, in devices or in biological systems. For this purpose, the system should be studied also far from equilibrium; energy storage may occur at higher energies under such conditions, and memory would be more effective if configurations at high energies could be utilized. Whether or not the possibility to utilize the effect is realistic depends on the details of the many-electron excitation spectrum. The main question is, how many configurations exist from which decay to the vicinity of the ground state must, at very low T, occur by many-electron transitions, so they can act as configurations for storage of energy or memory. Computer investigations addressing this problem are currently under way. (In this context it should be mentioned that related studies on spin-glass models for neural networks have been ongoing for some years (Little, 1974; Shaw and Vasudevan, 1974; Hopfield, 1982), with encouraging results.) Other important concerns are how such states could be reached selectively, how they can be detected (i.e. "input and output" considerations), and how effectively the memory can be reset by various possible means, e.g. a slight change of T, or a pulse of a strong electric field.

These problems are now under investigation in extensive computer studies. The main difficulty in these studies is the exponential increase in the number of configurations with system size. However, this very problem that makes computer studies difficult also provides hope that relatively small aperiodic units with interacting electrons, perhaps the size of macromolecules, or submicron semiconductor units attainable with modern technology, can sustain the long-lived nonequilibrium states. Conceivably, such small units could be coupled with each other to perform complex logical functions.

I am grateful to the editors of this volume for inviting me to contribute to a volume in honor of Professor Fröhlich. I also acknowledge the collaboration with my colleagues M. Ortuño, A. Hunt, and B. Hadley, on which this paper is based.

REFERENCES

V. Ambegoakar, B.I. Halperin, and J.S. Langer, Phys. Rev. B: Solid State **4**, 2612 (1971).

P.W. Anderson, Phys. Rev. **109**, 1492 (1958).

H. Böttger and V.V. Bryskin, *Hopping Conduction in Disordered Solids*, Berlin, 1985.

J.H. Davies, F.A. Lee, and M.A. Rice, Phys. Rev. B: Solid State **29**, 4260 (1982).

S.F. Edwards and P.W. Anderson, J. Phys. F: Metal Phys. **5**, 965 (1975).

A.L. Efros and B.I. Shklovskii, J. Phys. C: Solid State Phys. **8**, L49 (1975).

A.L. Efros and M. Pollak, *Electron-Electron Interactions in Disordered Systems,* North Holland, Amsterdam, 1985.

D. Emin, Adv. Phys. **24**, 305 (1974).

H. Fröhlich, Phil. Mag. Supplement **3**, 325 (1954).

H. Fröhlich, H. Pelzer and S. Zienan, Phil. Mag. **41**, 221 (1950).

P. Gosar, Physica A **120**, 339 (1983a).

P. Gosar, Phys. Stat. Sol. (b) **118**, 853 (1983b).

M. Grunewald, B. Pohlman, L. Schweitzer, and D. Wurtz, J. Phys. C: Solid State Phys. **15**, L1153 (1982).

T.D. Holstein, Ann. Phys. **8**, 325 (1959).

J.J. Hopfield, Proc. Natl. Acad. Sci. U.S.A. **79**, 2554 (1982).

A.L. Hunt, to be published.

A.K. Jonscher, Nature **267**, 673 (1977).

W.A. Little, Math. Biosci. **19**, 101 (1974).

J.M. Marshall, Rept. Progr. Phys. **46**, 1235 (1983).

A. Miller and E. Abrahams, Phys. Rev. **120**, 745 (1960).

K. Morigaki, Semicond. Semimet. **21C**, 155 (1985).

N.F. Mott, Adv. Phys. **16**, 49 (1967).

N.F. Mott, J. Non-Cryst. Solids **1**, 1 (1968).

M.A. Paalanen, T.F. Rosenbaum, and G.A. Thomas, Phys. Rev. Lett., in press (1986).

G. Pfister and H. Scher, Adv. Phys. **27**, 747 (1978).

M. Pollak, Disc. Faraday Soc. **50**, 17 (1970).

M. Pollak, J. Phys. C **14**, 2977 (1981).

M. Pollak, Phil. Mag. **50**, 265 (1984).

M. Pollak and T.H. Geballe, Phys. Rev. **122**, 1472 (1961).

M. Pollak and A. Hunt, Phil. Mag. **B52**, 391 (1985).

M. Pollak and M. Ortuno, Solar Energy Mat. **8**, 81 (1982).

M. Pollak and M. Ortuno, in *Electron-Electron Interaction in Disordered Systems*, A.L. Efros and M. Pollak, Eds., North Holland, Amsterdam, 1985a.

M. Pollak and M. Ortuno, J. Non-Cryst. Solids **77-78**, 33 (1985b).

M. Pollak and G.E. Pike, Phys. Rev. Lett. **28**, 1449 (1972).

N. Robertson and L. Friedman, Phil. Mag. **33**, 753 (1976).

J. Schnakenberg, Z. Phys. **208**, 165 (1968).

E. Schrödinger, *What is Life? The Physical Aspects of the Living Cell,* Cambridge University Press, Cambridge, U.K., 1944.

G.L. Shaw and R. Vasudevan, Math. Biosci. **21**, 207 (1974).

B.I. Shklovskii and A.L. Efros, *Electronic Properties of Doped Semiconductors,* Nauka, Moscow, 1979 (in Russian), Springer, Heidelberg, 1984 (in English).

G. Srinivasan, Phys. Rev. B: Solid State **4**, 2581 (1971).

D. Stauffer, Phys. Rept. **54**, 1 (1979).

R.A. Street, Adv. Phys. **30**, 593 (1982).

J. Tauc, Festkörperprobleme **22**, 85 (1982).

J.L. van Hemmen and I. Morgenstern, *Heidelberg Colloquium on Spin Glasses,* Springer, Heidelberg, 1983.

P. Young and R Bhatt, in press (1986).

9. Quantum Statistical Basis of Thermodynamics with Phase Structure

G. L. Sewell

With 5 Figures

ABSTRACT

We obtain a general quantum statistical derivation of thermodynamics with phase structure on the basis of the model in which many-particle systems are idealized as infinite. Macroscopic observables, corresponding to global densities of extensive conserved quantities, are explicitly introduced into the model to serve as thermodynamic variables. The resultant thermodynamics, expressed in terms of these variables and their conjugates, supports a phase structure characterized both by singularities in the Gibbs potential and by coexistent, macroscopically distinct, equilibrium states. It is noteworthy that, by contrast, the traditional statistical mechanics of finite systems could not support either of these characterizations of thermodynamic phase structure.

1. INTRODUCTION

This contribution to Professor Fröhlich's 80th birthday Festschrift is devoted to an approach to a very general and long-standing problem,[1] namely that of deriving classical thermodynamics, *with phase structure*, from the quantum theory of many-particle systems. I think it is fitting that this approach is based on a strategy that follows Fröhlich's precept that any theory purporting to extract macroscopic laws from quantum mechanics must contain, as a *primary* ingredient, some specification of the macroscopic variables in terms of which the relevant many-body problem is to be posed (cf. Fröhlich, 1969, 1973). Here, in fact, we formulate the theory of equilibrium states and phases in terms of macroscopic observables, corresponding to global densities of extensive conserved quantities, within the framework of the statistical mechanical model in which many-particle systems are idealized as infinite. The reason for this idealization will be discussed presently.

To see what the problems are, let us first recall the basic empirical fact that, at the thermodynamical level, the equilibrium states of a macroscopic system fall into phases, that depend on the temperature T and other 'control' variables (e.g. pressure), which we denote collectively by y. The situation is represented by Fig. 1, where the thermodynamical (T,y)-space is divided into regions I, II, III, ..., corresponding to single phases, separated by "surfaces" Γ of phase coexistence (Callen, 1960). Thus, in each single-phase region, there is precisely one equilibrium

[1]This problem is, of course, quite different from that of establishing the existence, and even some specific properties, of phase transitions in concrete models.

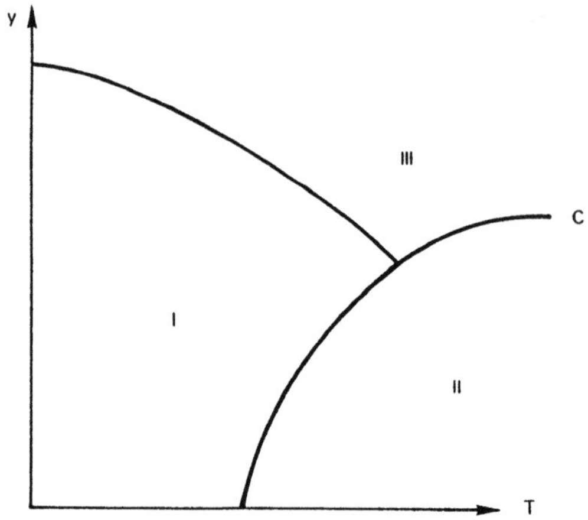

Fig. 1 — Example of a phase diagram. I, II and III are the single-phase regions, the boundaries separating them are the regions of coexistence of difference phases, and C is a critical point.

state corresponding to given values of (T, y), and the Gibbs potential, $\phi(T,y)$ is a smooth function of its arguments, while on the boundaries, Γ, separating these regions, different phases can coexist and the Gibbs potential has singularities.

Now the traditional statistical mechanics of finite systems cannot lead to a thermodynamics with such phase structure, for the following reasons. First, the Gibbs potentials of finite systems, as obtained from statistical mechanics, are smooth functions of their arguments everywhere; and consequently, they cannot possess the singularities required to characterize phase transitions (cf. Lebowitz, 1968). Second, for each value of the temperature T and of the effective Hamiltonian H_y corresponding to the external "forces" y, there is precisely one state (density matrix) ρ that minimizes the free energy, $Tr(\rho H_y + kT\rho \ln \rho)$; and that is the canonical state, given by const. exp $(-H_y/kT)$. In other words, there is but one equilibrium state corresponding to each (T, y), and that means that the finite system model cannot carry coexisting phases. Thus the model fails to support either thermodynamic singularities or coexisting phases; and indeed it suffers from the basic deficiency,[2] closely connected with these failings, of carrying no clear-cut *qualitative* distinction between micro scopic and macroscopic variables.

These considerations have led to a different approach to statistical mechanics, in which macroscopic systems are idealized as infinitely extended ones of finite density. This idealization, which has been extensively investigated over the past two decades,[3] has the great merit of exposing in sharp relief certain intrinsic properties of many-particle systems that would otherwise be masked by finite size-effects. Thus, for example, it represents as singularities in the Gibbs potential what for finite systems would merely be huge gradients, experimentally (but not mathematically) indistinguishable from singularities. Even more radically, it permits a sharp distinction between macroscopic and microscopic observables, the former being global, i.e., extending over all space, and the latter local. Correspondingly, it admits

[2]One can see, in fact, by comparing the models of finite and infinite systems, how the decisive advantage possessed by the latter model over the former one, with regard to phase structure, can be traced to the clear-cut distinction it permits between microscopic and macroscopic observables.

[3]See Ruelle (1969), Emch (1972), Dubin (1974), Thirring (1980), Bratteli and Robinson (1979, 1981), and Sewell (1986).

coexisting phases, i.e., macroscopically different equilibrium states under the same thermodynamical conditions.

The present article (cf. also Sewell, 1986, Chapter 4) consists of a new approach to statistical thermodynamics, based on the explicit introduction of a set of macroscopic variables, corresponding to global densities of extensive conserved quantities, into the infinite system model. The incorporation of these observables into the model enables us to describe it in both the macroscopic and microscopic terms needed for a statistical mechanical derivation of thermodynamics with phase structures. In particular, it enables us to label different coexisting phases by these intensive conserved quantities (e.g., magnetic polarization), as required by classical thermodynamics.

The material of this article will be organized as follows. In Section 2, we shall present a general formulation of the infinite system model. This involves a non-trivial extension of the corresponding formulation for finite systems. The equilibrium states of the infinite system are defined here by the stability condition that they minimize its global free energy density.

In Section 3, we shall introduce the extensive conserved quantities that form the basis of our macroscopic description of the model. There we shall explicitly specify an assumption that needs to be made in order that only a finite number of these variables suffice for a thermodynamic description of the model.

In Section 4, we shall combine the contents of the previous two Sections to pass from a microscopic to a macroscopic formulation of equilibrium states and to express the Gibbs potential in statistical mechanical terms. There we shall also introduce the key concept of *macroscopic degeneracy*, which signifies the existence of different equilibrium macrostates under the same thermodynamic conditions, and thus of phase coexistence.

In Section 5, we shall note some basic mathematical properties of the macroscopic description of the model that will be needed for the subsequent treatment of phase structure. These properties, whose proof will be given in the Appendix, stem from thermodynamical stability and concern the concavity[4] of the Gibbs potential and the entropy function.

In Section 6, we shall derive an important result from the properties of the macrostates and thermodynamic functions specified in Section 5, namely that macroscopic degeneracy occurs at precisely those values of the control variables (T, y) where the Gibbs potential has discontinuities in some first derivatives. This evidently means that such discontinuities correspond to the coexistence of different phases, as is tacitly assumed in classical thermodynamics.

In Section 7, we shall supplement the result of the previous Section with the assumption that the singularities of the Gibbs potential $\phi(T, y)$ from 'surfaces' Γ in the (T, y)-plane, as in Fig. 1. Such an assumption is needed because the general properties of the thermodynamics potentials obtained here tell us very little about the location of the singularities in the Gibbs potential. Here the assumption enables us to interpret both Landau's and Ehrenfest's classifications of phase transitions in statistical thermodynamical terms.

[4]Recall that a concave function $f(\nu)$ is one for which $f(\lambda\nu_1 + (1-\lambda)\nu_2) \geqslant \lambda f(\nu_1) + (1-\lambda) f(\nu_2)$ for $0 < \lambda < 1$.

In Section 8, we shall derive both the standard relation between entropy, internal energy, and mechanical work ($TdS = dE + dW$) for a single phase and the Clausius-Clapeyron formula for coexistent phases from the stability properties of equilibrium states.

We shall conclude, in Section 9, with a summary of the results of this article.

As regards the mathematics employed here, I have tried to keep it as simple as possible, without making unproven assertions. Thus, bearing in mind that the mathematics needed to formulate infinite systems is somewhat more complex than that required for finite ones, I have adopted the policy of merely sketching most of the proofs of relevant theorems, and referring to other works for more details. In fact, I would refer readers seeking a fuller version of the material of this article to my book (Sewell, 1986, especially Chapter 4).

2. THE MODEL

We consider a system, Σ, of particles of some given species in an infinitely extended space, X, which may be either a Euclidean continuum or a discrete lattice. We denote points of X by x, bounded regions of this space by Λ, the volume of Λ by $V(\Lambda)$, and the space translate of Λ through x by $\Lambda + x$.

The model[5] of Σ is constructed so that, in each bounded region Λ, it reduces to the finite system, $\Sigma(\Lambda)$, formed by particles of the given species, confined to Λ. This latter system is formulated in a standard way, its observables, $O(\Lambda)$, and its states, $S(\Lambda)$, corresponding to the self-adjoint operators and the density matrices, respectively, in a Hilbert space $H(\Lambda)$. The local observables are always defined in accordance with the consistency requirement that, if Λ is a subregion of Λ', then $O(\Lambda)$ is a subset of $O(\Lambda')$. The energy observable, i.e. the Hamiltonian, for $\Sigma(\Lambda)$ is denoted by H_Λ. We define O, the set of observables of the infinite system Σ, to comprise the total collection of those of the various finite systems $\Sigma(\Lambda)$, and we take the observables of Σ that are localized in Λ to be $O(\Lambda)$. Space translations of the observables of Σ, corresponding to spatial displacements through x, may be defined in a straightforward way[6] as transformations $A \rightarrow A(x)$ of O that preserve the algebraic structure of O and that map $O(\Lambda)$ into $O(\Lambda + x)$ for each bounded Λ. It will be assumed that the interactions in Σ are translationally invariant, so that $H_\Lambda(x) = H_{\Lambda+x}$.

An important class of observables is that of the extensive conserved quantities. Here, we define such a quantity, Q, to be a family of observables $\{Q_\Lambda\}$ for the respective bounded regions $\{\Lambda\}$, such that[7] (a) Q_Λ commutes with H_Λ; (b) Q_Λ has the extensivity property that, if Λ consists of two disjoint regions, Λ_1 and Λ_2, then $Q_\Lambda = Q_{\Lambda_1} + Q_{\Lambda_2}$; and (c) $Q_\Lambda(x) = Q_{\Lambda+x}$. The extensive conserved quantities, or their global densities, will play the role of macroscopic variables of the model.

---

[5]For more detailed formulations of this model, see, for example, Ruelle (1969), Emch (1972), and Sewell (1986).

[6]Cf. Ruelle (1969), Emch (1972), and Sewell (1986).

[7]More generally, the conditions (a) and (b) may be modified by surface corrections (cf. Sewell, 1986, Chapter 4, where further conditions on Q corrections (cf. Sewell, 1986, Chapter 4, where further conditions on Q needed for "good" thermodynamics behavior are specified).

The states of Σ are defined to be functionals of the observables which serve to yield their expectation values. Thus, if ρ is a state and A an observable, of Σ, then $\rho(A)$ is a number, representing the expectation value of A for the state ρ. The states of Σ are, in fact, defined so that they reduce, in each bounded region Λ, to states of $\Sigma(\Lambda)$, as given by density matrices in $H(\Lambda)$. Thus, each ρ of the infinite systems is given by a family of density matrices $\{\rho_\Lambda\}$ for the respective regions $\{\Lambda\}$; and if A is an observable localized in Λ, then its expectation value for the state ρ is

$$\rho(A) = Tr_\Lambda(\rho_\Lambda A), \tag{2.1}$$

the trace being taken over $H(\Lambda)$. Evidently, the density matrices $\{\rho_\Lambda\}$, representing ρ, must satisfy the consistency condition that, if A is an observable for Λ and hence for any region, Λ', containing Λ, then

$$Tr_\Lambda(\rho_\Lambda A) = Tr_{\Lambda'}(\rho_{\Lambda'} A).$$

The theory that follows will be centered on the translationally invariant states, i.e., those satisfying the condition that $\rho(A(x)) = \rho(A)$ for all observables A and spatial displacements x.

To pass to a macroscopic description of the model, we introduce global properties of the translationally invariant states, as given by their densities of entropy, energy, and other extensive conserved quantities. Thus we start by noting that it follows from our above definitions that the values of the entropy, energy, and extensive conserved quantity, Q, induced in a bounded region Λ by a state ρ of Σ are

$$-kTr_\Lambda(\rho_\Lambda \ln\rho_\Lambda), \quad Tr_\Lambda(\rho_\Lambda H_\Lambda) \text{ and } Tr_\Lambda(\rho_\Lambda Q_\Lambda),$$

respectively, k being Boltzmann's constant. Accordingly, we define the global densities of entropy, energy, and of an extensive conserved quantity Q, for the state ρ, to be the infinite volume limits of the ratios of these quantities to $V(\Lambda)$, i.e.

$$\hat{s}(\rho) = \lim_{\Lambda\to\infty} -kTr_\Lambda(\rho_\Lambda \ln\rho_\Lambda)/V(\Lambda) \tag{2.2}$$

$$\hat{e}(\rho) = \lim_{\Lambda\to\infty} Tr_\Lambda(\rho_\Lambda H_\Lambda)/V(\Lambda) \tag{2.3}$$

$$\text{and } \hat{q}(\rho) = \lim_{\Lambda\to\infty} Tr_\Lambda(\rho_\Lambda Q_\Lambda)/V(\Lambda), \tag{2.4}$$

the limits being taken over an increasing sequence of regions Λ. In fact, for translationally invariant states, these limits have been proved[8] to be well-defined and independent of the chosen sequence of Λ's, provided that these regions are suitably regular (e.g. cubic or spherical) and that, in the case of eq. 2.3, the interactions satisfy certain general realistic conditions. These are essentially that, on the one hand, the interactions fall off fast enough at long distances to insure that the energy is a good extensive variable and that, on the other hand, they are not sufficiently attractive at short distances to lead to a collapse of the system.

Thermal equilibrium states at temperatures T are now defined by the stability condition that they minimize the free energy density $\hat{e}(\rho) - T\hat{s}(\rho)$. Moreover,

[8]See Ruelle (1969), Robinson (1971) and Sewell (1986) and references contained there.

this thermodynamic functional, unlike its counterpart for finite systems,[9] may be minimised by a number of different states; and moreover, these are generally macroscopically different, in that they are mutually distinguishable by the values they yield for the densities of certain extensive conserved quantitis.[10] For example, in the case of an Ising ferromagnet, the equilibrium states at any temperature below the critical point are distinguishable by virtue of their different polarizations (Messager and Miracle-Sole, 1975).

3. THE MACROSCOPIC DESCRIPTION

We now seek to characterize a set of macroscopic observables that can serve to provide a thermodynamic description of the model Σ. To this end, we note again that the free energy density, $\hat{e} - T\hat{s}$, may be minimized by states that are macroscopically different, in that they are distinguishable by the values they yield for certain global intensive variables, $\hat{q}$. This may be equivalently[11] expressed in terms of the energy, rather than the temperature, by saying that the entropy density $\hat{s}$ of Σ may be minimized, subject to a constraint that fixes the value of its energy density, $\hat{e}$, by a number of macroscopically different states. Thus on the assumption that equilibrium states corresponding to maxima of $\hat{s}$, for given $\hat{e}$, this means that further global intensive variables, other than the energy density, are needed to specify these states.[12] Evidently, the number of such variables must be finite if the system is to enjoy "good" thermodynamical behavior.

In view of these considerations, we make the fundamental assumption that the system has a set $(Q_1,\ldots, Q_n)$ of extensive conserved quantities with the following properties, needed for a thermodynamical description of Σ.

($Q1$) *There is precisely one translationally invariant state of* Σ that maximizes its entropy density $\hat{s}(\rho)$ subject to constraints that hold the densities $\hat{e}(\rho)$, $\hat{q}_1(\rho),\ldots \hat{q}_n(\rho)$ of energy, E, and of the quantities $Q_1,\ldots Q_n$, respectively, to arbitrarily chosen values, e, $q_1,\ldots q_n$.

($Q2$) The functionals $\hat{q}_1(\rho),\ldots \hat{q}_n(\rho)$ are linearly independent of one another and of $\hat{e}(\rho)$.

Under these assumptions, we term the variables given by the energy and $Q_1,\ldots Q_n$ a *complete set of extensive conserved quantities.* For notational simplicity, we shall henceforth denote $(Q_1,\ldots Q_n)$ by Q, $(\hat{q}_1(\rho),\ldots \hat{q}_n(\rho)$ by $\hat{q}(\rho)$, and the values $(q_1,\ldots q_n)$ by q. We shall denote by $\rho_{e,q}$ the state, referred to in $Q1$, that maximizes $\hat{s}(\rho)$ subject to the condition that $\hat{e}(\rho) = e$ and $\hat{q}(\rho) = q$; and we shall denote by $s(e,q)$ the entropy density of this state, i.e.

$$s(e,q) = \hat{s}(\rho_{e,q}) = \max\{\hat{s}(\rho)\,|\,\hat{e}(\rho) = e;\ \hat{q}(\rho) = q\}\,. \tag{3.1}$$

[9] Cf. remark at the beginning of Section 1.

[10] See Sewell (1986).

[11] See Sewell (1986), Section 3.8.

[12] The assumption that equilibrium states should be specified in terms of a suitable set of extensive conserved quantities is, of course, basic to classical thermodynamics (cf. Callen, 1960), and ensues from elementary dynamical considerations in statistical mechanics (cf. Landau and Lifshitz (1959), Section 4).

To summarize, then, we describe the macroscopic properties of Σ in terms of the global intensive variables (e,q), representing the densities of the extensive conserved quantities (E,Q). Accordingly, we take the variables (e,q) to represent the macrostates of Σ. The equilibrium microstate corresponding to the macrostates (e,q) is then $\rho_{e,q}$ and the entropy density is $s(e,q)$, as defined by eq. 3.1.

4. GIBBS POTENTIAL AND EQUILIBRIUM STATES

Let $y \equiv (y_1, \ldots y_n)$ be the thermodynamical variables, i.e. external fields or chemical potentials, conjugate to the extensive conserved quantities Q. Then we may define the Gibbs potential $\phi(T,y)$ in terms of the entropy function $s(e,q)$ by the standard formula[13] (cf. Callen, 1960)

$$\phi(T,y) = \min_{e,q}(e - Ts(e,q) - y \cdot q) \tag{4.1}$$

where $y \cdot q \equiv \sum_1^n y_j q_j$.

Equivalently, by eqs. 3.1 and 4.1, ϕ may be expressed in statistical mechanical terms by the formula

$$\phi(T,y) = \min_{\rho} (\hat{e}(\rho) - T\hat{s}(\rho) - y \cdot \hat{q}(\rho)), \tag{4.2}$$

which is evidently the minimum free energy density of Σ at temperature T in the presence of the external 'force' y. Further, it has been proved[14] that this is identical to the infinite volume limit of the equilibrium free energy density of a finite version, $\Sigma(\Lambda)$, of Σ corresponding to given T and y, i.e.

$$\phi(T,y) = \lim_{\Lambda \to \infty} -\frac{kT}{V(\Lambda)} \ln Tr_{\Lambda} \exp-(H_{\Lambda} - y \cdot Q_{\Lambda})/kT. \tag{4.3}$$

Here we note a well-established fact, namely that, while the Gibbs potential for a finite system is a real analytic function of its arguments, its infinite volume limit $\phi(T,y)$ may have singularities.[15] This, of course, is of crucial importance for the theory of phase transitions.

As regards the equilibrium states of the infinite system Σ, for given (y,T), we assume that these are precisely the states that minimize the free energy density $(\hat{e} - T\hat{s} - y \cdot \hat{q})$. Thus, by eqs. 4.1 and 4.2, they are given at the microscopic and macroscopic levels by the values of ρ and (e,q) for which

$$\phi(T,y) = \hat{e}(\rho) - T\hat{s}(\rho) - y \cdot \hat{q}(\rho) \tag{4.4}$$

$$\text{and } \phi(T,y) = e - Ts(e,q) - y \cdot q, \tag{4.5}$$

---

[13] In fact, it has been proved (cf. Ruelle (1969), Sewell (1986)) that the minima referred to in Eqs. 4.1 and 4.2 are actually achieved by at least one state. The proof depends on continuity properties that have been established elsewhere for the functionals $\hat{e}$, $\hat{s}$, $\hat{q}$ and the function s.

[14] See Ruelle (1969), Chapter 7 for lattice systems, and Robinson (1971) for continuous ones.

[15] For a classic example of such singularities, see Onsager's (1944) treatment of the Ising model.

respectively. Further, it follows from our basic assumption $Q1$, of Section 3, and eq. 3.1 that there is a one-to-one correspondence between the equilibrium microstates and macrostates, $\rho_{e,q}$ being the solution of eq. 4.4 corresponding to the solution (e,q) of eq. 4.5. Hence, as Σ may support different equilibrium microstates for given (T,y), it follows that the macroscopic equilibrium condition eq. 4.5 may be satisfied by more than one macrostate. We shall refer to the existence of more than one macrostate as *macroscopic degeneracy*, which evidently corresponds to the coexistence of different phases. As we shall show in Section 6, the condition that macroscopic degeneracy occurs when the thermodynamic control variables take values (T,y) is precisely that some first derivative of the Gibbs potential is discontinuous there.

5. CONSEQUENCES OF THERMODYNAMICAL STABILITY: CONCAVITY OF s AND ϕ

As a preliminary to the formulation of the phase structure of Σ, we now note some simple mathematical properties of its macrostates, its entropy function, $s(e,q)$, and its Gibbs potential, $\phi(T,y)$, that stem from thermodynamical stability. The key concepts required here are those of convexity and concavity, which we shall now briefly summarize.[16]

Thus, a region K of a linear space V is termed convex if, for any two points v_1,v_2 in K and any real number λ between 0 and 1, the point $\lambda v_1 + (1-\lambda)v_2$ lies in K (cf. Fig. 2). Extremal points of K are then defined as those that cannot be expressed as combinations $\lambda v_1 + (1-\lambda)v_2$ of other points v_1,v_2 of K, with $0 < \lambda < 1$. Thus, if K has a boundary, then that consists of its extremal points.

A function $f(v)$, defined on a convex region K, is termed convex if, for any v_1,v_2 in K and $0 < \lambda < 1$, $f(\lambda v_1 + (1-\lambda)v_2) \leqslant \lambda f(v_1) + (1-\lambda)f(v_2)$; while it is termed concave if the inequality goes the other way, i.e., if (cf. Fig. 3)

$$f(\lambda v_1 + (1-\lambda)v_2) \geqslant \lambda f(v_1) + (1-\lambda)f(v_2) \tag{5.1}$$

It follows easily from this definition that functions of one variable, whose second derivatives are everywhere negative, must be concave.

We shall subsequently require the following general properties of concave functions.

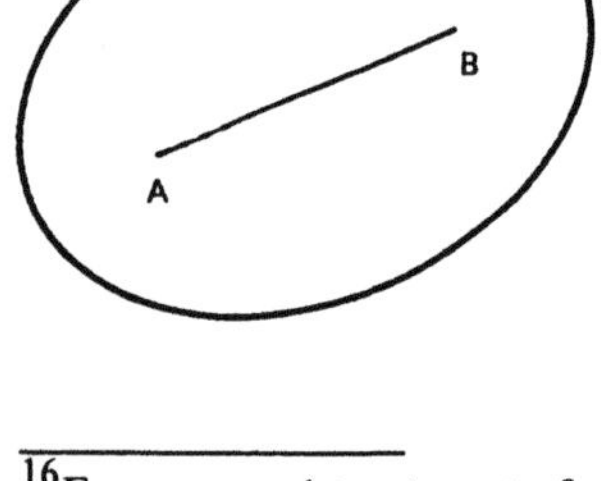

Fig. 2 — An oval-shaped region K of a plane, is convex, as the line joining two of its points, A and B, lies wholly in K. The extremal points of K form its boundary.

[16]For a general treatment of convexity and concavity, see Roberts and Varbert (1973).

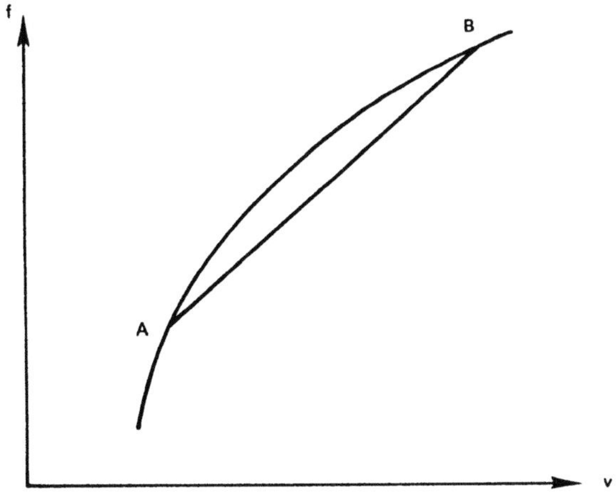

Fig. 3 — Graph of a concave function $f(v)$. Note that the chord AB lies below the graph between A and B.

(Cl) They are continuous in the interior of any convex regions in which they are defined (cf. Fig. 4 for an illustration of the incompatibility of a discontinuity with concavity).

(C2) A concave function $f(v)$ of one variable has so-called right and left derivatives at every point in the interior of its domain K, and these are defined to be

$$\frac{d^{(\pm)}f(v)}{dv} = \lim_{h \to \pm 0} \frac{f(v+h) - f(v)}{h} \tag{5.2}$$

respectively. These derivatives correspond to tangents to the graph of $f(v)$ at $v \pm 0$ (cf. Fig. 5). If they are equal then, of course, f is differentiable at v. if they are unequal, then the derivative of f is said to be discontinuous at v.

(C3) If, in the situation described by C2, a straight line L, of shape k through a point $(v_o, f(v_o))$ lies above the graph of f, i.e., if (cf. Fig. 5)

$$f(v_0 + h) - f(v_o) \leqslant hk \text{ for all real } h\,,$$

$$\text{then } \frac{d^{(+)}f(v_o)}{dv} \leqslant k \leqslant \frac{d^{(-)}f(v_o)}{dv}\,. \tag{5.3}$$

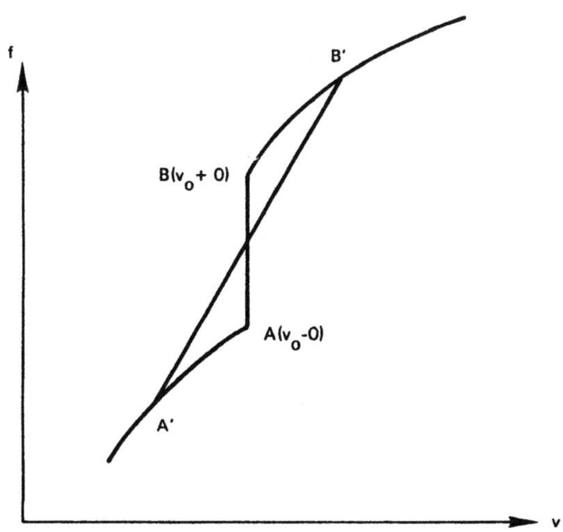

Fig. 4 — Function $f(v)$ with discontinuity at v_0 is *not* concave, since A lies below the chord A'B'.

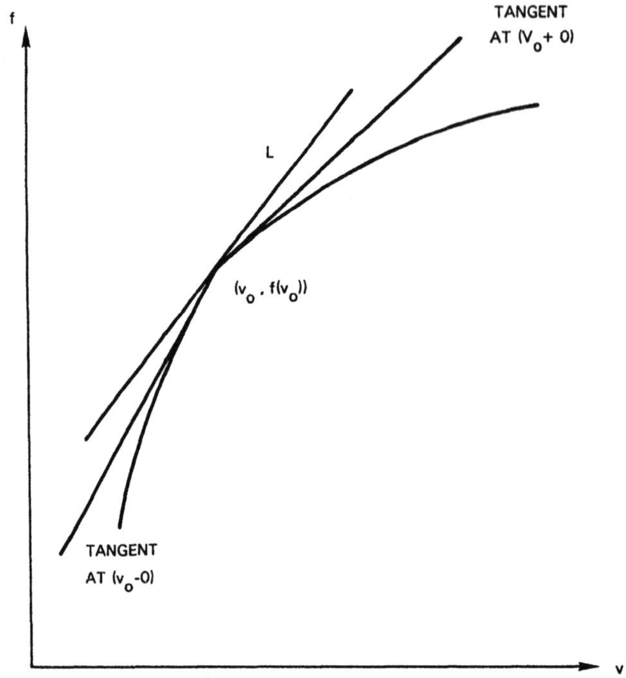

Fig. 5 — Graph of concave function with unequal left and right derivations at ν_0.

In particular, if f is differentiable at ν_o, then $\dfrac{df(\nu_o)}{d\nu} = k$.

(C4) In general, the right and left derivatives of a function $f(\nu)$, at a point ν_o, are limits of $\dfrac{df}{d\nu}$ at sequences of points $\{\nu_n^{(\pm)}\}$ at which f is differentiable and which converge to ν_o from above and below, respectively, as $n \to \infty$, i.e.

$$\frac{d^{(\pm)}f(\nu_o)}{d\nu} = \lim_{n\to\infty} \frac{df}{d\nu}(\nu_n^{(\pm)}) \text{ and } \lim_{n\to\infty} \nu_n^{(\pm)} = \nu_o \pm 0\,. \tag{5.4}$$

Returning to the macroscopic description of Σ, we now note the following facts, whose proof will be given in the Appendix.

(I) *The set M of all macrostates* (e, q) $(\equiv(e,q_1,\ldots, q_n))$ *form a convex region of an* $(n+1)$*-dimensional Euclidean space. Thus, if* (e, q) *and* (e', q') *are microstates, so too are "mixtures" of them as given by* $(\lambda e + (1-\lambda)e', \lambda q + (1-\lambda)q')$, with $0 < \lambda < 1$.

(II) *The set, E, of equilibrium macrostates corresponding to* (T, y) form a convex subset of M, i.e. if (e, q) and (e', q') are equilibrium macrostates, for given (T, y), then so too are their mixtures $(\lambda e + (1-\lambda)e', \lambda q + (1-\lambda)q')$.

(III) *The entropy function s and the Gibbs potential* ϕ *are both concave, i.e.*

$$s(\lambda e + (1-\lambda)e', \lambda q + (1-\lambda/q') \geqslant \lambda s(e, q) + (1-\lambda)s(e', q') \tag{5.5}$$

$$\text{and } \phi(\lambda T + (1-\lambda)T', \lambda y + (1-\lambda)y') \geqslant \lambda\phi(T, y) + (1-\lambda)\phi(T', y') \tag{5.6}$$

for $0 < \lambda < 1$. Note that these formulae imply that s and ϕ are separately convex in each of their arguments, e.g.

$$\phi(\lambda T + (1-\lambda) T', y) \geqslant \lambda\phi(T, y) + (1-\lambda)\phi T', y) \tag{5.7}$$

$$\text{and } \phi(T,\lambda y + (1-\lambda y') \geqslant \lambda\phi(T, y) + (1-\lambda)\phi(T, y') \tag{5.8}$$

the former inequality being obtained by putting $y = y'$ in eq. 5.6 and the latter by putting $T = T'$ in that equation.

The properties I-III of macrostates and thermodynamic potentials are of great significance for the theory of phase transitions, for the following reasons. Since the Gibbs potential, ϕ, is concave, it enjoys the above properties C1 and C2. Thus, this potential is continuous, though its derivatives are not necessarily so. Hence, any phase transitions corresponding to singularities in ϕ can occur only as discontinuities in derivatives of this potential but not in ϕ itself.

As regards the set E of equilibrium macrostates corresponding to given (T, y), the fact that E is convex signifies that its extremal elements are different from the rest in that they alone do not correspond to mixtures, $\lambda(e, q) + (1-\lambda)(e', q')$, of other equilibrium states. This suggests that we take the pure phases to be the external elements of E, and the mixed ones to be the other equilibrium macrostates at the same value of (T, y). Following Ruelle (1969), we shall in fact assume this to be the case, noting that this assumption is supported by the fact that the extremals are the only equilibrium states for which the global intensive variables $(\hat{e}, \hat{q})$ are all sharply defined, i.e. dispersion-free.[17]

MACROSCOPIC DEGENERACY AND THERMODYNAMIC DISCONTINUITY

In the classical theory of these transitions, it is tacitly assumed that the coexistence of different phases of a system (e.q. liquid and gaseous) arises precisely when its Gibbs potential has a discontinuous derivative. We shall now justify this assumption, on the basis of our statistical thermodynamical treatment of the model Σ, by showing that macroscopic degeneracy, i.e. phase coexistence, occurs at precisely those values of the thermodynamic variables (y, T) at which the Gibbs potential $\phi(T, y)$ has a discontinuous first derivative.

The argument[18] runs as follows. If (e, q) is an equilibrium macrostate corresponding to (T, y) then, by eq. 4.5

$$\phi(T, y) = e - Ts(e, q) - y \cdot q \tag{6.1}$$

while, by eq. 4.1

$$\phi(T', y') \leqslant e - T's(e, q) - y' \cdot q$$

[17] This was proved, at the microscopic level, by Ruelle (1969), and, at the macroscopic level, by the present author (Sewell, 1986). Note also that it was proved in those works that the set of equilibrium states possessed the required compactness property, at both microscopic and macroscopic levels, to insure that E has extremal points and that all other macrostates are "mixtures" of these.

[18] This is analogous to a more abstract argument, at the microscopic level, due to Lanford and Robinson (1968a).

for any other values (T',y') of the thermodynamic control variables. On subtracting eq. 6.1 from the two versions of this inequality obtained for $T' = T$ and for $y' = y$, we see that

$$\phi(T, y') - \phi(T, y) \leqslant - (y'-y)\cdot q \tag{6.2}$$

$$\text{and } \phi(T, y') - \phi(T, y) \leqslant - (T'-T)s(e, q) . \tag{6.3}$$

Suppose now that the first-order derivatives of ϕ are continuous at (T, y). Then, as ϕ is concave and so possesses the property C3, of Section 5, it follows from eqs. 6.2 and 6.3 that

$$\frac{\partial\phi}{\partial y}(T, y) = - q \tag{6.4}$$

$$\text{and } \frac{\partial\phi}{\partial T}(T, y) = - s(e, q) \tag{6.5}$$

and hence, by these equations and eq. 4.5, the equilibrium macrostate (e, q) is *uniquely* determined by the formulae

$$e = \phi(T, y) - T\frac{\partial\phi}{\partial T}(T, y) - y\cdot\frac{\partial\phi}{\partial T}(T, y) \tag{6.6}$$

$$\text{and } q = - \frac{\partial\phi}{\partial y}(T, y) \text{ (as in eq. 6.4)} .$$

Hence, there is no macroscopic degeneracy at (T, y) if ϕ is differentiable there.

To prove that, conversely, a discontinuity in a first derivative of ϕ implies a macroscopic degeneracy, we employ a construction which demonstrates that if, say, $\frac{\partial\phi}{\partial T}$ is discontinuous at (T, y), then are equilibrium macrostates there with different entropy densities, $\frac{\partial^{(\pm)}}{\partial T}\phi(T, y)$: a similar construction can be used to show that a discontinuity in $\frac{\partial\phi}{\partial y}$ implies equilibrium states with different values of q.

Thus, supposing that $\frac{\partial\phi}{\partial T}$ is discontinuous at (T, y), we note that, for fixed y, $\phi(T, y)$ is concave in T and therefore enjoys the property C4, of Section 5, with respect to this variable. Hence, there are sequences $\{T_n^{(\pm)}\}$, converging to T from above and below, such that ϕ is differentiable with respect to T at the points $(T_n^{(\pm)},y)$ and

$$\frac{\partial^{(\pm)}}{\partial T}\phi(T, y) = \lim_{n\to\infty}\frac{\partial\phi}{\partial T}(T_n^{(\pm)},y) . \tag{6.7}$$

Now let $(e_n^{(\pm)},q_n^{(\pm)})$ be equilibrium macrostates corresponding to $(T_n^{\pm)},y)$, respectively. Then the equilibrium condition eq. 4.5, as applied to these states, takes the form

$$\phi(T_n^{(\pm)},y) = e_n^{(\pm)} - T_n^{(\pm)}s(e_n^{(\pm)},q_n^{(\pm)}) - y\cdot q_n^{(\pm)}, \tag{6.8}$$

while the differentiability of ϕ with respect to T at $(T_n^{(\pm)}, y)$ ensures that eq. 6.5 is valid with e, q, T taking values $e_n^{(\pm)}$, $q_n^{(\pm)}$, $T_n^{(\pm)}$, respectively, i.e.

$$\frac{\partial}{\partial T}\,\phi(T_n^{(\pm)},y) = -\,s(e_n^{(\pm)},q_n^{(\pm)})\,. \tag{6.9}$$

Furthermore, by invoking the continuity properties of ϕ and s and the structure of the macrostate space M, one can show[19] that the sequences $\{T_n^{(\pm)}\}$ may be chosen so that $(e_n^{(\pm)},q_n^{(\pm)})$ converge to macrostates, which we denote by $(e^{(\pm)},q^{(\pm)})$, respectively, and that, correspondingly, the limiting forms of eqs. 6.8 and 6.9 are

$$\phi(T, y) = e^{(\pm)} - Ts(e^{(\pm)},q^{(\pm)}) - y\cdot q^{(\pm)} \tag{6.10}$$

$$\text{and } s(e^{(\pm)},q^{(\pm)} = -\lim_{n\to\infty}\frac{\partial^{(\pm)}}{\partial T}\,\phi(T, y) \equiv -\frac{\partial^{(\pm)}}{\partial T}\,\phi(T, y), \text{ by } (C4) \tag{6.11}$$

Equation 6.10 tells us that both $(e^{(+)},q^{(+)})$ and $(e^{(-)},q^{(-)})$ satisfy the equilibrium condition eq. 4.5, while eq. 6.11 implies that these are different macrostates, in view of the discontinuity in $\frac{\partial\phi}{\partial T}$ at (T, y). This establishes, then that such a discontinuity carries a macroscopic degeneracy of the model at (T, y). A parallel argument can be employed to show that a discontinuity in $\frac{\partial\phi}{\partial y}$ also implies such a degeneracy.

We conclude, therefore, that macroscopic degeneracy occurs at precisely those values of the thermodynamic variables (T, y) where the Gibbs potential has some discontinuous first derivative.

7. PHASE STRUCTURE

The result we have just discussed serves to relate the coexistence of different phases to discontinuities in first derivatives of the Gibbs potential. However the argument leading to this result was based on very general properties of concave functions and so tells us nothing about the location of the discontinuities. For example, an arbitrary concave function $f(T, y)$ could have discontinuous first derivatives at a set of points that formed no simple geometrical pattern, e.g., at all rational values of $T, y_1, \ldots y_n$![20]

Thus, in order to proceed further, we supplement the result of the previous Section with the classical thermodynamical assumption that the points of discontinuity of first derivatives of ϕ form a finite number of n-dimensional surfaces, Γ, in the $(n+1)$-dimensional (T, y)-space (cf. Fig. 1). This is an underlying assumption of the Gibbs phase rule, and is supported both by experimental evidence and by treatments of solvable models.[21]

Under this assumption, we may summarize the results obtained so far as follows.

(1) The Gibbs potential $\phi(T, y)$ is continuous, but its derivatives may have discontinuities.

[19] See Sewell (1986), Chapter 4.
[20] See Roberts and Varberg (1973), p. 7.
[21] See Lebowitz (1977) for a treatment of the Ising model.

(2) The region of discontinuities in the first derivatives of ϕ form a finite set of surfaces, Γ, in (T, y)-space. Thus, by equations eqs. 6.4 and 6.5, the values of the macrovariables (e, q) changes abruptly as the phase point (T, y) crosses Γ, but nowhere else.

(3) The surfaces Γ comprise the regions of coexisting phases, and the rest of (T, y)-space consists of single-phase regions.

Thus, the (T, y)-space is divided into single phase-regions separated by surfaces Γ, where the thermodynamic variables (e, q) change abruptly (cf. Fig. 1). These surfaces comprise the regions of *first-order* phase transitions, in Ehrenfest's sense, as they correspond to discontinuities in the Gibbs potential.

The other type of phase transitions correspond to higher-order singularities in the Gibbs potential, and were termed phase transitions of the *second kind* by Landau. According to classical thermodynamics,[22] these transitions occur at the so-called critical points which mark the termination of the boundaries Γ separating different phases (cf. Fig. 1) and this assumption is also supported by exact solutions of solvable models.[23]

8. THERMODYNAMICAL RELATIONS

In the scheme we have just described, the thermodynamic (T, y)-space consists of single-phase regions, separated by boundaries, Γ. We shall now provide statistical thermodynamical derivations of both the standard formula relating changes in entropy and internal energy to mechanical work in a single phase and the Clausius-Clapeyron equation for coexistent phases.

Thus, considering first a single phase, we see from eqs. 6.4 and 6.5 that

$$d\phi = -\, sdT - q \cdot dy$$

and hence, as

$$\phi = e - Ts - y \cdot q$$

$$Tds = de + dw \tag{8.1}$$

with

$$dw = -\, y \cdot dq\,, \tag{8.2}$$

the work per unit volume done by the external forces, y, in an infinitesimal change dq in q. Equation 8.1 is therefore the standard thermodynamical formula relating changes in entropy to changes in internal energy and to mechanical work done by external forces for a single phase.

Next we consider two existing phases, I and II, say. If Γ denotes the surface separating the regions of (T, y)-space corresponding to these phases, it follows from the continuity of the Gibbs potential that if $\phi_I(T, y)$ and $\phi_{II}(T, y)$ are the forms taken by this potential in phases I and II, respectively, then $\phi_I = \phi_{II}$ on Γ. Further, as ϕ_I and ϕ_{II} satisfy the versions of the formulae 8.1 and 8.2 appropriate to the respective phases, it follows that

[22] See Tisza (1966), and Landau and Lifshitz (1959).

[23] Cf. Lebowitz (1977) for a treatment of the Ising model.

$$d(\phi_I - \phi_{II}) = -(s_I - s_{II})\, dT - (q_I - q_{II}) \cdot dy \text{ on } \Gamma .$$

Hence, as the left-hand side of this formula vanishes on Γ,

$$(s_{II} - s_I) = (q_I - q_{II}) \frac{dy}{dT} \text{ on } \Gamma .$$

Therefore, as $T(s_{II} - s_I)$ is the *latent heat*, L, for a transition from I to II.

$$L = T(q_I - q_{II}) \frac{dy}{dT}, \tag{8.3}$$

which is the Clausius-Clapeyron equation.

CONCLUSION

We have constructed a general statistical mechanical framework for thermodynamics with phase structure by incorporating macroscopic observables, corresponding to global intensive conserved quantities, into the infinite system model. The macroscopic observables here must form a complete set, in the sense defined in Section 3, in order that they may serve to classify equilibrium states at both the macroscopic and microscopic levels.

The model supports characterizations of phase transitions both by singularities in the Gibbs potential and by macroscopic degeneracy of equilibrium states, i.e., coexistent phases. Although these two characterizations are a priori independent of one another, we established an essential connection between them, stemming from thermodynamical stability, by showing that macroscopic degeneracy occurs at precisely those values of the control variables (T, y) at which the first derivatives of the Gibbs potential are not all continuous. This result justifies a standard tacit assumption of classical thermodynamics, namely that phase transitions of the first order, in Ehrenfest's sense, occur at precisely those values of (T, y) where the system supports coexistent phases. For such values of these control variables, the pure phases are then naturally interpreted as the extremal equilibrium states (cf. discussion at the end of Section 5).

Other results obtained here, on the basis of stability, were the standard relation 8.1 relating entropy, internal energy and mechanical work in a single phase, and the Clausius-Clapeyron equation 8.3 for coexistent phases.

On the debit side, it should be emphasized that the arguments centered on the concavity of the Gibbs potential ϕ and the entropy function s, that were the basis of the above results, tell us virtually nothing about the location of the singularities of the Gibbs potential. For that reason we needed to assume, in Section 7, that the discontinuities in the first derivatives of this potential formed "surfaces" in the thermodynamic (T, y)-space, as is assumed for the derivation of the Gibbs phase rule. Thus, there remains the fundamental problem of what further general statistical mechanical properties of the model need to be unearthed in order to obtain this rule.

Appendix

We shall now sketch the proofs of the assertions I, III and II, of Section 5, in that order.

Proof of I. The definitions of Sections 2 imply that, if ρ and ρ' are translationally invariant states, then so too is $\bar{\rho} = \lambda\rho + (1-\lambda)\rho'$ (i.e. $\bar{\rho}(A) \equiv \lambda\rho(A) + (1-\lambda)\rho'(A)$ for $0 < \lambda < 1$). Further, by 2.3 and 2.4,

$$\hat{e}(\lambda\rho + (1-\lambda)\rho') = \lambda\hat{e}(\rho) + (1-\lambda)\hat{e}(\rho') \tag{A.1}$$

$$\text{and } \hat{q}(\lambda\rho + (1-\lambda)\rho') = \lambda\hat{q}(\rho) + (1-\lambda)\hat{q}(\rho')\,. \tag{A.2}$$

Suppose now that (e, q) and (e', q') are macrostates of Σ, and that ρ,ρ' are corresponding microstates, i.e.

$$\hat{e}(\rho) = e,\ \hat{q}(\rho) = q;\ \hat{e}(\rho') = e',\ \hat{q}(\rho') = q'\,.$$

Then, by eqs. A.1 and A.2, the macrostate corresponding to $\lambda\rho + (1-\lambda)\rho'$ is $(\lambda e + (1-\lambda)e', \lambda q + (1-\lambda)q')$. Hence, if (e, q) and (e', q') are macrostates, then so too is $(\lambda e + (1-\lambda)e', \lambda q + (1-\lambda)q')$, for $0 < \lambda < 1$, which means that the set M of all macrostates of Σ is convex.

Proof of III. We start by noting that the second derivative of the function $f(v) = v\ ln\ v$ (with $v > 0$) is positive and therefore that this function is concave. From this and the definition (eq. 2.2) of $\hat{s}(\rho)$ one can infer[24] that $\hat{s}$ is also concave, i.e.

$$\hat{s}(\lambda\rho + (1-\lambda)\rho') \geqslant \lambda\hat{s}(\rho) + (1-\lambda)\hat{s}(\rho')\,, \tag{A.3}$$

for $0 < \lambda < 1$. Suppose now that ρ and ρ' are the states that maximize $\hat{s}$ subject to the condition that the values of $(\hat{e}, \hat{q})$ are (e, q) and (e',q'), respectively. Then, by the definition of the entropy function s, in Section 3,

$$s(e, q) = \hat{s}(\rho) \text{ and } s(e',q') = \hat{s}(\rho')$$

and therefore, by eq. A3,

$$\hat{s}(\lambda\rho + (1-\lambda)\rho') \geqslant \lambda s(e, q) + (1-\lambda)s(e', q')\,. \tag{A.4}$$

Moreover, by our above specifications of ρ,ρ' and the definitions (eqs. 2.3 and 2.4) of $\hat{e}$, $\hat{q}$, we see that

$$\hat{e}(\lambda\rho + (1-\lambda)\rho') = \lambda e + (1-\lambda)e';\text{ and } \hat{q}(\lambda\rho + (1-\lambda)\rho') = \lambda q + (1-\lambda)q'\,.$$

Consequently, by eq. 3.1, the left-hand side of eq. A.4 cannot exceed $s(\lambda e + (1-\lambda)e', \lambda q + (1-\lambda)q')$, which implies that s is concave.

To prove the concavity of ϕ, let (e, q) be an equilibrium macrostate corresponding to values $(\lambda T + (1-\lambda)T', \lambda y + (1-\lambda)y')$ of the thermodynamic control parameters, i.e. by eq. 4.5,

[24] See Lanford and Robinson (1968a,b).

$$\phi(\lambda T + (1-\lambda)T', \lambda y + (1-\lambda)y') = e - (\lambda T + (1-\lambda)T')s(e,q)$$
$$- (\lambda y + (1-\lambda)y') \cdot q$$
$$\equiv \lambda(e - Ts(e, q) - y \cdot q)$$
$$+ (1-\lambda)(e - T's(e, q) - y' \cdot q) .$$

By eq. 4.1, the coefficients of λ, $(1-\lambda)$ on the right-hand side of this equation cannot exceed $\phi(y, T)$ and $\phi(y', T')$, respectively, from which it follows that ϕ is concave.

Proof of II. Suppose that (e, q) and (e', q') are two equilibrium macrostates corresponding to (T, y). Then these both satisfy the condition given by eq. 4.5, and consequently

$$\phi(T, y) = \lambda e + (1-\lambda)e' - T(\lambda s(e, q) + (1-\lambda)s(e', q'))$$
$$- y \cdot (\lambda q + (1-\lambda)q') .$$

By the concavity of s, the coefficient of T in this equation cannot be less than $s(\lambda e + (1-\lambda)e', \lambda q + (1-\lambda)q')$, and therefore

$$\phi(T, y) \geqslant \lambda e + (1-\lambda)e' - Ts(\lambda e + (1-\lambda)e', \lambda q + (1-\lambda)q') \quad (A.5)$$
$$- y \cdot (\lambda q + (1-\lambda)q')$$

moreover, by eq. 4.1, $\phi(T, y) \leqslant \bar{e} - Ts(\bar{e},\bar{q}) - y \cdot \bar{q}$ for any macrostate $(\bar{e},\bar{q})$ and therefore for the one given by $(\lambda e + (1-\lambda)e', \lambda q + (1-\lambda)q')$. This implies that the equality sign must be operatve in eq. A.5, i.e., that the state $(\lambda e + (1-\lambda)e', \lambda q + (1-\lambda)q')$ satisfies the equilibrium condition given by eq. 4.5. Hence the set, E, of macrostates corresponding to (T, y) is convex.

REFERENCES

O. Bratteli and D.W. Robinson, *Operator Algebras and Quantum Statistical Mechanics*, Springer, Heidelberg/New York, Vol. 1, 1979; Vol. 2, 1981.

H. Callen, *Thermodynamics*, Wiley, New York, 1960.

D.A. Dubin, *Solvable Models in Algebraic Statistical Mechanics*, Clarendon Press, Oxford, 1974.

G.G. Emch, *Algebraic Methods in Statistical Mechanics and Quantum Field Theory*, Wiley, New York/London, 1972.

H. Fröhlich, in Proc. 1st Int. Conf. Theor. Phys. Biol., H. Maurois, Ed., North Holland, Amsterdam, 1969, pp. 13-22.

H. Fröhlich, Nuovo Cimento **3**, 490 (1973).

L.D. Landau and E.M. Lifshitz, *Statistical Physics*, Pergamon, London/New York, 1959.

O.E. Lanford and D.W. Robinson, Commun. Math. Phys. **9**, 327 (1968a).

O.E. Lanford and D.W. Robinson, J. Math. Phys. **9**, 1120 (1968b).

J.L. Lebowitz, Ann. Rev. Phys. Chem. **19**, 389 (1968).

J.L. Lebowitz, J. Stat. Phys. **16**, 463 (1977).

A. Messager and S. Miracle-Sole, Commun. Math. Phys. **40**, 187 (1975).

L. Onsager, Phys. Rev. **65**, 117 (1944).

A.W. Roberts and D.E. Varberg, *Convex Functions*, Academic, London/New York, 1973.

D.W. Robinson, *The Thermodynamical Pressure in Quantum Statistical Mechanics*, Lecture Notes in Physics, Vol. 9, Springer, Berlin/Heidelberg/New York, 1971.

D. Ruelle, *Statistical Mechanics*, W.A. Benjamin, New York, 1969.

G.L. Sewell, *Quantum Theory of Collective Phenomena*, Clarendon Press, Oxford, 1986 (to appear).

W. Thirring, *Quantum Theory of Large Systems*, Springer, New York/Vienna, 1980.

L. Tisza, (and others), *Generalized Thermodynamics*, M.I.T. Press, Cambridge, MA, and London, 1966.

10. On Molecular Control

T.W. Barrett

Molecular control is the efficient transfer of energy to a molecular system for purposes of useful work, whether the energy originates from electromagnetic field or photon, electron, or phonon. The transferred energy may be used in any purposeful way and may serve, for example, in increasing biocatalysis in reactors dedicated to bulk chemical production, in the formation of proton gradients across membranes, or in radiation-induced reactions. It is thus evident that, given this goal, one requirement is an understanding of the periodic and quasiperiodic behavior of molecular systems to ensure that the transferred energy is not dissipated. In contrast, the molecular system should not be of the kind that treats the energy stochastically by intramolecular randomization (Noid et al., 1981), because the transferred energy is then lost for useful work purposes.

There are two limiting forms of unimolecular dynamics. The first, the RRKM (Rice, Ramsperger, Kassel, Marcus) theory, is based on the assumption that the energy can be partitioned among all available molecular states (Marcus, 1952). In the second form, it is assumed that:

(1) the normal modes of a molecule are anharmonic;
(2) the KAM (Kolmogorov, Arnol'd, Moser) theorem for low perturbations applies (Kolmogorov, 1954; Arnol'd, 1963, 1978; Moser, 1962, 1973);
(3) the conditions exist as described in the Fermi-Pasta-Ulam problem (Fermi et al., 1955); and
(4) the classical motion is quasiperiodic.

In the case of a quantal system, the first limit also requires the preparation of a group of states and is described by KAM at high perturbation. Furthermore, there are current attempts at stochastic approaches to Brownian motion of a quantum system (cf. Dattagupta, 1984; Gardiner and Collett, 1985). With the goal of molecular control and efficient transfer of energy in mind, the first limit describes those conditions to be avoided, and the second limit, those to be attained.

Therefore, either to "craft" (engineer) energy to be compatible with a molecular system or to "craft" (engineer) a molecule to be compatible with available energy, so that in both cases the energy is efficiently transferred for useful work, means that the second limiting form is being considered. In the case of phonon energies, the method of seeking the second limit in energy-molecule interactions and avoiding the first, or chaotic, stochastic, ergodic or irregular form of interaction, is beginning to be understood conceptually (Eilbeck et al., 1984; Takeno, 1983, 1984). In the case of quantum systems and photon energies, it is more difficult. There are no simple correspondences between classical and quantum mechanical chaos and classical and quantum mechanical quasiperiodicity. Future theoretical work is required to define precisely the distinction between quantum mechanical chaos and quasiperiodicity (Cf Reichl and Lin, 1986; Eidson and Fox, 1986; Fox and Eidson, 1986), so that with the present purposes in mind, the first may be avoided and the second achieved. The conceptualization must also include, besides resonance interaction, the possibility of parametric pumping of energy transfer, as in the well-known case of Fermi resonance. An understanding of parametric excitation in energy transfer will permit a greater choice of mechanisms and greater versatility.

There are two competing tendencies in systems with two degrees of freedom. The competition arises between the number of resonances becoming increasingly dense in the phase space and the width of the resonances tending to become small. As the number of degrees of freedom increases, either complete stochasticity or complete harmonicity can be achieved, depending upon the average ratio of the resonance width to distance between resonances (Lichtenberg and Lieberman, 1983).

The determination of resonance energies in systems of coupled oscillators must also address the problem of resonance states of the individual oscillators with widely different lifetimes. Recent advances in the theory of resonances have led to an understanding of resonance states in terms of complex energies and associating them with "quantum momenta" (Lefebvre, 1985).

Multiphoton excitation, in which a single polyatomic molecule absorbs many infrared photons (Lyman, Galbraith and Ackerhalt, 1982), is one prototype form of molecular control. Multiphoton excitation experiments exhibit many features that are independent of the particular molecular species. It appears that multiphoton excitation depends upon a number of effects including absorption line broadening, multiphoton absorption, higher-order interactions among normal-mode states, and intramolecular energy flow between states. Theoretical treatment and prediction is thus difficult. Furthermore, the radiation is not simply a probe, because it severely perturbs the absorbing sample. A new picture has recently emerged in which the polyatomic molecule possesses:

(1) discrete low-energy level states permitting resonant absorption described by conventional spectroscopic theory;
(2) a quasicontinuum of higher energy states due to mixing with other "background" states and described by multiphoton absorption theory (Lyman, Galbraith and Ackerhalt, 1982); and
(3) a region of higher energy states described by RRKM theory (Robinson and Holbrook, 1972).

Therefore, in considering energy transfer to a polyatomic system, defining the boundaries of (1), (2) and (3) is of fundamental importance. At the present time, such definitions for most systems remain both an experimental and a theoretical challenge.

Because of the complexities of defining the resonance and stochastic excitation conditions of a complicated polyatomic system, a model system may offer rule-of-thumb behavior. The dc superconducting quantum interference device

(SQUID) offers itself in this regard as a simple example of a four-dimensional autonomous nonlinear dynamical system (Ketoja et al., 1984). The stability problems of the SQUID are susceptible to analysis in terms of Hill's equation and Mathieu's equation (Magnus and Winkler, 1966). The Feigenbaum sequence to chaos (Feigenbaum, 1978, 1979) has been found in the dc SQUID equations in a large number of narrow marginal regions of the phase space between different types of solution (Ketoja et al., 1984).

The way to efficiently transfer or harvest energy is presently a theoretical or analytical problem. But even if its solution is presupposed, there are then two other problems bearing on the use of that energy presumed captured and stored, which may not lend themselves to general solutions.

The first concerns the function of enzymes. Many of the uses on which the energy will be spent involve an increase in enzymatic reaction rates. Unfortunately, there is no agreement on how enzymes function at the present time. In fact, there may be many mechanisms, many different forms of energy may be used, and how the enzyme recognizes the substrate may be a critical concern. We have to know many precise attributes of the excited molecule to use the captured energy.

The second concerns the realization that membranes have an organizing purpose essentially required by statistically mechanical arguments (Fox, 1982; Lavenda, 1985). Membranes are required in conjunction with chemical reactions to bring about the translocation of molecules, ions, or chemical groups. The result of this symbiosis of chemical reactions and membranes is a separation of charge whose recombination underlies the performance of osmotic, chemical, and mechanical work. We have to know even more attributes of the excited molecule to utilize this symbiosis and obtain the result desired.

We have barely begun to attack these two problems. The recent successful demonstration of frequency-specific effects (resonances) in DNA of uniform length from exposure to low-level (producing nonsignificant thermal induction) microwave or radio frequency radiation (Edwards et al., 1984) indicates that radio frequency energies of radiation can be efficiently transferred to molecular systems. However, how such transferred energy is usefully employed is not yet understood.

Another example of energy harvesting is light transduction in rhodopsin and bacteriorhodopsin (Birge, 1981). This example can be placed in the context of thermodynamic law. A strict observance of the second law of thermodynamics (Atkins, 1984) tells one that *heat* engines cannot produce work, unless in some stage of the energy cycle heat is discarded into a cold sink. In other words, heat cannot be completely converted into work except at zero temperature of the cold sink—which is unobtainable. Similarly, a strict observance of the second law in the case of molecular *entropy* engines tells one that no work can be produced unless in some stage of the energy cycle entropy is discarded into an entropy sink, which is the molecular system. In other words, transferred energy cannot be completely converted into work except at zero entropy of the molecular system—which also is unobtainable. Another type of engine, the molecular *internal energy* engine, loses internal energy (or enthalpy) in the transformation from initial state to final state. External energy sources must be used to reconstitute the internal energy for another cycle.

To return to the light-harvesting examples, a recent study of the enthalpy differences between light-adapted bacteriorhodopsin and its primary photoproduct demonstrated that about 30% of the absorbed photon energy is stored at an absorption maximum (568 nm) (Birge and Cooper, 1983). On the other hand, the primary step in rhodopsin excitation stores 60% of the absorbed photon energy at an absorption maximum (500 nm) (Cooper, 1979). However, the primary photochemical events associated with both rhodopsin and bacteriorhodopsin involve similar isomerization of the retinyl chromophore. Thus, the difference between the two protein-chromophore systems is remarkable and needs explanation. It may be relevant, therefore, within the context of the preceding remarks on the thermodynamic requirements of energy transfer, to notice the energy cycle differences in the two systems. On the one hand, bacteriorhodopsin uses part of the photon energy to perform work and part to complete the energy cycle to be in the initial state to receive another photon. It resembles an entropy engine. Rhodopsin, on the other hand, in the absence of energy input from respiratory enzymes, resembles more an internal-energy engine and cannot complete an energy cycle without energy transfer from those enzymes. Such considerations should dictate choices of molecular systems for energy harvesting. The second law would state that no process is possible in which the *sole result* is the absorption of quantum energy from a reservoir and its *complete conversion* into work. A tax on transferred quantum (photon) energy is necessary if useful work is to be performed.

The field of chemical synthesis and biosynthesis is very empirical and must remain so because research interests, eventually, have to be justified by products and usable results. In the field of specialty chemical and biomedical and pharmaceutical chemical production, there is little concern for scale-up to bulk chemical production because in this field of endeavor, small production often gives profit enough. But theory and experiment are complementary and there is an imbalance at the present time, as in the past, in the biosynthesis field toward experiment. An investment in theoretical analysis now may create vast areas for empirical use and development.

REFERENCES

V.I. Arnol'd, Sov. Math. Dokl. **3**, 136 (1963).

V.I. Arnol'd, *Mathematical Methods of Classical Mechanics*, New York, Springer-Verlag, 1978.

P.W. Atkins, *The Second Law*, Freeman, New York, 1984.

R.R. Birge, Ann. Rev. Biophys. Bioeng. **10**, 315 (1981).

R.R. Birge and T.M. Cooper, Biophys. J. **42**, 61 (1983).

T.M. Cooper, Nature **282**, 531 (1979).

S. Dattagupta, Phys. Rev. **A30**, 1525 (1984).

G.S. Edwards, C.C. Davis, J.D. Saffer, and M.L. Swicord, Phys. Rev. Lett. **53**, 1284 (1984).

J. Eidson and R.F. Fox, Phys. Rev. A, in press (1986).

J.C. Eilbeck, P. Lomdahl and A.C. Scott, "The Discrete Self-Trapping Equation," Preprint LA-UR-84-3115, Los Alamos National Laboratory, NM, 1984.

M.J. Feigenbaum, J. Stat. Phys. **19**, 25 (1978).

M.J. Feigenbaum, J. Stat. Phys. **21**, 669 (1979).

E. Fermi, J. Pasta, and S. Ulam, "Studies of Non-Linear Problems," Sci. Rep. LA-140, Los Alamos Scientific Laboratory, Los Alamos, NM, 1955.

R.F. Fox, *Biological Energy Transduction*, Wiley, New York, 1982.

R.F. Fox and J. Eidson, Phys. Rev. **A34**, 482 (1986).

C.W. Gardiner and M.J. Collett, Phys. Rev. A: Gen. Phys., 31, 3761 (1985).

J.A. Ketoja, J. Kurkijarvi, and R.K. Ritala, Phys. Rev. **B30**, 3757 (1984).

A.N. Kolmogorov, Dokl. Akad. Nauk SSSR, **98**, 527 (1954).

B.H. Lavenda, Sci. Am. **252**, No. 2, 70 (Feb. 1985).

R. Lefebvre. Europhys. News **16**, 4 (1985).

A.L. Lichtenberg and M.A. Lieberman, *Regular and Stochastic Motion*, Springer, New York, 1983.

J.L. Lyman, H.L. Galbraith, and J.R. Ackerhalt, Los Alamos Sci., **3**, 66 (1982).

W. Magnus and S. Winkler, *Hill's Equation*, Dover, New York, 1966.

R.A. Marcus, J. Chem. Phys. **20**, 359 (1952).

J. Moser, Nachr. Akad. Wiss. Göttingen II. Math. Phys. Kl. **1962**, 1 (1962).

J. Moser, *Stable and Random Motions in Dynamic Systems,* Ann. Math. Studies No. 77, Princeton University Press, Princeton, NJ, 1973.

D.W. Noid, M.L. Koszykowski, and R.A. Marcus, Ann. Rev. Phys. Chem. **32**, 267 (1981).

L.E. Reichl and W.A. Liw, Phys. Rev. A: Gen. Phys. **33**, 3598 (1986).

P.J. Robinson and K.A. Holbrook, *Unimolecular Reactions*, Wiley-Interscience, New York, 1972.

S. Takeno, Progr. Theor. Phys. **69**, 1798 (1983).

S. Takeno, Progr. Theor. Phys. **71**, 395 (1984).

11. Structure of Liquid 1-Alkanols and Some Isomers*

W. G. Scaife

With 7 Figures

For a large class of fluid dielectrics, that is for nondipolar substances, the Clausius-Mosotti, (CM), relationship between density, ρ, and relative dielectric permittivity, ϵ, holds very well. It can be written as

$$M(\epsilon-1)/(\epsilon+2)\rho = 4\pi N_a \alpha/3 \tag{1}$$

where M is the molecular weight, N_a is Avogadro's number, and α is a molecular, or displacement polarizability. Measurements on members of the n-alkane family, for example, over a temperature range from 248 K to 373 K, and at pressures up to 350 MPa, show that the left-hand side of eq. 1 varies by not more than 1.5 to 2% (Scaife and Lyons, 1980).

In the case of fluids that are composed of molecules with permanent dipole moments, μ, the CM equation is not successful, especially when temperature is varied. Debye (1929) went a considerable distance towards remedying the deficiency. In one model he pictured a molecule subject to random, Brownian, reorientations induced by the thermal fluctuations of its neighbors. The externally applied field has the effect of making it slightly more probable that the permanent dipole will align with the applied field rather than against it. Debye appreciated that other models could lead to the same mathematical expressions for time-dependent behavior. In particular he analyzed the behavior of a dipole that has two possible antiparallel orientations, either of which is equally probable. In order to jump from one to the other the molecule must possess sufficient energy to surmount a potential hill. The existence of an applied field will favor one orientation slightly more than the other.

The permittivity $\epsilon(w)$ in Debye's theory ranges between two extremes, $\epsilon(0)$ the equilibrium value at low frequencies, and $\epsilon(\infty)$, the limiting value at high frequencies. Between these extremes the dielectric relaxation process absorbs energy from the measuring field and $\epsilon(w) = \epsilon' - i\epsilon''$, or, to follow Debye,

$$\epsilon(w) = \epsilon(\infty) + (\epsilon(0) - \epsilon(\infty))/(1 - iw\tau). \tag{2}$$

By using the concept of time constant or relaxation time τ, the time-dependent behavior of ϵ is well reproduced for a number of materials. Typical of these are the n-alkanols. Using Mizushima's (1927) data for methanol at radio frequencies, Debye showed that there is one process which dominates the changes of

*The substance of this paper was first given at a meeting of the European Molecular Liquids Group at the University of Bielefeld in September 1985.

polarization on a relatively long time scale. Later work has shown that in all cases such measurements when taken over a wider frequency range reveal the existence of other processes with smaller time constants. To interpret the behavior of a molecule that is subject to Brownian fluctuations, Debye proposed that the molecule be treated as a sphere of radius a, turning against a drag due to its neighbors characterized by a viscosity β. This leads to an expression for the relaxation time τ:

$$\tau = 4\pi\beta_a^3/kT\,. \tag{3}$$

It is expected that β will respond to changes of temperature and pressure about as the macroscopic viscosity does. It was pointed out by Fröhlich (1949) that for this model to be valid it must relate to dilute solutions of an axisymmetric molecule in a nonpolar solvent of relatively low viscosity. In the alternative model of a dipole switching its orientation by hopping over a relatively high potential hill, the height of the hill, Q, does not affect the equilibrium permittivity. However, it does affect the rate at which the system approaches equilibrium. The probability of making a hop, and therefore the rate of its occurrence, is proportional to $\exp(-Q/kT)$. Debye (1929) considered this to be an appropriate model to apply to the strongly pronounced relaxation observed in ice. Fröhlich and Sack (1944) also applied it to solid polar hydrocarbons. The behavior of τ can be discussed by an extension of the theory of absolute rate processes (Glasstone et al., 1941). The reorientation of a molecule is supposed to require that it enters an activated state, symbolized by ‡. The state is characterized by extra volume and altered entropy, and exists in equilibrium with molecules in their normal state. The appropriate function to describe two states in contact is the Gibbs free energy G. The theory of absolute rate processes leads to

$$K = 1/\tau = k'(kT/h)\exp(-\Delta G^{\ddagger}/RT) \tag{4}$$

where h is Planck's constant, k is Boltzmann's constant, and for want of better information k' is made equal to unity. Standard thermodynamic relations can be applied to eq. 4 to yield values of $\Delta H^{\ddagger}$, $\Delta V^{\ddagger}$, and $\Delta S^{\ddagger}$. In practice experimental results do not fit eq. 4 exactly, but it is a useful basis for examining the effects of pressure and temperature.

Debye's theory was modified by Onsager (1936). He realized that the local field which acts on the molecule cannot be related to the external field by the Lorentz-Lorenz factor used in the CM formalism. His equation for a polar fluid is

$$\epsilon - n^2 = [3\epsilon/(2\epsilon + n^2)]\ [4\pi\mu_0^2 N/3kT]\ [(n^2+2)^2/9] \tag{5}$$

where N is the number density and μ_o is the dipole moment of an isolated molecule. When applying eq. 5 to measured data various workers have taken the value of n measured for the sodium D line and written $\epsilon(\infty)$ in place of n^2 in eq. 5, where $\epsilon(\infty) = 1.05$ or $1.10\ n_D^2$. When applied to many polar liquids, Onsager's equation relates ϵ to ρ quite well. This is best seen when high pressures as well as temperature have been used to vary ρ. For example, in the case of eugenol, which is a phenol, in which the hydroxyl group is intramolecularly bonded, the equilibrium permittivity can be predicted from a knowledge of density and a set of measurements at ambient conditions, with an accuracy of 1.5% between 273 K and 323 K. The effect of pressure is somewhat greater. At 293 K an increase of pressure from ambient up to 400 MPa produces a change in permittivity some 5% less than is predicted by eq. 5 (Scaife, 1976). There are, however, groups of materials for which Onsager's equation proves quite inadequate, notably hydrogen-bonded liquids such as water and the alkanol family.

Kirkwood (1939) modified the Onsager analysis by considering a small sphere of material with the molecule of interest at its center. The sphere is large enough to allow the behavior of the dielectric that surrounds it to be calculated by macroscopic electrostatic theory. Yet it is small enough to allow, in principle at least, account to be taken of the interactions between the central molecule and each of its neighbors. Fröhlich (1949) analyzed this model and arrived at the now generally accepted equation,

$$9(\epsilon - n^2)(2\epsilon + n^2) / \epsilon(n^2+2)^2 = 4\pi N g \mu_0^2 / kT . \tag{6}$$

The correlation parameter $g = \mu \cdot \mu'$, where μ is the dipole moment of a molecule at the center of the sphere and μ' is the average dipole moment of a spherical region embedded in its own medium if one of its dipoles is kept in a fixed orientation. The product $\mu \cdot \mu'$ can be written:

$$\mu \cdot \mu' = \mu^2(1 + z < \cos\theta >) \tag{7}$$

where z is the number of nearest neighbors, and θ is the angle between the directions of a pair of dipoles, with $< \cos\theta >$ being an average over all orientations of pairs formed by the central molecule and every other molecule of the spherical region. The computation of $< \cos\theta >$ requires a knowledge of the energy of interaction between neighboring dipoles in the liquid in terms of the angular orientations of their axes. In most cases sufficient information is not available to carry out this calculation. However, a generalization can be made, that molecular interactions which favor parallel alignments should yield $g > 1$, while antiparallel alignments should give values of $g < 1$.

When eq. 6 is applied, various factors must be considered. Experiments measure the product $Ng\mu_0^2$, where N, g, and μ_0 refer to a single "unit". Should a complex of two or more dipoles be formed, for example a dimer, then μ_0 should be replaced by the net moment of the complex and the number density N should be reduced by the size of the complex. Equation 6 is derived on the basis that every dipole within the spherical region contributes equally. There may be factors that disable some molecules from responding to the electric field. Even though each molecule eventually has an opportunity to respond, at any given instant the fraction so disabled would remain the same. This again amounts to a reduction in the number density N. Finally, there is always the possibility of more than one species existing at the same time, e.g., a mixture of single molecules and complexes of one kind or another. The value of g calculated from eq. 6 will then be an average.

In discussing the behavior of the n-alkanols and isomeric alkanols the precision of the estimate of μ_0 and $\epsilon(\infty)$ is not critical. As was found by Smyth and Stoops (1929), Dannhauser and Cole (1955), and Johari and Dannhauser (1968), these families exhibit values of g ranging down from 3 to one tenth of this, even though the same values of $\epsilon(\infty)$ and μ_0 are used in interpreting measurements on each member. Before briefly reviewing the experimental evidence, it is useful to distinguish four different categories among the family of isomeric alkanols. Data for a representative of each category is given as a plot of g versus density in Figs. 4 to 7. The first of these, (a), consists of the 1-alkanols, molecules in which the hydroxyl group is attached to a carbon atom at an end of the carbon chain. The second, (b), comprises isomers in which the hydroxyl group is attached to some other carbon atom. A third, (c), consists of those isomers which have methyl branching of the carbon chain, but at a position that does not cause steric interference with the hydroxyl group. In the last, (d), there are those isomers in which branching does cause steric interference. In this connection attention must be directed to the fact that except for category a, nearly all the isomers can exist as enantiomeric variants; that is, the hydroxyl group may be attached to either of two different bonds on a given carbon atom. Moreover, the hydrogen bond formed with the lone pair electrons of the oxygen atom of the hydroxyl group may take two different, roughly tetragonal bond directions. Given that most specimens consist of racemic mixtures, these facts increase the number of possible structural arrangements in chains that must be considered.

Early speculation about the nature of the structure created by hydrogen bonds in alcohols was based on X-ray diffraction measurements of liquid samples. Data on methanol and 1-nonanol led Zachariasen (1935), for example, to analyze a structure in which the carbon chains in the case of 1-nonanol, and the methyl groups in the case of methanol, are disposed on alternate sides of a hydrogen-bonded chain of oxygen atoms. The O H..O link is considered to be linear; adjacent links are angled at approximately the tetrahedral angle, and free rotation about the bond axis is assumed to be possible. Oster and Kirkwood (1943) applied Kirkwood's (1939) model, assuming that the average configuration of a chain is spherical. The value for g that they obtained was 2.57, which is much smaller than the values of 4 to 4.5 observed at low temperatures by, for example, Bordewijk (1968).

The dipole orientation of a molecule forming part of a chain can be correlated with that of its neighbors in one of two ways. The most obvious is that the chain moves, for a period at least, as an entity, each segment preserving its orientation relative to its neighbors. Except in states of low density and for chains of relatively small numbers of elements, it is probable that a chain will be only partially free to move as an entity, because it will be pinned by its neighbors. Freedom for a dipole to respond to an applied field will therefore extend at a particular instant to only a limited distance along the chain. In order to explain the variation of g with temperature, Brot (1957) assumed that chains would have various lengths, each with its own average g. He proposed to calculate the equilibrium constants regulating the populations of chains of different lengths and therefore contributing different amounts to the average, measured, g by analyzing the effect of temperature on g. In order to explain values of $g<1$ Dannhauser (1968) extended this analysis by also postulating the existence of rings, such as dimers and trimers. This amounts to assuming smaller values of N, and a different value of μ_0 for the units. While dimers with very small moments do form in carboxylic acids, where bond angles are favorable to such arrangements, their existence in alkanols would require a large degree of bending in the bonds, and is therefore much less likely to be found. Nevertheless small rings of low dipole moment could perhaps form in favorable circumstances.

The theories that have been outlined above may perhaps explain dielectric behavior at relatively high temperatures and low densities, as for example in methanol close to its critical temperature, 513 K, where g is still about two (Dannhauser and Bahe, 1964; Franck and Deul, 1978). Even so, the theory is not considered entirely satisfactory. Certainly in conditions of high density and low temperature a different explanation is called for. We therefore suggest a different model drawn from a consideration of the results of measurements at high pressures and low temperatures. Our model is based on the implications of two important facts.

(a) The highest values of g are observed under conditions of high density.

(b) Hydrogen bonding, as has been mentioned, persists even to temperatures as high as the critical temperature of methanol.

It seems reasonable therefore to assume that hydrogen-bonded chains persist under most conditions, regardless of the apparent value of g. What determines this apparent value of g, then, is the structural environment in which the chains exist. It is suggested that two conditions are necessary for values of g to exceed unity. On the one hand, chains must exist. On the other, the degree of local order must be sufficient for one chain to lie parallel to a neighboring chain, at least for an appreciable distance. It is asserted that such proximity is required both to permit the chain to sunder and so allow for dipolar reorientation to occur, and also to ensure the possibility of correlated motion by neighboring molecules when they have received sufficient thermal energy to break free from their chains.

In searching for an understanding of the structure that gives rise to large values of g, we should realize that it will be found in material which has a density close to that of the solid at ambient pressure, as for example in 1-heptanol at 283 K and 350 MPa, where $g=3$ (Fig. 4). Under these conditions molecules will be forced into mainly parallel orientations; spherical conformations are most unlikely. In the equivalent alkane, molecules would tend to form layers separated by the length of one molecule. In the case of 1-alkanols, molecules will array themselves on either side of the hydrogen-bonded, zigzag chain of oxygen atoms, as shown in Fig. 1a. The picture is essentially that of Zachariasen (1935), but with successive chains being laid down alongside each other until they form a two-dimensional sheet of chains. Tauer et al. (1952) described such a structure in solid methanol. The layers will now be rather more than double the length of the alkyl group in width. The molecules in a given layer will be laced together by the sheet containing the chains. The dipoles forming this sheet will tend to lie in one direction, while the direction in the next sheet will be antilikeness here to this. There is a likeness here with solid ketones, for which Fröhlich (1946) suggested that although of course there is no hydrogen bonding in ketones, the dipoles would be aligned in a two-dimensional sheet, and adjacent sheets would have an antiparallel dipole orientation. It is suggested that individual molecules move chiefly by rotation about their long axes, though some small lateral movement will also be necessary. The result of such movements will be the breaking of two hydrogen bonds and the pos-

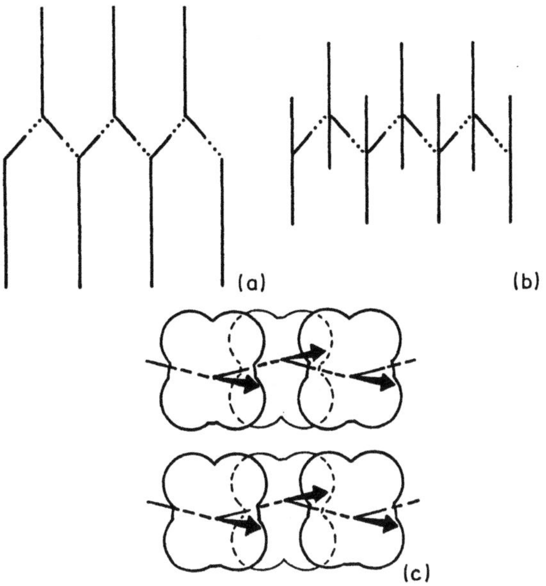

Fig. 1 — Arrangement of hydrogen-bonded chains for isomers. (a) 1-alkanols. (b) unbranched alkanols with hydroxyl group not attached to the end of the carbon backbone; the hydrogen bonds are shown dotted. (c) End view of molecules, showing chain stacking.

sible making of two new bonds, so providing for the possibility of that correlation of reorientation which is required to ensure $z < \cos\theta >$ in eq. 7 is greater than zero. In fact, since alternate dipoles in a chain are parallel, and adjacent dipoles are inclined at an angle of approximately 70° (Fig. 1c), we can calculate that $1 + z < \cos\theta >$, assuming four nearest neighbors, is 3.68. This is close to the maximum observed value of g. It can be seen that in effect such rotations cause the orientation of a chain to be switched from one direction to another mutually perpendicular to it. It is probable that the path of individual chains across the two-dimensional sheet is continually changing. The picture is not unlike that proposed for water by Bernal and Fowler (1933), but with linkages in two rather than three dimensions. The fact that z is four and that g has an upper limit implies that although many such rotations are required to establish a new orientation for a chain, each one is statistically independent of the others. It will be noted that a feature of this model is the relative uniformity of motion normally performed by individual molecules. This is a prerequisite for the existence of a single relaxation time governing the slow dielectric response. There is some evidence that in solid methanol at least, dipoles retain some freedom to rotate at temperatures close to the melting point (Denney and Cole, 1955; Tauer and Lipscomb, 1952). Certainly in the cycloalkanols the value of g changes relatively little on freezing (Poser et al., 1985; Pingel et al., 1984). It is suggested that as temperature rises and density decreases, the planes of chains persist, but become increasingly disordered. Gradually the planes will extend over smaller areas and tend to lose their structure, but nevertheless reorientation of individual molecules in chains will continue to require the existence of some order on a local scale. Without this they cannot respond to the field.

The exact way in which chains build up in isomers with the hydroxyl group located away from the end of the molecule is more uncertain because of the existence of enantiomers. However, it is clear that at high densities there must be a parallel orientation of the alkyl parts of the molecules. As with the 1-alkanols it is expected that chains will stack one above another to form a sheet of hydrogen-bonded oxygen atoms. But in this case the ends of the alkyl groups will protrude by different amounts at either side of the layers, (Fig. 1b). At high densities there will have to be a degree of alignment between the structures in adjacent layers. This interlocking represents a significantly greater degree of order than that in 1-alkanols and it may also give the chains themselves some protection from a tendency to be disrupted by high pressures.

As temperature increases, chains will continue to exist for all classes of alkanols, but they will increasingly often fail to align with neighbors in a way that allows them to break free and become involved in correlated motion. It may be said that N, the average number density of molecules suitably positioned to reorient if given the necessary thermal impetus, is reduced. Molecules forming part of a chain that has not been aligned will continue to find difficulty in breaking free, and will respond only weakly to an electric field. The average value of $Ng\mu_o^2$ will fall as smaller numbers of chains align favorably. Eventually chains will break up at higher temperatures, and the apparent value of g will approach unity.

This picture can also readily explain the behavior of isomers in category d. Steric hindrance in such isomers will add to the difficulty of alignment of chains, even at low temperatures, and thus lead to a reduction in the value of N, and also in the apparent value of g, even to less than unity. But the application of high pressure will tend to overcome the steric barriers to packing, as can be seen, for example, in the case of 2-methyl-2-hexanol (Fig. 7). In general, chains seem very tolerant to steric variability. Thus mixtures of methanol and 1-propanol (Denney and Cole, 1955), and 1-heptanol and 4-heptanol (Bordewijk, 1968), show no tendency for a reduction in g as a result of intermixing.

The behavior of solutions of alkanols in nonpolar solvents can also be understood in terms of our model. At very low concentrations it is found that g rises as alkanol concentration increases (Bordewijk et al., 1973). This is explained as due to the formation of very short chains containing perhaps only two molecules, which are capable of moving as an entity. However, as soon as these chains grow in size they effectively trap the dipoles. Because solvent molecules get in the way, chains cannot easily align. Once again we can say that N has been reduced, and the apparent g falls. As the concentration of alkanol increases, the fraction of obstructing, solvent molecules is diminished, more and more chains are found in favorable alignment with neighbors, and g begins to rise. This trend will continue until the liquid is all alkanol.

Turning from consideration of equilibrium permittivity $\epsilon(0)$, let us consider the time-variant behavior. Many of the alkanols share with ice and water a complex permittivity plot that is dominated by a single, Debye-like relaxation process, (Fig. 2). If this is interpreted as the consequence of molecules jumping over a potential hill, then a single relaxation time implies a uniformity in the height of such barriers. This is just what our model provides. The magnitude of τ depends to only a small extent on the size of the molecule (Böttcher and Bordewijk, 1978). The principal influence is the availability of extra energy necessary to jump the potential barrier. As the temperature rises, the ordered stacking of chains becomes less and less complete, but reorientation can take place only when chains do align,

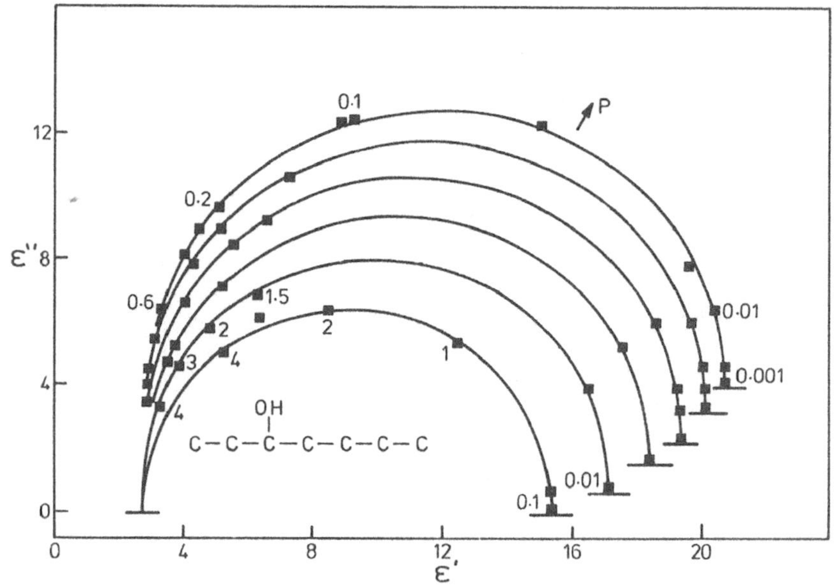

Fig. 2 — Plot of complex permittivity for 3-heptanol at 253.2 K and pressures of 0.1, 49, 98, 147, 196 and 245 MPa. Frequencies shown in kHz. (Taken from Vij et al. (1981), with permission.)

at least in a localized region, to permit molecules to break free. It is important to remember that under similar conditions alkanols have values of τ that are significantly larger than similar molecules which do not have hydrogen bonding. At 273 K and ambient pressure, 1-heptanol has $\tau = 0.48\ 10^{-8}\,s$, while heptyl bromide has $\tau_1 = 0.02\,10^{-8}\ s$ and $\tau_2 = 0.0005\,10^{-8}\,s$, (Higasi, 1961). The larger τ is no doubt associated with bodily movement of the whole molecule. Dielectric relaxation in alkanols is clearly a much slower process than in equivalent alkyl halides.

In situations in which g is < 1, as for example in 2-methyl-2-hexanol at low pressures, the complex permittivity plot (Fig. 3), is no longer almost entirely a semicircle. The amplitude of the dispersion has shrunk as the main Debye-like process, which is possible only in an ordered environment, progressively disappears, and the higher-frequency processes in turn become relatively more important. The value of τ, however, remains if anything somewhat larger than in 1-heptanol, for which $g = 3.5$ at the same temperature and pressure. In other words, the process of reorientation remains essentially the same.

Consider now the parameters of the activated state as determined by measurements using pressure as well as temperature as a variable (Table 1). The height of the potential barrier can be associated with $\Delta H^{\ddagger}$. The activation entropy $\Delta S^{\ddagger}$ can best be understood if we appreciate that the entropy needed by molecules of alkanols of whatever category, in the activated state, is roughly the same. Therefore the larger the value of $\Delta S^{\ddagger}$, the greater is the relative degree of order in the structure from which the molecule breaks free when it reorients. It can be seen that $\Delta H^{\ddagger}$ is least for 1-alkanols. Other categories have values two or three times as high. This reflects the greater degree of influence of neighbors on the ability of a molecule to rotate. In all these isomers there will be a degree of interlocking of

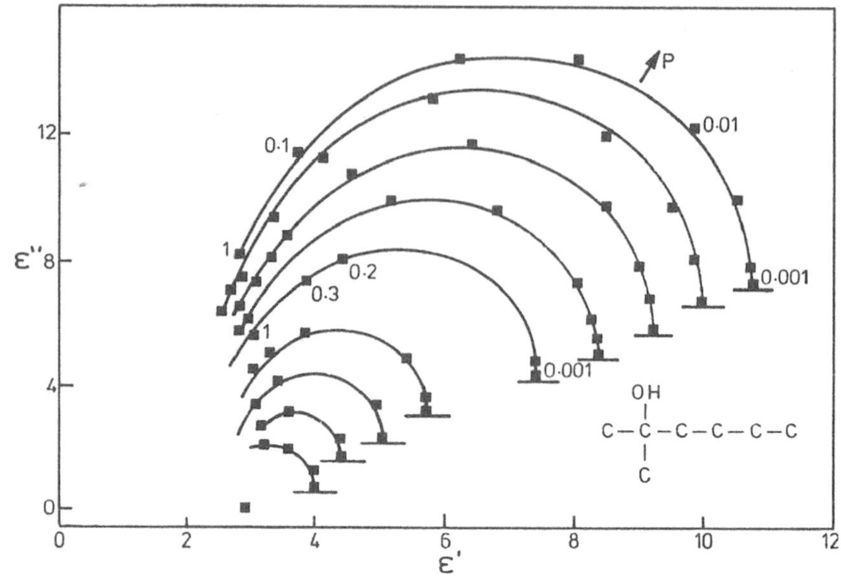

Fig. 3 — Plot of complex permittivity for 2-methyl-2-hexanol at 248.7 K and pressures of 0.1, 98, 108, 147, 186, 206, 226, 245 and 265 MPa. Frequencies shown in kHz. (Taken from Vij et al. (1981), with permission.)

adjacent layers. Likewise the trend in the size of activation volume reflects the increased space required by branched isomers especially, to allow rotation. The activation volume for 1-heptanol is only 10% or so of the molar volume. This compares with almost 30% in the case of eugenol, which is compatible with the assumption that eugenol reorients as a single molecule in a disordered environment, while 1-heptanol rotates about its long axis in a cage formed by its neighbors. The values of $\Delta\nu^{\ddagger}$ for isomers in categories (c) and (d) on the other hand are comparable with that of eugenol. But it is in the values of activation entropy that the distinction between 1-alkanols and other alkanols is most evident. Referring to Figs. 4-7 shows that application of pressure is capable of causing g to increase even in cases where it is well below unity at ambient pressure. It may be that at the greater densities produced by high pressures, the racemic mixture of enantiomers is segregated on a microscopic scale, and this facilites a greater perfection of structure, and with this, increased density.

Data on the molar volume of both n-heptane and some heptanols is available (Vij et al., 1978; Scaife and Lyons, 1980), and comparisions can be made. At ambient pressure heptanols have smaller molar volumes than n-heptane at all temperatures. The difference is greatest at 100°C, reaching 7% for 1-heptanol, and rather less for other isomers. This reflects the effect of hydrogen-bonded chains in maintaining some degree of structure, even at elevated temperatures. At 100°C the molar volume of 1-heptanol is 1.5% less than that of 3-heptanol, but at −20°C the values are almost identical. At high pressures the differences in molar volume are greatly reduced, and lie within ±1% or so of that of n-heptane. Finally, in order of compressibility, we can rank n-heptane highest, followed by the branched alkanols, with 3-heptanol and 1-heptanol the least affected by pressure.

Table 1. Activation parameters for reorientation in several categories of isomeric heptanols and octanols at 200 MPa[a]

Alkanol	g	$10^8\tau$, (s)	$\Delta H^{\ddagger}$, $\left(\frac{\text{kcal}}{\text{mole}}\right)$	$\Delta V^{\ddagger}$ $\frac{\text{cm}^3}{\text{mole}}$	$\Delta S^{\ddagger}$, $\frac{\text{cal}}{\text{mole.K}}$	T, K
Category a						
1-heptanol	3.4[b]	4.5[b]	8.3[b]	14[b]	9[b]	258
Category b						
2-octanol[c]	3.2	25	17	23	40	259
3-heptanol	3.7[b]	57[b]	17[b]	26[b]	35[b]	258
3-octanol[c]	3.2	93	18	22	39	258
Category c						
6-methyl-3-heptanol[c]	2.8	160	21	33	50	259
5-methyl-3-heptanol[c]	1.7	205	26	38	68	257
Category d						
3-methyl-2-hexanol	2.0	12	20	31	49	259
2-methyl-2-hexanol	1.0	23	25	43	68	258
2-methyl-3-heptanol[c]	0.4	32	23	35	60	260

[a]Taken from Vij et al. (1981) with permission.
[b]Estimated by interpolation.
[c]Johari and Dannhauser (1969).

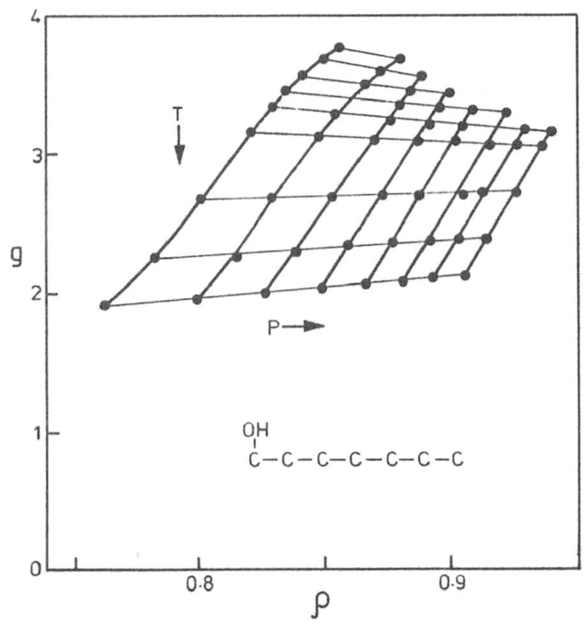

Fig. 4 — Plot of correlation parameter g against density ($g.cm^{-3}$), for 1-heptanol. Isobars at 0.1, 50, 100, 150, 200, 250, 300, and 350 MPa. Isotherms at −29, −20, −11, 0, 10, 20, 50, 75, and 101°C. (Taken from Vij et al. (1978), with permission.)

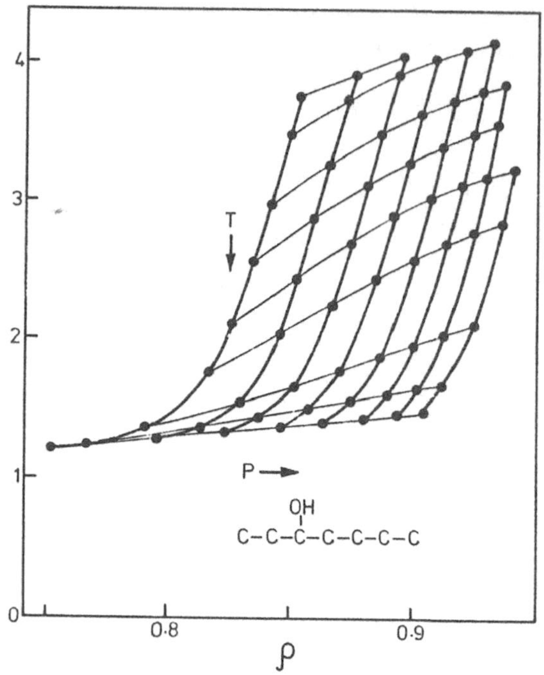

Fig. 5 — Plot of correlation parameter g against density (g.cm^{-3}), for 3-heptanol. Isobars at 0.1, 50, 100, 150, 200, 250, 300, and 350 MPa. Isotherms at −20, −10, 0, 10, 23, 50, 60, and 100°C. (Taken from Vij et al. (1978), with permission.)

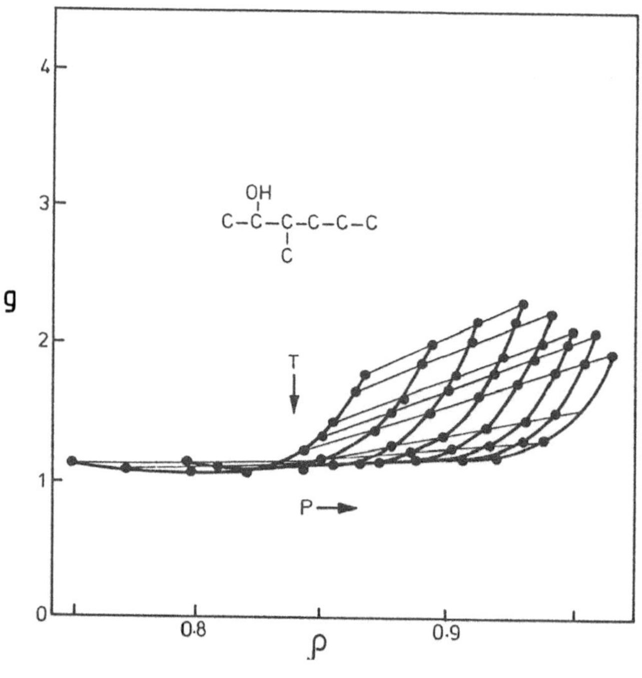

Fig. 6 — Plot of correlation parameter g against density (g.cm^{-3}), for 3-methyl-2-hexanol. Isobars at 0.1, 50, 100, 150, 200, 250, 300, and 350 MPa. Isotherms at −29, −25, −17, −10, −1, 24, 50, and 99°C. (Taken from Vij et al. (1978), with permission.)

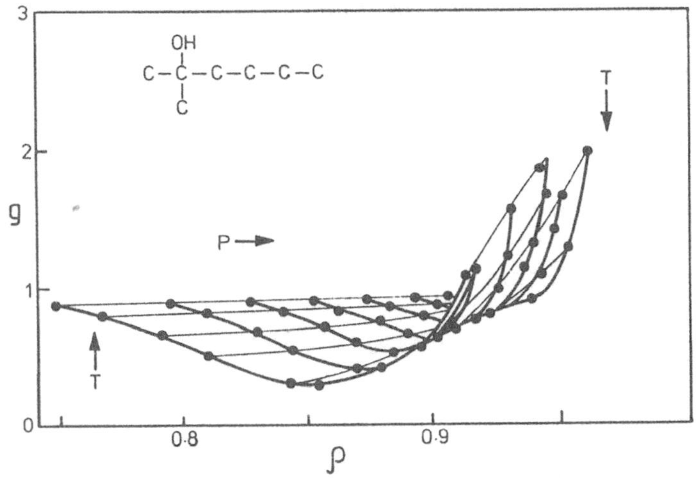

Fig. 7 — Plot of correlation parameter g against density ($g.cm^{-3}$), for 2-methyl-2-hexanol. Isobars at 0.1, 50, 100, 150, 200, 250, 300, and 350 MPa. Isotherms at −29, −25, −1, 24, 50, 75, and 99°C. (Taken from Vij et al. (1978), with permission.)

SUMMARY

(a) Most molecules in liquid alkanols exist in hydrogen-bonded chains, even when the correlation factor is <1.

(b) The number density N of molecules active in reorientation depends on the extent of structure in the arrangement of these chains.

(c) The main variable should not be considered to be the correlation factor g, which tends to a maximum value of 3.5 to 4, but rather the number density N as defined above.

REFERENCES

J.D. Bernal and R.H. Fowler, J. Chem. Phys. **1**, 515 (1933).

C.J.F. Böttcher and P. Bordewijk, *Theory of Electric Polarization, Vol. 2: Dielectric in Time-Dependent Fields*, Elsevier, Amsterdam, 1978, p. 254.

P. Bordewijk, *Fluid structure and dielectric behavior of the mono-alcohols*, Doctoral Thesis, Univ. Leiden, 1968.

P. Bordewijk, M. Kunst, and A. Rip, J. Phys. Chem. **77**, 548 (1973).

P.C. Brot, Ann. Phys. (Paris) **2**, 714 (1957).

W. Dannhauser, J. Chem. Phys. **48**, 1911 (1968).

W. Dannhauser and L.W. Bahe, J. Chem. Phys. **40**, 3058 (1964).

W. Dannhauser and R.H. Cole, J. Chem. Phys. **23**, 1762 (1955).

P. Debye, *Polar Molecules*, Chemical Catalog Co., New York, 1929.

D.J. Denney and R.H. Cole, J. Chem. Phys. **23**, 1767 (1955).

E.U. Franck and R. Deul, Faraday Discuss. Chem. Soc. **66**, 191 (1978).

H. Fröhlich and R. Sack, Proc. R. Soc. London **A182**, 388 (1944).

H. Fröhlich, Proc. R. Soc. London **A185**, 399 (1946).

H. Fröhlich, *Theory of Dielectrics*, Clarendon Press, Oxford, 1949.

S. Glasstone, K.J. Laidler, and H. Eyring, *Theory of Rate Processes*, McGraw-Hill, New York and London, 1941.

K. Higasi, *Dielectric Relaxation and Molecular Structure*, Monograph No. 9, Res. Inst. of Applied Electricity, Hokkaido Univ., 1961.

G.P. Johari and W. Dannhauser, J. Phys. Chem. **72**, 3273 (1968).

G.P. Johari and W. Dannhauser, J. Chem. Phys. **50**, 1862 (1969).

J.G. Kirkwood, J. Chem. Phys. **7**, 911 (1939).

S. Mizushima, Bull. Chem. Soc. Jpn. **1**, 47, 83, 115, 143 and 163 (1926).

L. Onsager, J. Am. Chem. Soc. **58**, 1486 (1936).

G. Oster and J.G. Kirkwood, J. Chem. Phys. **11**, 175 (1943).

N. Pingel, U. Poser, and A. Würflinger, J. Chem. Soc., Faraday Trans. **80**, 3221 (1984).

U. Poser, L. Schulte, and A. Würflinger, Ber. Bunsenges. Phys. Chem. **89**, 1275 (1985).

W.G. Scaife, J. Phys. D: Appl. Phys. **9**, 1489 (1976).

W.G. Scaife and C.G. Lyons, Proc. R. Soc. London **A370**, 193 (1980).

C.P. Smyth and W.N. Stoops, J. Am. Chem. Soc. **51**, 3330 (1929).

K.J. Tauer and W.N. Lipscomb, Acta Cryst. **5**, 606 (1952).

J.K. Vij, W.G. Scaife, and J.H. Calderwood, J. Phys. D: Appl. Phys. **11**, 545 (1978).

J.K. Vij, W.G. Scaife, and J.H. Calderwood, J. Phys. D: Appl. Phys. **14**, 733 (1981).

W.H. Zachariasen, J. Chem. Phys. **3**, 158 (1935).

12. Surface Transport in Time and Frequency Domains

A. K. Jonscher

With 13 Figures

ABSTRACT

Recent experimental studies are described of the time- and frequency- dependence of conduction processes on humid surfaces of insulating and semi-insulating materials. The two-dimensional conduction by slowly mobile ionic carriers is shown to lead to characteristic "signatures" in time and in frequency, which are associated with nonlinear behavior with respect to the applied voltage and to nonuniform spatial patterns of current flow. It is shown that some of these processes are associated with electrochemical interactions leading to storage of charge and energy in chemical reactions. Novel features of the dependence of the admittance on the geometrical dimensions of the samples are related to proposed mechanisms involving contact and "bulk" surface flow. Experimental results are quoted for humid zeolites, sand, mica paper, glass, and mica single crystals. The relevance of this novel approach to other areas of transport is outlined.

INTRODUCTION

Surface transport on insulators in the presence of humidity and other conditions giving rise to the production of mobile ionic species represents an important form of charge transport which affects many practical situations and which has a considerable fundamental significance. In most general terms, one is dealing here with the two-dimensional analog of the familiar three-dimensional conduction processes and, as is often the case, the change from three to two dimensions introduces into the observed behavior changes of kind rather than merely of degree. Proceeding by analogy, we may distinguish two-dimensional electronic and ionic transport processes, and the former may, in principle, involve free electrons moving in two-dimensional bands of allowed states, or localized electrons moving by some form of hopping motion along the surface.

The present paper is concerned with ionic transport but it is advisable to mention briefly the electronic situations so as to clarify the picture. Truly two-dimensional free-carrier transport has been considered in the context of bands of surface states that were believed to exist on some solids (Shockley, 1939). More important in present-day context is the motion of electrons in strongly inverted bands in semiconductors next to insulator interfaces in metal-oxide-semiconductor devices, which gives rise to a quasi-two-dimensional transport with important consequences, especially in the limit of strong quantization of electronic states in the direction normal to the surface. The corresponding localized hopping transport would arise in situations in which the inversion was insufficient to cause free-carrier densities to be present but defect-related localized levels might facilitate hopping transport near the surface. It is possible to envisage a truly surface mode of electronic hopping conduction between localized levels on the surface or interface between two media, e.g. at an insulator-vacuum interface, but these will not enter into our discussion since there is insufficient evidence for the existence of this form of transport.

Concentrating, therefore, on ionic surface transport, we note that one of the fundamental boundary conditions is given by the contacts to the surface, which are typically metallic. Just as in the three-dimensional situation (Fig. 1) contacts are two-dimensional, in the case of two-dimensional surface flow contact must be regarded as subject to one-dimensional boundary conditions. This raises interesting questions regarding the nature of such contacts, since concepts such as capacitance and contact resistance have to be redefined for this process, which differs in important respects from the corresponding three-dimensional situation.

One of the most important questions here is the nature of the processes that mediate the transfer of charges from the metallic contact to the ionically conducting surface. In the metal one expects electronic conduction and the charge transfer process must involve some form of ionization of neutral species which then are transported away from the interface, or alternatively the neutralization of ionic species arriving at the line contact, with the generation of an electron (or a hole) which then continues the transport in the metallic electrode. It is possible to envisage ionic transport across the contact interface, but experience with three-dimensional ionic conductors suggests that transport of ions across interfaces is energetically unfavorable and we have no reason to think that it would be easier in two-dimensional conduction processes.

Whatever detailed physical and chemical processes determine the supply of carriers to and from the electrodes in two-dimensional surface transport, we expect

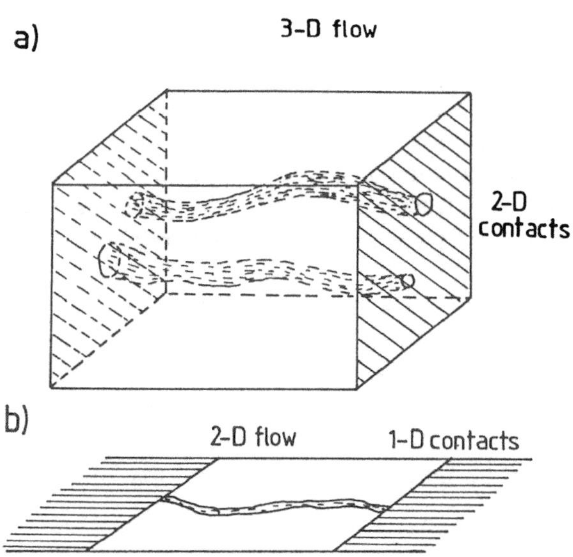

Fig. 1 — A schematic representation of 3-dimensional flow with 2-dimensional contacts and with the formation of filaments of high current density arising from the presence of inhomogeneities, diagram (a). 2-dimensional flow with 1-dimensional contacts and with the formation of filaments is shown in diagram (b)

the formation of some one-dimensional analog of the conventional charge double layer arising at metal-electrolyte contacts with both liquid and solid electrolytes. The field configuration would be very different emanating from the edge of the metallic conductor and it is evident that the detailed nature of this edge, its "sharpness" on the atomic scale and its uniformity, might play an important role in determining the properties of such contacts. The high fields involved, with the very high space and surface charge densities, would give rise to ionization phenomena that are necessary to supply the charge carriers.

A very important concept entering into our discussion of two-dimensional flow on surfaces concerns the *spatial uniformity* of this transport. In three-dimensional flow it is possible to distinguish longitudinal nonuniformities, arising from the presence of some barriers that limit the flow of current along the direction of the applied field, and transverse nonuniformities resulting in the formation of filamentary paths that concentrate the flow of current in discrete points or small areas on the cross-section of the conducting region normal to the direction of the applied field, as shown in Fig. 1. Filamentary flow may be particularly important in hopping conduction in disordered materials where there is evidence for this type of behavior arising from the statistical distribution of favorable paths.

The corresponding phenomena of nonuniformity in two dimensions are potentially even more critical, since the lower dimensionality restricts the number of alternative paths in the event of one of them being forbidden. We expect both filamentary flow and nonuniform field distribution along the lines of flow in two-dimensional conduction. What is not easily determined is the spatial scale attaching to this nonuniformity, since this would require making visible the paths of flow and there are not many sensitive methods of achieving this end. Nevertheless, some progress may be expected in this direction in the near future.

One specific form of spatial nonuniformity found in both three- and two-dimensional systems arises from the nonuniform emission and absorption of charge carriers at electrode boundaries. Planar electrodes are often "patchy" in their characteristics, whether the charges are electrons or ions, and this is due to sometimes very slight nonuniformities of the interface. The same situation would be expected, *a fortiori*, in one-dimensional edge contacts to surface conduction. If the material adjoining to the patchy electrodes is itself assumed to be uniform, the effects of electrode nonuniformities would be expected to be smoothed out over a "spreading" distance related to the mean distance between emitting areas. In this limit, the effect would only add the spreading resistance of the emission sites to the resistance of the volume material, and in extreme situations this spreading resistance may, indeed, limit the behavior of the system, which is then independent of the geometrical spacing between the electrodes.

One limiting example of surface transport in a quasi-three-dimensional setting is provided by humid zeolited (Haidar and Jonscher, 1986; Jonscher and Haidar, 1986) in which a regular three-dimensional network of interconnected cavities is capable of absorbing large quantities of water and other substances that profoundly affect its conducting properties. Examples of this behavior will be given in the present paper since they are considered to be highly relevant to the topic of surface transport.

A very important aspect of nonuniform transport in disordered three- and two-dimensional systems concerns their dependence on time and frequency; this will form the principal subject of the present paper. In the case of three-dimensional systems the main features of this transient behavior are at least reasonably well established, even if their interpretation is not necessarily generally agreed. In two-dimensional surface transport, by contrast, very little is known about time- and frequency-dependent behavior and consequently no attempts have been made, to our knowledge, to develop models that might elucidate the phenomena in question. We hope to be able to show in the following that a considerable amount of information may be obtained from the study of such transient behavior. As against this, the hitherto generally adopted attitude is that surface transport is tantamount to the superposition of an ohmic time- and frequency-independent conductor over the surface of an insulator.

FREQUENCY- AND TIME-DOMAIN REPRESENTATIONS

When dealing with system responses that are not instantaneous with respect to the exciting forces one has to assume, first of all, that the response is linear in the applied force and that it is additive, i.e., the response to consecutive excitations is equal to the sum of the responses that would have arisen from each separate excitation. With this assumption of linearity and superposition it is possible to define the response in terms of either a delta-function stress in time or a sinusoidally variable stress at an angular frequency ω. The total displacement of charge, i.e. the polarization arising from a delta-function stress $E\Delta t$ is defined as (Jonscher, 1983a, referred to here as DRS)

$$\Delta P(t) = \epsilon_0 f(t)\ (E\Delta t) \tag{1}$$

where the product of the field amplitude e and time duration Δt defines the strength of the delta function and remains constant as $\Delta t \rightarrow 0$. ϵ_0 is the permittivity of free space. The superposition principle leads to the following response to an arbitrarily time-varying field $E(t)$ in the form of a convolution integral with the characteristic response function $f(t)$

$$P(t) = \epsilon_0 \int_0^\infty f(\tau) E(t-\tau)\, d\tau\,. \tag{2}$$

In the particular case where the field is a step function of amplitude E_0 starting at $t = 0$, the convolution integral simplifies and its derivative, which gives the current flowing in the system under step-function field excitation, is then

$$i(t) \doteq \epsilon_0 E_0\, f(t)\,. \tag{3}$$

This gives a very simple interpretation of the physical sense of the response function $f(t)$ in terms of the current response of the system.

It is easily shown by applying the Fourier transformation to both sides of eq. 2 that the harmonic response to an alternating driving field at a frequency ω is

$$P(\omega) = \epsilon_0 \chi(\omega)\, E(\omega) \tag{4}$$

where the Fourier components $P(\omega)$ and $E(\omega)$ are to be understood in the sense of vectorial notation in the complex time plane and the susceptibility $\chi(\omega)$ is defined as the complex Fourier transform of the function $f(t)$

$$\chi(\omega) = \chi'(\omega) - i\chi''(\omega) = \int_0^\infty f(t) \exp(-i\omega t)\,dt . \tag{5}$$

The real and imaginary components of $P(\omega)$ relate to the components of the polarization $\chi(\omega)$ that are, respectively, in phase and in lagging quadrature with respect to the driving field $E(\omega)$.

The above formulation in terms of volume polarization and electric field in the system is, strictly speaking, applicable to situations such as plane-parallel dielectric samples in which the field can easily be defined and the geometry is unambiguous. If this is not the case, and most surface conduction situations are a good illustration of this, it is possible to redefine our concepts in terms of the complex capacitance $C(\omega)$ and the voltage $V(\omega)$ applied to the system—geometrical considerations do not enter into consideration explicitly then but they become subsumed implicitly in the response function $f(t)$. On the other hand, the time and frequency dependences remain valid quite generally.

The presentation of our experimental results is in the form of logarithmic plots either of the real and imaginary components of the capacitance against frequency, or of the charging and discharging currents against time. In some cases we may adopt the technique of normalization of data with respect to temperature or relative humidity or some other relevant parameter (DRS; Hill and Jonscher, 1983), which leads to a better appreciation of the dependence of the response on these parameters and also enhances the reliability of the data obtained.

Since the complex susceptibility $\chi(\omega)$ is a Fourier transform of the characteristic function $f(t)$ it follows that the charging and discharging currents are Fourier transforms of the corresponding frequency-domain capacitive responses. It is instructive to list a few simple transformations to be able to appreciate the type of result to be expected in any given situation. These pairs of Fourier transforms are given in Table 1 and their time and frequency characteristics are shown in Fig. 2. The power-law type of response is seen almost universally in most situations, and in the case of surface transport it will be found that the most common type of behavior corresponds, at least at low frequencies, to the values of the exponent $n << 1$. In that case the time-domain response is almost independent of time, and we shall see many examples of this.

Another convenient method of presentation of dielectric data is in the form of the complex impedance plot. For the simple case of a parallel combination of capacitance C_0 and conductance G_0 this impedance is

$$Z(\omega) = \frac{1/G_0}{1 + i\omega/\omega_p} \text{ with } \omega_p = G_0/C_0 . \tag{6}$$

The plot of the real part against the imaginary part in the complex Z plane is a semicircle with diameter $1/G_0$ and with its highest point at $\omega = \omega_p$.

A very important aspect of time-to-frequency-domain transformation and vice versa is the question of linearity of the response in terms of the amplitude of the exciting signal. Experience shows that, while the high-frequency dielectric properties are very closely linear at all but the highest electric fields, approaching breakdown values, the position is very different in the case of transport over longer distances, the dispersive effects are involved and where linearity is no longer preserved; examples of this are given by DRS. This raises the question of the validity of frequency-domain measurements in nonlinear situations.

Two points may be made immediately. First, the application of an alternating signal at a given frequency ω to a nonlinear system results in the response at the fundamental frequency and at an infinite series of either odd or even harmonics, depending on whether the system is symmetric or asymmetric with respect to the origin. All instruments for measuring frequency-dependent responses, be they alternating current (ac) bridges or frequency response analyzers, are designed to reject rigorously all harmonics, unless they are specifically programmed not to do so, e.g. in the case of harmonics analyzers. Thus, without resort to special measures, the results of ac measurements give only the fundamental frequency response, which represents a suitably linearized first term in the Fourier series expansion of the complete nonlinear response. In the absence of detailed knowledge not only of the amplitude of consecutive harmonics but also of their phase relationship with respect to the driving fundamental waveform, it is impossible to obtain a complete answer to the question of the response of a nonlinear system. The conclusion must be that ac measurements on nonlinear systems are, at best, very approximate and may be seriously misleading in extreme cases.

Second, nonlinear systems do not admit linear integral transformations such as the Fourier transformation from the frequency-to-time domain, or the Kramers-Kronig (Hilbert) transformation between the real and imaginary components of the complex $C(\omega)$. One may still get an approximate result, but its validity remains in doubt.

The only route to measurements of the dynamic properties of nonlinear systems lies in time-domain measurement under step-function excitation, since this reflects the elementary fact that (DRS)

Mother Nature works in the time domain

We are going to see in the course of the following presentation that the charging and discharging currents are no longer related by the simple relation applicable in the case of dipolar systems

$$i_c(t) = |i_d(t)| + i_0 \tag{7}$$

where we take account of the fact that the discharging current is of the opposite polarity, but has the same time dependence, while i_0 is the direct current arising from the dc conductivity. Many samples of various types of departures from this simple relationship are given by DRS.

PHYSICAL ORIGINS OF TIME DEPENDENCE AND FREQUENCY DEPENDENCE OF TRANSPORT

It is instructive to reflect on the physical principles underlying any frequency and time dependence of transport phenomena, since this will serve to emphasize the causes of some of the observations that will be discussed.

Consider first a completely homogeneous medium, which may be solid, liquid or gaseous, containing some charge carriers of an electronic or ionic nature. The application of a step-function electric field to such a system will cause different types of response according to the nature of charge carriers in question. If these carriers are completely free, as would be the case in highly rarefied plasmas, with no collisions with any neutral, i.e. immobile, species, the response would be determined totally by inertial processes limiting the acceleration of the carriers in the applied field. In the case of a sinusoidally time-varying field, such ideally free charges would execute sinusoidal motions whose amplitudes would be in complete antiphase with the driving field, implying that their velocities would be in phase quadrature. This implies that no work is being done by the alternating field on these charges, energy input during one half cycle resulting in an enhanced kinetic energy which is then given back without any loss to the driving system in the other half cycle. The system exhibits zero current in phase with the driving field, i.e. zero conductivity. It is a paradox that the amount of in-phase current, i.e. conductivity, increases with increasing scattering with immobile species in the system, since only such collisions can lead to an exchange of energy between the system and the driving force. Electrically this type of response is illustrated by a perfect inductor-capacitor combination where energy is being exchanged between magnetic and electric storage form.

A physical example of such a system is represented by metals at ultraviolet frequencies that are well above any collision frequencies of the conduction electrons. In these conditions metals become transparent to radiation and damping is very low.

The other extreme form of time-dependence behavior is found in systems in which carriers are so severely damped by collisions that inertial effects are totally absent and the velocity of these carriers, i.e. the current, is wholly in phase with the driving force, independently of the frequency of this driving force. This is the limit of purely direct current (dc) conduction that is found in most solids conducting by heavily scattered free carriers at frequencies below the scattering or collision frequency. The same dc response would be obtained also in systems in which carriers are not free, in the sense of electrons or holes moving in quasi-free bands in crystalline solids, but are moving with very rapid Brownian or random motions under thermal excitation. Such examples may be found in liquid and solid electrolytes and generally in ionic conductors, also in molten salts and in liquid metals. In all these, the ionic charges, if not the electronic ones, have very rapid thermally activated random motions and any externally applied electric fields give rise to a velocity component that is practically instantaneous, i.e., dc response.

The most elementary example of an inhomogeneity-induced frequency dependence in a system consisting of rapidly scattered free carriers is afforded by a $p-n$ junction. Here, two essentially conducting regions are separated, within the same single crystal lattice, by a region depleted of free carriers which therefore behaves like an insulator and the $p-n$ junction may be modelled, to a first approximation, by a series combination of a perfect capacitor C_0 and a perfect conductor G_0, the latter only becoming effective in influencing the behavior at angular frequencies of the driving field in excess of $\omega_p = G_0/C_0$. The dielectric response of a system consisting of a series combination of C_0 and G_0 is given by the Debye form of frequency response in which the effective complex capacitance is given by

$$C(\omega) = C'(\omega) - C''(\omega) = C_0\,(1 + i\omega/\omega_p)^{-1}. \tag{8}$$

The actual behavior of $p-n$ junctions and Schottky diodes reveals considerable departures from this idealized behavior (DRS; Jonscher et al., 1986) arising from the presence in the space charge region of deep trapping levels, and also from interfacial processes in the case of Schottky barriers. The fundamental point remains, however, that the presence of an inhomogeneity is capable of profoundly affecting the transient response of a conducting medium involving free charge carriers.

It is evident that the special case of surface conductance would give rise to a frequency-independent dc behavior if the conducting fluid could be regarded as a perfectly homogeneous medium in which ionic charge carriers were free to move as in an ordinary electrolyte. We shall see from the experimental evidence to be presented later in this paper that this is far from being the case and that this model is certainly not applicable to the situations studied by us.

The prevailing form of frequency and time dependence that will be found in surface transport on insulators follows the "universal" power law t^{-n} and its counterpart ω^{n-1} shown in Table 1 and in Fig. 2 with values of the exponent n either close to unity or close to zero. As shown in Table 1, the characteristic feature of the power-law frequency spectra is that the real and imaginary components of the complex capacitance are in a frequency-independent ratio $C''(\omega)/C'(\omega) = \cot(n\pi/2)$ which is smaller than 1 for $n > \frac{1}{2}$ and larger than 1 for $n < \frac{1}{2}$. The resulting spectra consist of two power-law components shown schematically in Fig. 3, in which the high-frequency region has relatively low loss and the low-frequency branch is very lossy and strongly dispersive and is therefore known as Low-Frequency Dispersion (LFD) (DRS; Jonscher, 1978, 1983b).

In the context of *volume* transport in the presence of slowly mobile carriers of ionic or electronic nature (hopping electrons) LFD may be explained in terms of

Table 1. Fourier transforms of typical response functions

Time Domain	Frequency Domain
$\exp(-\omega_p t)$	$\dfrac{1}{1 + i\omega/\omega_p}$
t^{-n}	$[\sin(n\pi/2) - i\cos(n\pi/2)]\omega^{n-1}$

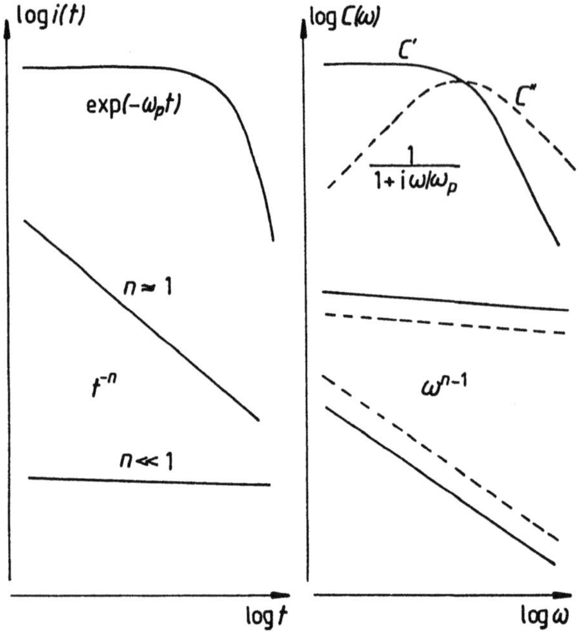

Fig. 2 — Corresponding pairs of Fourier transforms of typical functions, represented as the logarithmic plots of the time dependence of the charging currents and the frequency dependence of the real (—————) and imaginary (– – – – –) components of the complex capacitance. The exponential function leads to the Debye frequency response, the fractional power law in time gives the complementary fractional power law in frequency, in which the real and imaginary components are in a constant ratio $C'(\omega)/C''(\omega) = \tan(n\pi/2)$.

many-body interactions (Dissado and Hill, 1984) which also leads to the expectation of spatially nonuniform flow of the type shown in Fig. 1. While no theory exists of the corresponding *surface* transport, we suggest that an analogous situation arises and that the presence of surface LFD is to be associated with nonuniformity of charge flow, as in Fig. 1. We shall see later that there is good evidence that electrochemical interactions play an essential role in surface LFD, in complete contrast with the volume situation envisaged by Dissado and Hill.

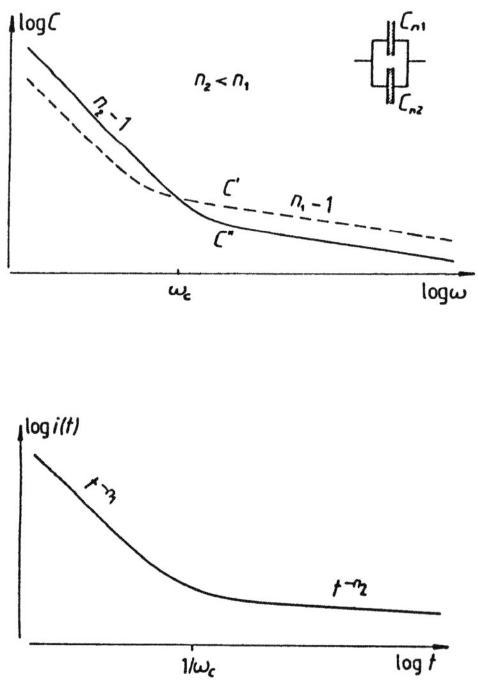

Fig. 3 — The upper diagram (a) shows the frequency dependence of the real (– – – – –) and imaginary (———————) components of the complex capacitance of a dielectric medium dominated by hopping charge carriers. The smaller exponent, $n_2 < \frac{1}{2}$, corresponds to the low-frequency branch with LFD, the larger exponent $n_1 > \frac{1}{2}$, corresponds to the high-frequency less lossy branch. The transition frequency is denoted by ω_c. The inset shows the equivalent circuit diagram consisting of two dispersive capacitors in parallel, which may or may not correspond to the simultaneous presence in a given medium of two separate physical processes. The lower diagram (b) shows the corresponding time dependence of the charging current under step-function excitation. Note the very slow, quasi-dc decay of current at long times, corresponding to LFD in diagram (a).

It may be mentioned here for the sake of completeness that another LFD process with very similar time and frequency dependence, and yet with different physical mechanisms *vis-a-vis* the volume LFD, is to be found in transverse flow *across* insulator-metal and semiconductor-metal interfaces (Jonscher et al., 1986; Jonscher, 1986). These situations may have their counterpart in the 1-D contact regions to surface flow, but we have no firm evidence for this at the present time.

PRESENTATION OF EXPERIMENTAL RESULTS

We present our experimental evidence for the time- and frequency-dependence of surface currents in the presence of humidity under the headings of zeolites, sand, glass, and mica—the substances on which most of our experiments were carried out.

(a) Zeolites

Zeolites are crystalline aluminosilicates with Group I and Group II cations; they consist of large molecular cavities connected by short ring-like structures into three-dimensional regular networks which have the remarkable property of being able to absorb large volumes of guest substances such as water or, at sufficiently low temperatures, various gases. In the case of water, it is possible to add up to 30% by weight. Dielectric measurements were carried out by various groups (Morris, 1969; Fedorov et al., 1964, 1966; Glazun et al., 1967a, 1967b;, Jones, 1975) and here we propose to give the most recent data from our own group (Haidar and Jonscher, 1986; Jonscher and Haidar, 1986).

As pointed out in the introduction, we are regarding zeolites as a special case of materials composed entirely of "internal surfaces" where transport takes place in the internal cavities by means of ions solvated by the absorbed water. For our present purpose it is less important to know the precise mechanism by which conduction takes place, a point on which there is a lack of complete agreement, than to note that the characteristics of this conduction process are very similar to those seen in transport on humid solid surfaces.

The dielectric spectra of three types of zeolites, known as Na-X, Na-Y and Ca-A, are shown in Fig. 4 in the form of logarithmic frequency dependence of the real and imaginary components of the complex capacitance. Several observations are in order.

At low relative humidity (RH) there is relatively little dispersion in the low-frequency capacitance and loss; Na-X and Ca-A even show loss peaks, indicating strictly localized to-and-fro motion of charges, probably within a single molecular cavity. At 12% RH there is already a measurable increase of low-frequency loss and capacitance and this proceeds rapidly with a further increase of RH until the low-frequency values of $C'(\omega)$ and $C''(\omega)$ exceed the dry-zeolite values by five or more orders of magnitude. Without going into a detailed analysis, it is possible to show that these results are not Kramers-Kronig-compatible between C'' and C', a point which will be taken up later in connection with the time-domain results. It is interesting to note that none of the zeolites shows any evidence of dc conduction at the lowest frequencies and highest humidities; in Na-X and Na-Y both the loss component $C''(\omega) = G(\omega)/\omega$ and $C'(\omega)$ *saturate* while in Ca-A *both* $C''(\omega)$ and $C'(\omega)$ continue to rise towards low frequencies.

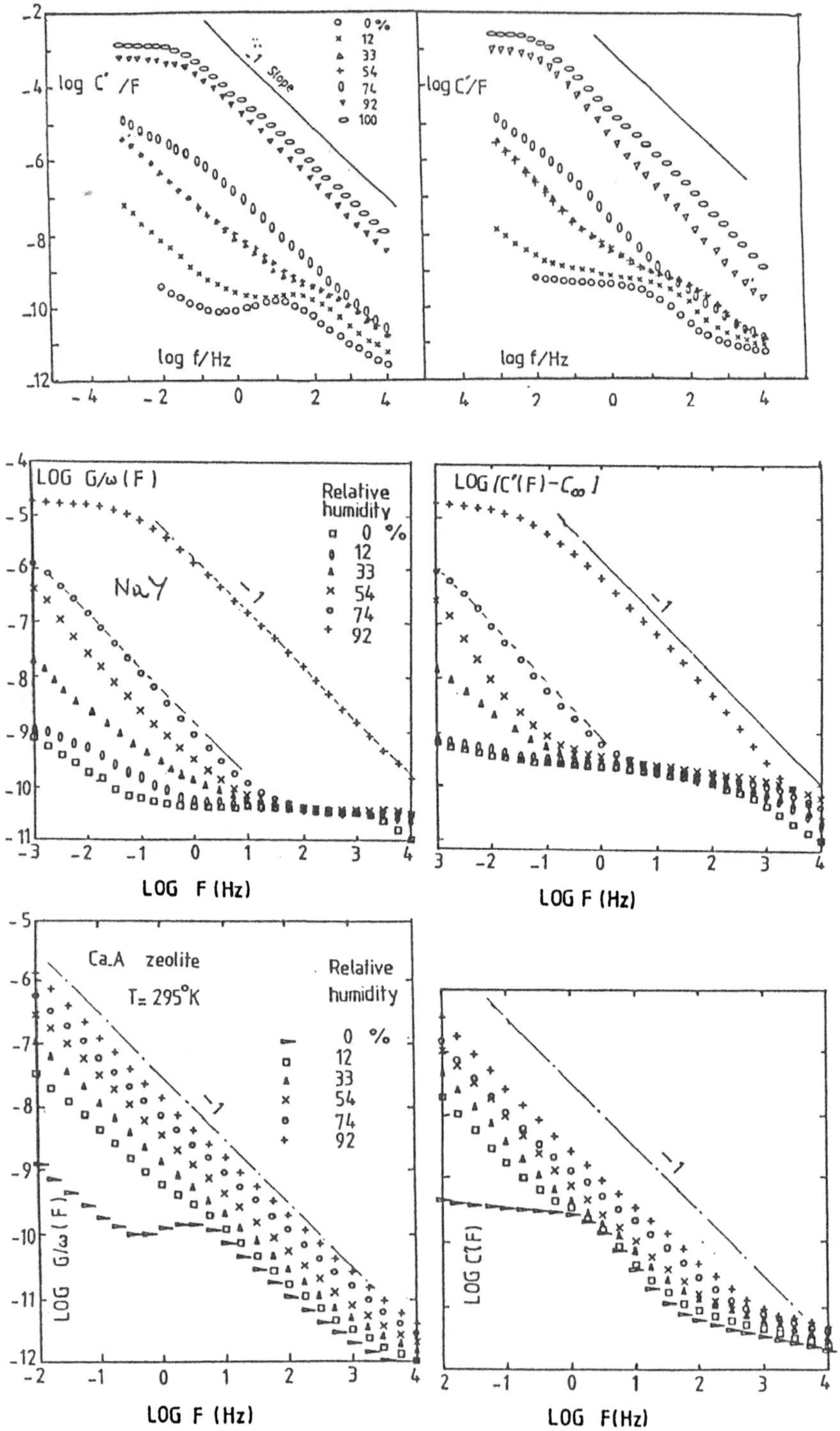

Fig. 4 — Logarithmic plots of the frequency dependence of the real component of capacitance, $C'(\omega)$ or $C'(\omega) - C_\infty$ (right-hand diagrams) and the imaginary component $C''(\omega) = G(\omega)/\omega$ (left-hand diagrams) for three types of zeolites with variable relative humidity of the ambient air. From Haidar and Jonscher 1986.

The increase of permittivity with increasing humidity, expressed as a percentage weight increase ΔW, is shown in Fig. 5a in the form of a semilogarithmic plot that indicates that the increase is *exponential* in ΔW; this may be regarded as surprising in the light of ordinary conductivity theory, which indicates a linear dependence of conductance on the number of carriers. The inference, which is also supported by other examples to be discussed later, must be either that the number of available ionic carriers is an exponential function of the humidity content, which is rather implausible, or that the conduction process is not a uniformly distributed one but relies on some form of percolation along paths of least resistance which should be strongly nonlinear in humidity.

An important aspect of strongly dispersive behavior concerns its linearity in the amplitude of the applied signal. This is illustrated in Fig. 5b, where both $C'(\omega)$ and $G(\omega)/\omega$ rise with rising signal amplitude (at low frequencies in the region of saturation for Ca-A in this example). The effect becomes negligible at higher frequencies—a common form of behavior in these situations, suggesting that short-distance transport, probably dominated by movements within the molecular cavities, is linear in the applied field, while longer-range transport only is affected.

In view of the strongly dispersive frequency-domain behavior of humid zeolites it was considered essential to look at the time-domain response under step-

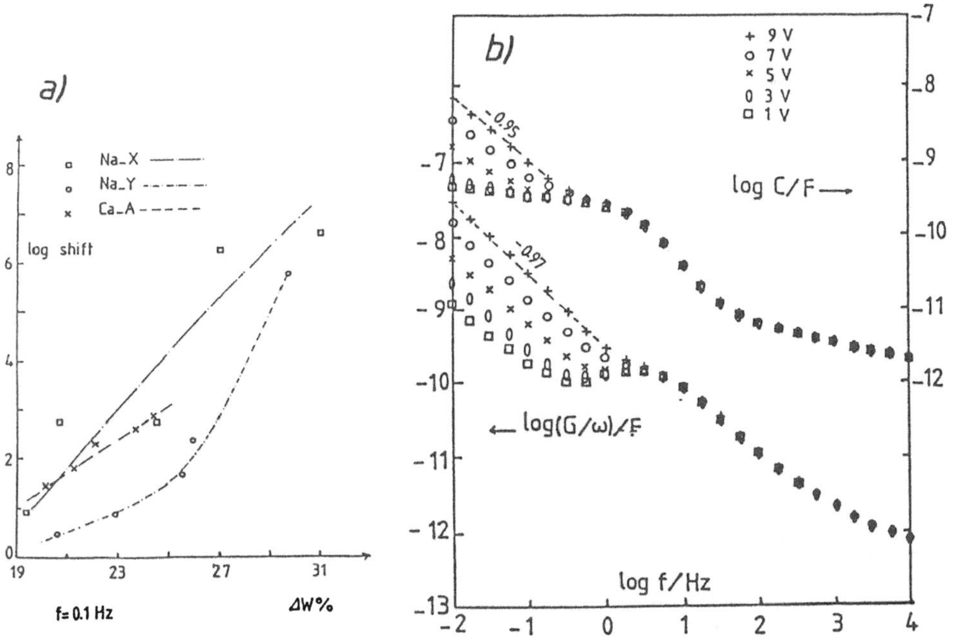

Fig. 5(a) — The dependence of the amplitude of the complex capacitance, plotted logarithmically against the fractional water intake (by weight) ΔW, for the three types of zeolites shown in Fig. 3. Note the approximately exponential dependence on ΔW. From Haidar and Jonscher 1986. (b) — Evidence for the non-linear response of the humid zeolite Ca-A with respect to the amplitude of the applied voltage signal. Only the low-frequency part of the spectrum is seen to be affected. From Haidar and Jonscher 1986.

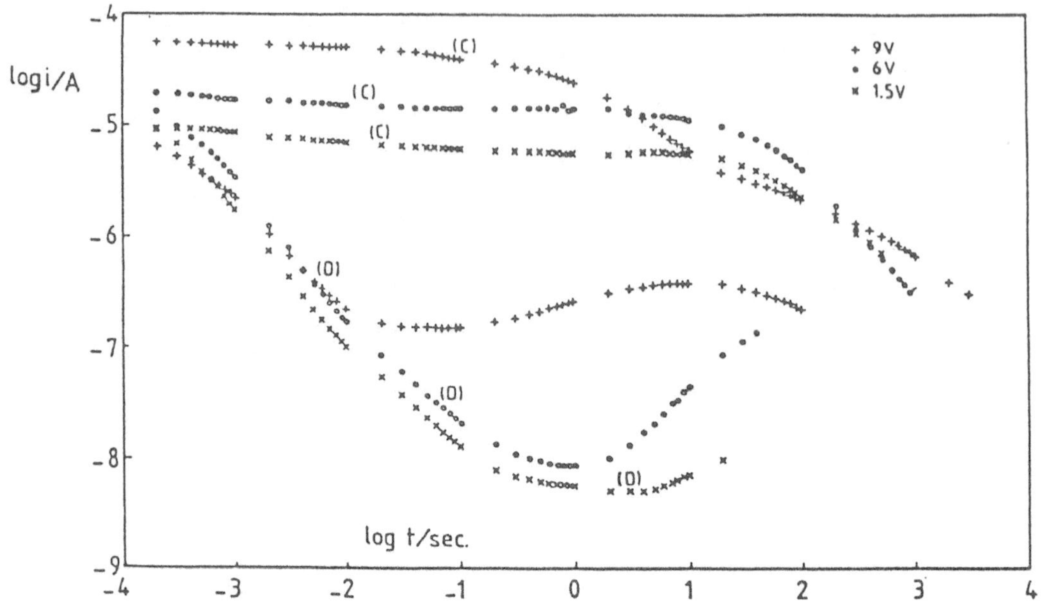

Fig. 6 — The time dependence of the charging (C) and discharging (D) currents in a Na-Y zeolite pellet at 33% RH, for three amplitudes of the applied voltage step. The sample was kept in a desiccator at the stated humidity for the entire period of measurement. Note the recovery of discharging currents after the initial drop. From Jonscher and Haidar 1986.

function excitation. Bearing in mind the relations given in Table 1, we expect a nearly time-independent current response. Figure 6 shows the response for Na-Y, which behaves very similarly to Na-X, and Fig. 7 for Ca-A, which is rather different in that the discharging current is so small that it could not be recorded on the available equipment. In both cases, the charging current curve shows a very large plateau in the logarithmic time variable in which the variation is very slight and a definite drop sets in at longer times. Ca-A, which showed the strongly dispersive behavior down to the lowest frequencies, gives a much longer time-independent current with just a trace of drop at the end. Na-X and Na-Y show the drop at much shorter times.

The amplitude of the initial currents in Fig. 6 shows at least a monotonic increase with the charging voltage, while in Fig. 7 the 6-V and 1.5-V responses are actually transposed, suggesting that there are important influences beyond our control affecting the magnitude of the currents.

The feature of the data in Fig. 6 most unexpected and difficult to interpret is the nature of the discharging currents, which initially slope faster than at t^{-1}, subsequently go through a minimum to come up to within a decade or so of the charging currents, and finally show a tendency to decay again. It may be noted that in samples which were being discharged in vacuum after charging at the given RH, the initial fall is identical but there is no subsequent recovery of the discharging current.

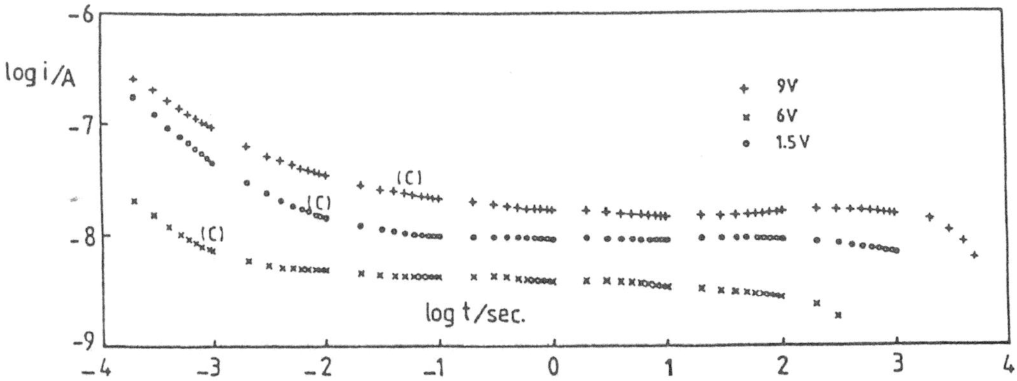

Fig. 7 — The charging currents at the Ca-A zeolite for three amplitudes of the step voltage. Note that the reversal of the magnitudes for the 1.5 and 6V run. The charging current are much flatter than in Fig. 6 and the discharge currents were too small to be recorded on our equipment.

The interpretation of the time dependence of $i_c(t)$ is qualitatively in agreement with the strongly dispersive nature of the frequency-domain response, but the quantitative agreement is much less satisfactory. As regards $i_d(t)$, on the other hand, the only possible inference is that somehow the system becomes "exhausted" during the initial passage of the relatively high current and only "recovers" gradually as the prolonged exposure to ambient humidity helps to restore the current by supplying more charge carriers. This is confirmed by the lack of recovery in discharging in vacuum. The most intriguing corollary to this is that the system somehow retains a "memory" of the past events which enables the current to go up again after the initial dip. A somewhat similar memory effect will be seen later in surface conduction on mica and glass.

This behavior of the charging and discharging currents illustrates the difficulty of applying Kramers-Kronig relations to nonlinear systems; quite clearly the transformation of the discharging currents would have led nowhere near to the observed frequency-domain response.

(b) Humid sand

Many granular media show a remarkable sensitivity of their dielectric properties to moisture, largely on account of their high specific area. Measurements on sand have given clear evidence of very strong low-frequency dispersion (Shahidi et al., 1975; Jonscher, 1978; DRS) and also show the property that data for different RH may be "normalized" by lateral shifts along the frequency axis to obtain a single master curve (Fig. 8). The logarithmic plot of shift against linear RH gives a straight line suggesting that the surface transport depends exponentially on RH. One interpretation of this result may be that each individual sand grain, or whatever particle may be involved, forms a "giant dipole" which makes only very poor contact with neighboring grains for the purpose of dc conduction—there is no evi-

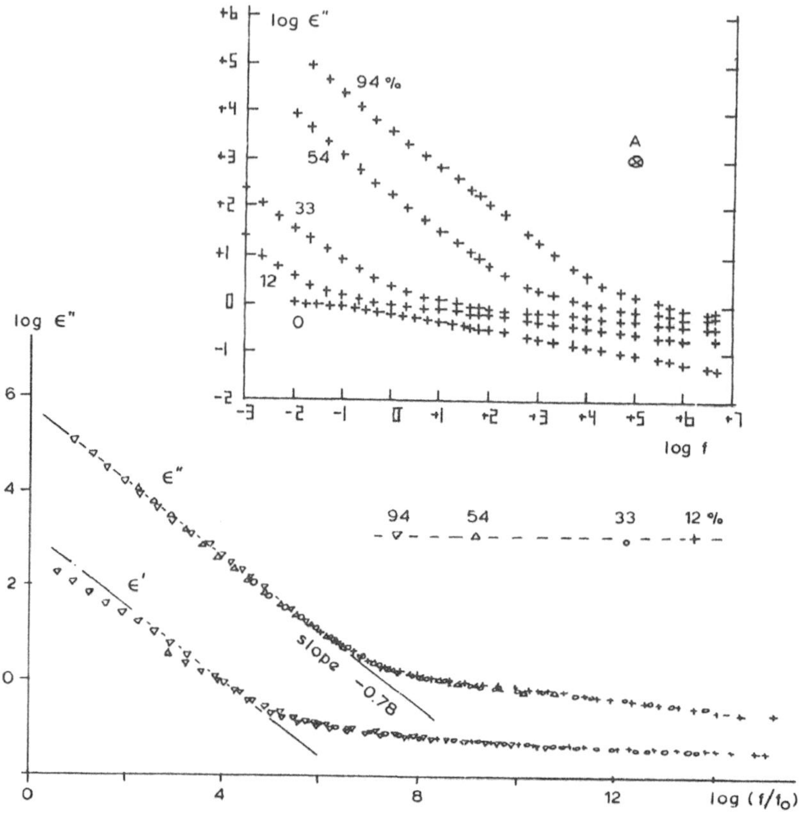

Fig. 8 – The dielectric response of humid sand at various levels of relative humidity (RH) of the ambient air circulating round the sample. The upper diagram gives the individual loss curves, showing the complete absence of any low frequency rise in zero RH, and rapidly rising loss with higher RH. The lower diagram gives the normalisation of these data for various RH, both for the real and the imaginary components of the effective permittivity. Both normalisations were effected with the same displacement locus of a representative point "A", shown as the horizontal chain-dotted line. The ϵ' data are displaced vertically for clarity, but the straight line is drawn in Kramers-Kronig compatible position with the slope of the loss data. The zero RH data cannot be normalised in the same way. The logarithmic shift in frequency given by the displacement locus shows an approximately linear dependence on RH, in agreement with the corresponding result for zeolites. From Jonscher 1983, from data by Shahidi et al. 1975.

dence of dc conductivity in these systems. It is possible to regard humid sand as "zeolite inside out" in that conduction takes place on the outside of grains rather than on the inside of molecular cavities. The end results appear to be remarkably similar.

(c) Mica paper

Measurements of surface conduction on mica in the frequency domain were first reported by Chaudhry and Jonscher (1985) for the case of mica paper, which consists of pure mica flakes held together by van der Waals forces without any binder or other form of glue. It is possible to envisage the electrical behavior of

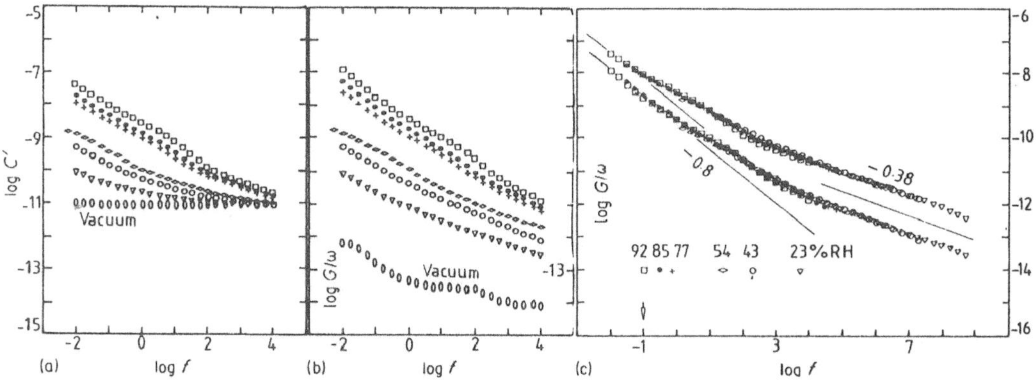

Fig. 9 — The humidity dependence of the complex capacitance of mica paper, with evaporated Al contacts capped by Cu foil, for a range of values of RH. The normalisation shown relates to $C'(\omega) - C_\infty$, but omitting the vacuum data as not normalisable with the rest. The two slopes of -0.8 at low frequencies are drawn in Kramers-Kronig compatible positions. From Chaudhry and Jonscher 1985.

mica paper along its surface, which is also the dominant plane of the individual mica flakes, as being dominated by the "giant dipoles" of individual mica flakes that are making only very small "point" contact with one another. In this respect the conditions are very similar to those encountered in humid sand, and the frequency-domain spectra are likewise very similar. Figure 9 shows the individual data for a range of RH values and also their "normalization" with respect to RH, indicating clearly the presence of two frequency regimes, both following power laws but with different exponents. In the case of the real part, we are plotting $C'(\omega) - C_\infty$, where the subtracted value corresponds to the limit of $C'(\omega)$ at high frequencies. The advantage of this form of representation is that it is more conspicuously Kramers-Kronig-compatible with the loss data. The amplitude of the dielectric response in mica paper again shows a good exponential dependence on RH, prompting conclusions like those for zeolites.

(d) Mica sheet

A more recent and a more systematic study of the conduction of mica along the planes of cleavage was reported by Jonscher and Ramdeen (1986), who measured the time-domain response of a "raft" of mica strips in order to enhance the total surface area. Aluminum electrodes were applied to both sides of the raft, with the planes of cleavage running normal to the electrodes. Charging and discharging currents were measured for a range of step voltage amplitudes and the results for 92% RH are presented in Fig. 10. The charging and discharging currents follow similar trends but their relative magnitudes vary according to the amplitude of the charging voltage step. In fact, the discharge currents, while persisting for very long times, are virtually independent of the amplitude of the voltage, while the charging currents are linearly dependent on voltage with an intercept V_0 on the voltage axis and with a slope corresponding to a series resistance R_s that depends on RH and on

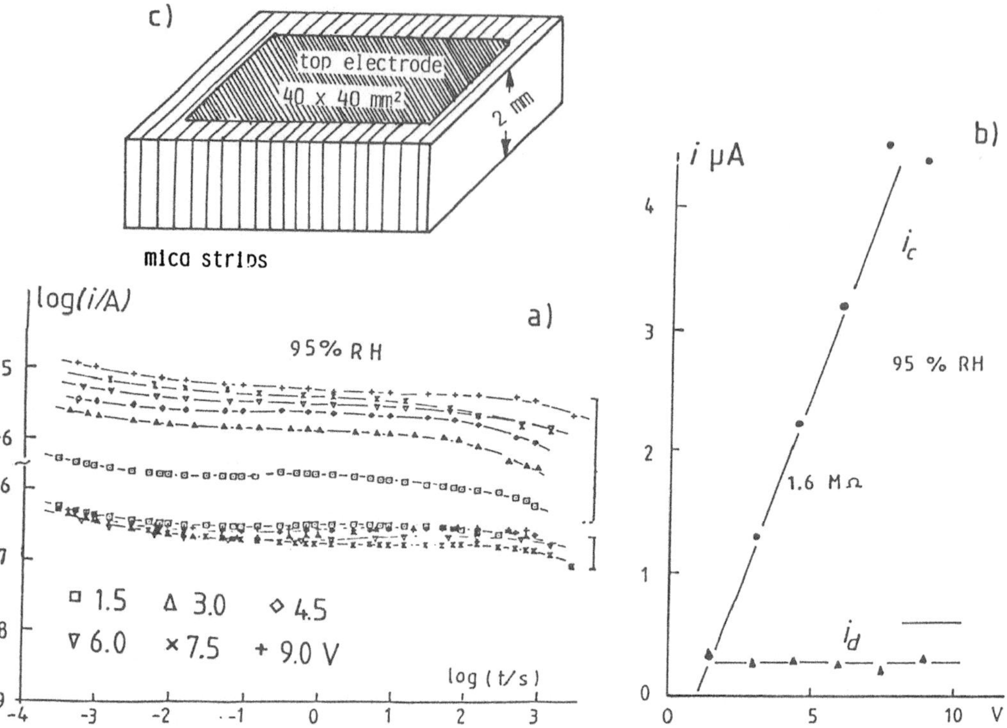

Fig. 10 – The time dependence of the charging and discharging currents on the surface of mica at 95% relative humidity, for a range of amplitudes of the exciting step voltage, diagram (a). The two sets of current data are displaced vertically by one decade, to avoid overlapping. In diagram (b) the amplitudes of $i_c(t_1)$ and $i_d(t_1)$ for $t_1 = 0.1$ s are plotted against the voltage amplitude, showing that the charging current amplitude increases linearly with the voltage, except for an intercept of approximately 1 V, and with a resistance which depends on the relative humidity. The discharging currents are practically independent of the voltage. The short horizontal line indicates the level of current that would flow if V_0 = 1 V were applied to the resistance of 1.6 MΩ which applies to the charging current. Diagram (c) shows the arrangement of mica strips to form a large area sample with current flow between the parallel top and bottom electrodes parallel top and bottom electrodes parallel to the surfaces of the strips.

the geometry of the sample. It is evident that the behavior of this surface system may be interpreted in terms of a battery of voltage V_0 with the series resistance R_s. The charging condition corresponds to a current

$$i_c = (V - V_0) / R_s \tag{9}$$

while the discharge current is independent of the previous charging voltage and is given by the ratio

$$i_d = V_0 / R_s \tag{10}$$

These relations are very different from the classical behavior of linear dipolar systems, described by eq. 7, and they illustrate clearly the problems arising in non-linear systems in which LFD, and therefore electrochemical processes, are playing an important role. We believe that this notion of "battery action" in low-frequency dispersive systems is fundamental to the understanding of this type of response, as will be described later in this paper.

(e) *Glass surfaces*

Very similar time-domain behavior has been observed in glass surfaces by Owede and Johnscher (1987), with electrode configurations giving a large range of aspect ratios s/l, where s is the spacing between the electrodes and l is their length, (Fig. 11). This figure shows the complex impedance plots, which appear as fairly regular semicircles with diameters R_s that are identical with the corresponding resistance values determined from the transient $I-V$ plots of the type shown in Fig. 10. Some irregularities appear at the lowest frequencies and their significance may be better understood from the frequency dependence of the real and imaginary components of the complex capacitance in logarithmic representation, (Fig. 12). These emphasize strongly the low-frequency end of the spectrum, while the linear Z plots emphasize the high-frequency end. The last three decades of the plots in Fig. 12 show almost constant $C'(\omega) \cong C_0$ and $G(\omega)/\omega \propto \omega^{-1}$, corresponding to constant G_0. This parallel combination of C_0 and G_0 implies a semicircular complex Z plot, as seen in Fig. 11. A very intriguing feature of the data in Fig. 12 is that at large separations s and high RH values there appears a negative capacitance, which in our logarithmic representation is plotted as $|C'(\omega)|$, different symbols being used to distinguish them from positive values. This physical significance of negative values of $C'(\omega)$ must be sought in the corresponding time-domain behavior (Jonscher, 1986), which should show charging and discharging currents that *rise* over a certain range of times, instead of falling monotonically as in the classical charging and discharging of ordinary capacitors. The physical reason for this partly rising current must be some form of instability or variability of the system under investigation, as if the very passage of current was tending to alter the physical properties of the medium. At the present moment we do not understand the reasons for this strange behavior in the particular case of our surface conduction experiments, and its confinement to large separations and high RH values. We must state that these results were well reproducible and there is no doubt about their validity.

The dependence of the values of R_s obtained from the time-domain charging experiments, (Fig. 10), or the equivalent values of $G_0 = 1/R_s$ from complex impedance plots, are plotted against the aspect ratio s/l and give the very remarkable behavior shown in Fig. 13. For $s/l > 0.2$ approximately, the dependence of the resistance is exponential:

$$R_s \propto \exp(\alpha s/l) \tag{11}$$

where the coefficient α depends also exponentially on relative humidity. For smaller values of the aspect ratio, below 0.1 or so, the dependence becomes a power law

$$R_s \propto (s/l)^k \tag{12}$$

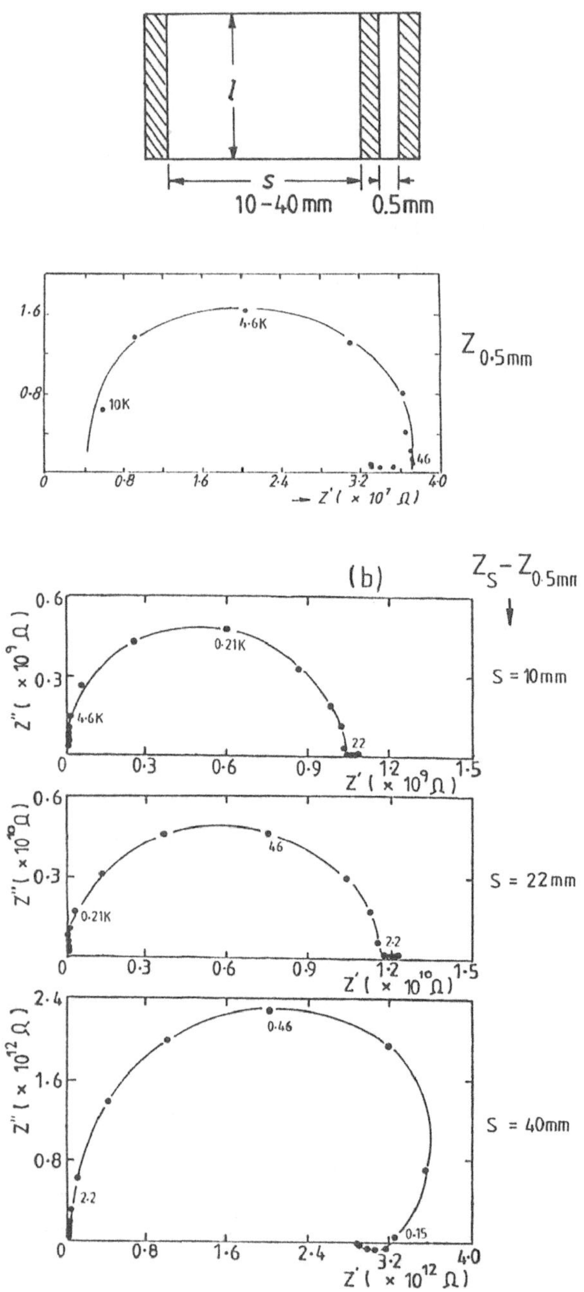

Fig. 11 — The complex surface impedance diagrams of samples of glass of dimensions shown in the inset, at 97% RH. The top diagram refers to the closely spaced electrodes at s = 0.5 mm; the following diagrams give the differences between the impedances for the stated s values and that for 0.5 mm. The diagrams are substantially semicircular, with some complications at the lowest frequencies and, in the case of 40 mm spacing, a negative overshoot into the fourth quadrant. Note the rapidly increasing diameter of the consecutive plots, corresponding to the effective parallel resistance $R_s = 1/G_0$. From Owede and Jonscher 1987.

where the exponent k is greater than unity and is also strongly dependent on RH. This type of behavior is completely at variance with normally accepted laws of resistance for uniform sheet conductors, for which the exponent k should be equal to 1. Further discussion of the consequences of this result will be given later in this paper.

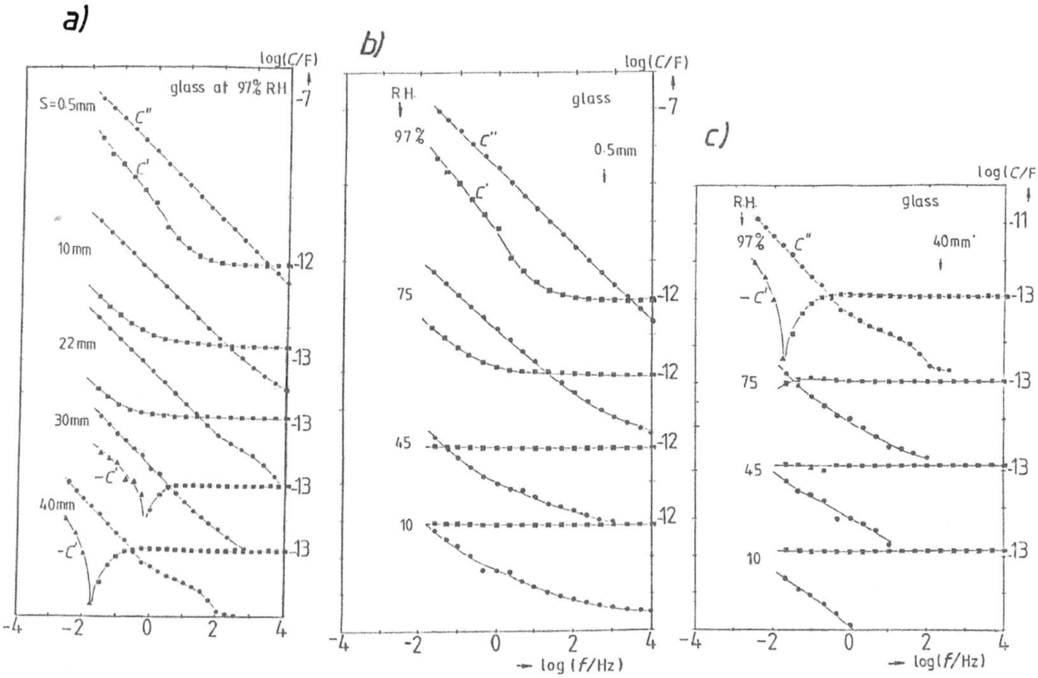

Fig. 12 — The frequency dependence of the real and imaginary components of the effective capacitance of the samples shown in Fig. 11, plotted logarithmically. Diagrams (a) refer to variable spacing *s* at 97% RH, diagrams (b) to variable RH at *s* = 0.5 mm, and diagram (c) to variable RH at *s* = 40 mm. Note the clear transition to negative capacitance at the highest RH and longest *s* values. From Owede and Jonscher 1987.

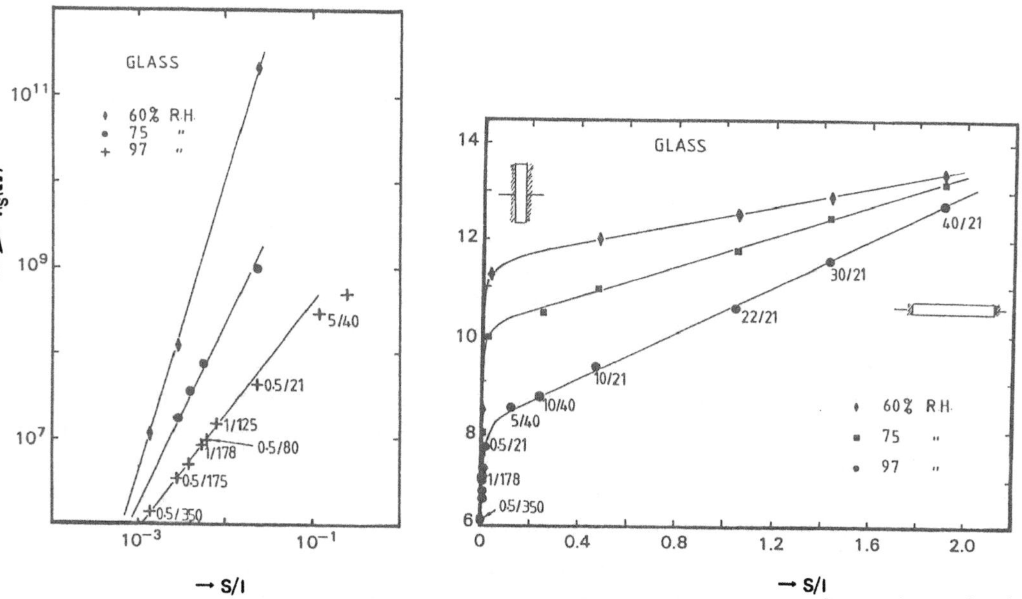

Fig. 13 — The dependence of the series resistance R_s of humid glass surfaces, determined by the impedance method, as in Fig. 11, or from the current-voltage plots, as in Fig. 10, both being equivalent to one another. The diagram on the left gives the date for small values of the aspect ratio s/l, plotted logarithmically against log s/l. The diagram on the right gives the plot against linear s/l for larger values of the ratio. The pairs of figures at the experimental points give the dimensions s/l in mm.

It should be stressed that the relationships 11 and 12 remain valid, including the two ranges of dependence, regardless of the individual values of s or of l—only their ratio is relevant. A similar dependence on s/l is also found for surface conduction on mica sheets and for Si_3N_4 films covering a glass substrate.

DISCUSSION AND CONCLUSIONS

The totality of experimental results relating to the dynamic aspects of surface conduction in the presence of humidity open up, we believe for the first time, some very important questions of interpretation and of understanding of the physical and chemical significance of what is being seen. It is clear that surface conduction is a very common phenomenon easily seen on any material that is sufficiently insulating to show the effect of surface conduction and that absorbs water and thus releases mobile ionic species which may take part in conduction. Similar effects have been seen on semi-insulating gallium arsenide (Owede, private communication). The chemistry and electrochemistry of the reactions involved have been the subject of extensive investigations and will not be reviewed here. Our principal concern is with the way in which the experimental evidence relating to the dynamic response of surface conductance may contribute to a better understanding of the nature of this surface transport.

The salient points of the evidence presented here may be summarized as follows.

1. Many different materials, such as mica, glass, silicon nitride, sand, zeolites, and GaAs, show very similar behavior in the presence of humidity, characterized by strong low-frequency dispersion in the frequency domain.
2. The time-domain equivalent of this is the presence of very slowly time-varying charging and discharging currents which may be understood in terms of the storage of charge and of energy by electrochemical processes, rather than, as in the case of the classical Maxwell-Wagner processes, by electrostatic processes (Jonscher, 1983b).
3. Clear experimental evidence shows that the value of the effective resistance R_s in the system depends on the *aspect ratio* s/l, rather than on the value of either dimension separately, and that this dependence falls into two well-defined ranges. For small $s/l < 0.1$, the relation follows a power law with an exponent approaching unity at the highest humidities and rising rapidly at lower humidities. For $s/l > 0.2$ the relation follows an exponential law.
4. In all cases, the conductance $1/R_s$ or the equivalent admittance increases exponentially with RH.
5. A permanent open-circuit voltage V_0 appears on samples that have been subjected to a polarizing bias. The magnitude of V_0 appears to depend on the aspect ratio s/l, increasing as s/l increases. Once impressed on the system, its polarity may be reversed only by the application of the opposite polarising bias exceeding the magnitude of the original one.

These general properties of surface conduction in the presence of humidity suggest that the classical concept of a surface sheet of uniform conductivity cannot apply and the existence of two regimes in the aspect ratio s/l suggests that two different types of processes obtain. The first of these, applicable at large s/l values, is evidently dominated by the properties of the space between the electrodes, the equivalent of "bulk" conduction in three-dimensional processes. This transport, being exponentially dependent on the spacing between the electrodes, would appear to be limited by some form of percolation processes, but certain difficulties arise with this interpretation. One of these is the evident linearity of the processes in the applied voltage, as seen in Fig. 10. If percolation through more conducting regions separated by insulating regions were the process, then nonlinear behavior might be expected. The detailed mechanism is not yet known with any certainty and further studies are required, but it is evident that the observed behavior is completely inconsistent with the simple notion of a conducting sheet representing the surface transport on a humid insulator. The strong nonlinearity of the effective resistance with the spatial dimensions and the LFD character of the spectral response are being tentatively associated with the presence of filamentary flow, and there is also clear evidence for the electrochemical nature of the processes responsible for this behavior.

It follows, by contrast, that the different behavior at small values of s/l is very likely dominated by the response of the 1-D contacts to the surface conduction process. The presence of these two different regimes of behavior, both nonlinear in the spatial dimensions, makes it impossible to carry out the normal division of the total response into two impedances in series, one representing the contacts and the other the "space" in between. This aspect, too, is not sufficiently well understood at the present time, but it is clear that a proper understanding of the background picture will be essential for correct interpretation of many phenomena in which processes take place similar to the ones described above.

In the light of the above discussion of mica and glass surfaces, it is possible to regard the behavior of mica paper and of sand as corresponding to very large s/l ratios, with approximately point contacts separated by the relatively very large distances. On the other hand, sand and mica do not involve metallic contacts, the dominant influence being insulator-insulator contacts.

Central to the understanding of the behavior of humid surface conduction is the notion of the electrochemical storage of charge and of energy in such systems, with the resulting LFD type of response. Without this new approach, it is very difficult to see how the observed behavior can be interpreted in terms of any simple "equivalent circuit" model. What remains to be investigated in more detail is the nature of the battery action that is seen and the quantitative dependence of its emf on the geometry, the RH, and the nature of the substances in question.

RELEVANCE TO OTHER AREAS OF TRANSPORT

While all our experimental materials have humid surfaces, including such "internal surfaces" as zeolites, many of these considerations may be applicable to other systems, two of which will be mentioned here. One is the behavior of biological systems, which consist of porous or cellular structures filled with liquid electrolytes, with the transport across the cellular boundaries being more restricted than in the volume the cells. It is known that these systems show LFD behavior of very like the one described in the present paper; the only open question is whether the physical mechanisms are similar or not.

The second area in which the ideas developed in the context of humid surfaces may be directly applicable is vacuum transport of ions or electrons on insulating surfaces, e.g., under the action of corona discharges or electron beam irradiation. The "electrochemical" processes at the electrodes may be very different in these situations in comparison with our humid surfaces, but the inhomogeneities of flow and the resulting spectral and temporal dependence of the response might turn out to be similar.

A special example of this type of transport process would be surface flashover resulting from an extremely high voltage applied to the sample. The presence of nonuniform filamentary flow may give rise to excessive heating and this is likely to result in a degradation of the material and subsequent catastrophic breakdown.

REFERENCES

M.A. Chaudhry and A.K. Jonscher, J. Materials Sci. **20**, 3581 (1985).

L.A. Dissado and R.M. Hill, J. Chem. Soc., Faraday Trans. 2, **80**, 291 (1984).

V.M. Fedorov, B.A. Glazun, I.V. Zhilenkov, and M.M. Dubinin, Bull. Acad. Sci. USSR, Div. Chem. Sci. 1839 (1964).

V.M. Fedorov, B.A. Glazun, M.M. Dubinin, and I.V. Zhilenkov, Bull. Acad. Sci. USSR, Div. Chem. Sci. 1087 (1966).

B.A. Glazun, M.M. Dubinin, and I.V. Zhilenkov, Bull. Acad. Sci. USSR, Div. Chem. Sci. 956 (1967).

B.A. Glazun, M.M. Dubinin, I.V. Zhilenkov, and M.F. Rakityauskaya, Bull. Acad. Sci. USSR, Div. Chem. Sci. 1156 (1967b).

A.R. Haidar and A.K. Jonscher, J. Chem. Soc., Faraday Trans. 2, in press (1986).

R.M. Hill and A.K. Jonscher, Contemporary Phys. **24**, 75 (1983).

G. Jones, J. Chem. Soc., Faraday Trans. 1 **71**, 2085 (1975).

A.K. Jonscher, Phil. Mag. **B38**, 587 (1978).

A.K. Jonscher, *Dielectric Relaxation in Solids*, Chelsea Dielectrics Press Ltd., London, 1983a.

A.K. Jonscher, Annual Report, IEEE 1983 Conference on Electrical Insulation and Dielectric Phenomena, IEEE, New York, 1983b, p. 479.

A.K. Jonscher, J. Chem. Soc., Faraday Trans. 2 **82**, 75 (1986).

A.K. Jonscher and A.R. Haidar, J. Chem. Soc., Faraday Trans. 2, in press (1986).

A.K. Jonscher, C. Pickup, and S. Zaidi, Semiconductor Sci. Technol. **1**, (June 1986).

A.K. Jonscher and T. Ramdeen, Elec. Insul. IEEE Trans. **E1**, (1986), in press.

B. Morris, J. Phys. Chem. Solids **30**, 73 (1969).

E. Owede and A.K. Jonscher, to be published (1987).

M. Shahidi, J.B. Hasted, and A.K. Jonscher, Nature **258**, 595 (1975).

W. Shockley, Phys. Rev. **56**, 317 (1939).

13. Absolute Entropy and the Problem of Microscopic Measurement

L. Tisza

INTRODUCTION

I first met Herbert Fröhlich at conferences on low temperature physics almost half a century ago. I am happy to remember these encounters on the occasion of his 80th birthday with a contribution on entropy and the *Nernst Heat Theorem.* This choice gives me also the opportunity to express my indebtedness to Franz Simon, later Sir Francis, who gave me at about the same time an insider's introduction to these ideas.

Simon was one of the foremost students of Nernst who continued and modernized his work, and developed the Third Law into the guiding principle of all of low temperature physics.

However, in spite of its impressive success, the Third Law came under severe attack (Fowler and Sterne, 1932) which aimed to discredit the usefulness of its central ingredient, the "absolute entropy." This elusive concept, often used improperly and obscured by careless semantics, proved to be a plausible target for adversary criticism.

My thesis is that the shortcomings of the Nernst Heat Theorem stem primarily from the historical context of its origin. As outlined in the next section, Nernst discovered a deep-seated connection between the thermodynamics of high temperature gas reactions and the low temperature processes in condensed matter. His main interest was in the former, which would have been much simpler to handle, had the necessary quantum statistics been available. However, this was in 1906 and Nernst had no choice but to start at the "wrong" end.

This was bound to lead to a certain confusion, which was partially rectified in the quantum statistical reformulation of the theory of about 1930. However, a residual conceptual quandary persisted, as the above-mentioned controversy shows. This is resolved in an almost embarrassingly simple fashion in the analysis of the section: "The Anatomy of a Controversy." I refer to the new version of the Heat Theorem, as the "Nernst Principle." As shown in the section on the absolute entropy, it enables us to extract microscopic information from macroscopic measurements on a class of condensed systems.

Experiments of this sort had been performed already in the 1930's and are reviewed here. This interesting line of research has not received the attention it deserves, presumably because of the uncertain status of the underlying theory. The present reformulation removes these doubts, and at the same time sharpens the conclusions.

In the final discussion a case is made for a more general practice of the "consolidation" of theories as demonstrated in this paper.

THE NERNST HEAT THEOREM

Nernst's discovery of the Heat Theorem is a fascinating story (Mendelssohn, 1964, 1973; Hiebert, 1979). However, I have to confine myself to the bare essentials.

Nernst was led to "his theorem" while attempting to predict thermodynamic criteria for chemical equilibria in gas reactions (Nernst, 1906). This was particularly significant in instances of reactions which were vitally important for practical reasons, such as the production of ammonia from atmospheric nitrogen, but were not yet "operational" at that time. He exploited the fact which is quite obvious in the chemical context, that the problem naturally divides into one of the *equilibrium constant* and another of the *rate* of reaction. His great achievement was to establish the problem of equilibrium as an instance of law, whereas the rate problem involved contingent engineering inventions, such as the choice of the proper catalyst. The latter was indispensable for industrial realization; it was the Haber-Bosch process that led in 1913 to the large scale production of ammonia, and hence of fertilizers and explosives. It overcame the exclusive reliance on Chilean nitrates. But let us return to Nernst's pursuit of his theorem.

Equilibrium in chemical reactions would seem to be governed by the thermodynamic properties of the participating substances. However, the classical theory leads to an arbitrary constant of integration in the entropy function of each substance. The existence of well-determined equilibria in gas reactions suggested to Nernst that these constants must be related.

At present we know that this incompleteness of classical thermodynamics is remedied by quantum statistics, and this is the basis for the formulation of the Nernst Principle in the section, "Absolute Entropy and the Nernst Principle." However, Nernst grappled with this problem at a time when the quantum theory of matter was just about to start.

Using his wide knowledge of the experimental material he conjectured that the differences of entropy constant associated with reactions would tend to zero near absolute zero, a bold assumption which turned out to be amazingly productive. Without even involving Planck's constant, this "poor man's quantum theory" implied many results developed only later in quantum statistics, such as the low temperature behavior of specific heats, and even the subtle prediction of "gas degeneracy." Most important it provided Nernst with the reference point for an absolute entropy, the concept that enables one to determine chemical equilibria from the physical properties of the individual substances.

However, there was a price. The "hidden" introduction of quantum statistics, while appearing only to supplement phenomenological thermodynamics, resulted in a conceptual fuzziness. The stage was set for a controversy which could not be resolved without removing the fuzziness.

THE ANATOMY OF A CONTROVERSY

Insofar as physicists were concerned, the Third Law entered the scene when Planck translated the chemical context into physical terms. It was justified to confine the discussion to a single substance and consider phase transitions instead of reactions. Accordingly, Planck postulated the vanishing of the entropy in the limit of absolute zero for each individual substance (Planck, 1912, 1932).

For Planck the meaning of the Nernst theorem was its complete harmony with the many changes in the counting of states which quantization introduced into classical statistical mechanics. His calculations referred to idealized models, and a discrepancy developed with respect to the experimentalist who found that his real substances often exhibit a frozen-in disorder where the entropy does *not* vanish at absolute zero.

This difficulty did not show up in Nernst's formulation, because the frozen-in disordered states do not manifest themselves in chemical reactions.

It was plausible that the required ordered behavior might be tied to some sort of "ideality" which was not easy to define. Subsequently it was often required that the validity of the postulate be restricted to *pure crystalline substances.* (Lewis and Randall, 1923, 1961). This qualification became standard for a number of years, but its insufficiency became evident with the rapidly expanding range of experimentation.

On the one hand, there are pure crystalline substances, such as ice, that have finite zero-point entropies due to frozen-in disorder (Pauling, 1935). On the other hand, the entropy tends to zero for the well-documented case of liquid helium and even for the mixture of the two helium isotopes which "unmix" around 0.5 K into two liquid phases separated by a meniscus, one richer in the heavy, the other in the light isotope (Peshkov and Zinovieva, 1957).

It is not surprising that in the case of such a theory with obvious merits but lacking in precise definitions, a polarization of opinions should evolve. The negative view was expressed most forcefully in a much quoted passage (Fowler and Sterne, 1932):

> "We reach therefore the rather ruthless conclusion that *Nernst's Heat Theorem strictly applied may or may not be true, but is always irrelevant and useless....applied to "ideal solid states" at absolute zero, which are physically useful concepts. The theorem, though often true is sometimes false, and failing in generality, must be rejected altogether.* It is no disparagement to Nernst's great idea that it proved ultimately of limited generality. The part that it has played in stimulating deeper understanding of all these constants, and its reaction on the development of the quantum theory itself cannot be overrated. But its usefulness is past and it should now be eliminated."

This total rejection is based on the conviction that the defects of the theory are inherent and irreducible. It is at the opposite pole from Simon's pragmatic conviction that the Nernst theorem is indeed a valid law, although in need of a more precise formulation (Simon, 1930, 1937). Her suggested that the postulate is valid for those "*degrees of freedom*" *for which internal thermodynamic equilibrium is established.* In addition, Simon pointed out that all quantitative results of thermodynamics depend on this condition. Thus equations of state and specific heats have no precise meaning in the presence of frozen-in disorder.

Simon's emphasis on the role of thermodynamic equilibrium in the interpretation of the Third Law was an important insight, and ensured the continued use of this principle within low temperature physics. It also provided a guideline for those physico-chemists, who had an intuitive feeling for avoiding its pitfalls and produced original results in the study of molecular crystals. However, as discussed in the section on absolute entropy, many of these seem now forgotten, maybe because Fowler's critique was never effectively countered.

In order to make up for this deficiency, we have to add two more ideas to the above analysis. Simon seems to say: "If your system is in equilibrium near absolute zero, then its entropy vanishes." However, this is not a verifiable statement, because we cannot tell whether the internal degrees of freedom in a condensed system are in equilibrium.

This situation can be remedied by turning around the logic: instead of *predicting* that the entropy of a system tends to zero as $T \to 0$, we consider this property as the much needed *criterion that equilibrium is indeed established.* (Tisza, 1966). This criterion is subject to an exacting experimental verification for those condensed systems which possess a vapor phase the entropy of which is rigorously computable by the methods of quantum statistics. This condition is satisfied for a vast number of interesting cases.

It was an extraordinary *tour de force* when Nernst in 1906 established a connection between the entropies of the high temperature gaseous and the low temperature condensed states, respectively. It should be obvious to us that the proper way to proceed is from the weak coupling to the strong coupling states, and this idea is developed in the next section. However, the inverse procedure was the only one open at the particular historical junction.

"ABSOLUTE ENTROPY" AND THE NERNST PRINCIPLE

Planck illuminated the Nernst Heat Theorem in terms of quantum statistics (Planck, 1932), by computing the absolute entropy of a monatomic ideal gas, and the vanishing of the entropy of an ideal solid according to the Debye theory. However, the difficulties discussed in Fowler and Stern (1932) refer to molecular crystals, where the building blocks have an internal structure.

We consider now a model for the gas phase which is only a little more complicated, but enables us to deal with the essence of the situation. By contrast, for the condensed phase we consider caloric experiments, rather than a theoretical approximation.

Let us consider a gas in the state of reference P_r, T_r where it is sufficiently close to ideality. For simplicity, we keep the pressure constant: $P_r = P$. Also we confine ourselves first to the case of a monatomic gas with a degree of degeneracy g stemming, say, from the electronic and nuclear spin. We have then the well-known statistical expression for the entropy:

$$S_{stat}\,(P,\, T_r;g) - Nk\ ln\ \frac{gkT_r}{P\lambda_r^3}\,, \tag{1}$$

where

$$\lambda_r = \frac{h}{(2\pi mkT_r)^{1/2}}\,, \tag{1.a}$$

and the symbols have their standard meaning.

Let us now consider a path of constant pressure in the P, T diagram and write for the experimental entropy difference between T_r and an experimentally attainable low temperature, T_0, the so-called *caloric entropy*:

$$S_{\mathrm{cal}}(P, T_0, T_r) = \int_{T_0}^{T_r} \frac{C_p}{T}\, dT\,. \tag{2}$$

This expression is directly measurable. The specific heat, C_p, is meant to include δ-function terms to account for latent heats.

By combining (1) and (2) we obtain

$$S(P,T_0;g) = S_{\mathrm{stat}}(P,T_r;g) - S_{\mathrm{cal}}(P,T_o,T_r)\,. \tag{3}$$

This equation has a number of noteworthy properties.

(i) The second term is "operational" in the usual experimental sense. We emphasize that it contains only observable temperatures, and *does not* involve the absolute zero.

(ii) The first terms is statistical, but the calculation refers to a reasonably realistic model and is rigorous.

(iii) The left hand side of the equation does not depend on the reference state P, T_r which can be varied within wide limits. The entropy difference between, say, T_r and $T_{r'}$ can be both measured and calculated, and the difference cancels in Eq (3).

We sum up the aforementioned properties by saying that Eq (3) is "operational in the generalized sense": a combination of experimental and mathematical operations which are free from questionable assumptions. (See Bridgman (1959) and p. 334 of Tisza (1966)).

The only parameter which is problematic is the degree of degeneracy, g, and the purpose of the present discussion in terms of the Nernst Principle is to enable us to deal with the specific problems associated with it.

However, before formulating the principle, we wish to point out that the essential features of Eq (3) have a much wider scope than the derivation from the monatomic model might suggest.

The second term, the "statistical entropy" becomes really interesting for diatomic molecules. This case was first established in an entirely satisfactory manner for the diatomic iodine molecule (Gibson and Heitler, 1928). It is often called "spectroscopic entropy," because the statistical method is brought to bear on a model, the parameters of which are obtained from precise spectroscopy data, and not from an approximate solution of the Schrödinger equation. The case of diatomic molecules leads to a much larger variety of degrees of freedom connected with two nuclear spins, isotopic distribution and the like. The problem arising in the condensed state is the measure of the resulting order or disorder. We can formally think of the problem in terms of the parameter g in the expression (1) and replace it with the appropriate parameters in each concrete instance of application.

We may say that the entropy is "relative" with respect to the definition of the internal parameter, g, while it is "absolute" in the sense that it does not contain an arbitrary integration constant. This clarification takes care of the objection against "absolute entropy" on p. 708 of Fowler and Sterne (1932).

The case of polyatomic molecules can become very complicated, but the conceptual situation is basically the same, and, in order to demonstrate the essence of an application of the Nernst Principle, we can even confine ourselves to the case of $g = 2$.

How does this degeneracy of the lowest state of the building blocks affect the properties of the structure? In this simplest case the problem is analogous to the Ising model and we expect that at high temperatures the contribution to the entropy is

$$S = Nk\ln 2. \tag{4}$$

This degeneracy is split by interaction, and the lowest ordered state has an entropy $S_0 = k\ln 2$, that is, the entropy per molecule is practically zero. Of course, a given system may not reach this state in a finite amount of time. We have a competition between the ordering tendency and the resistance to the rate-process due to activation energy. This competition is particularly interesting if the ordering tendency is furthered by the quantum mechanical zero-point energy. This leads to very visible effects if it involves translational motion as in liquid helium and superconductivity. However, we then deal with the internal modes of motion, where direct observation is more elusive. It is here that the Nernst Principle provides information that is of limited scope, but is precise and reliable.

Application of the Nernst Principle in evaluating the low temperature entropy by Eq (3), must take into account three considerations:

(a) If $S(P,T_0) \approx 0$, we infer that the degrees of freedom built into the formalism assumed an ordered state and no significant caloric effects can be expected form a further lowering of the temperature. This is the case in which the validity of the Nernst theorem is uncontested.

(b) If $S(P,T_0) > 0$, the system has a frozen-in disorder, and it is up to further experiments, maybe combined with theoretical considerations, to decide whether a lowering of the temperature might help, or whether the rate of ordering is inherently too slow. It is convenient to distinguish between *active* and *inactive degrees of* freedom in this case.

(c) If $S(P,T_0) < 0$, this is one of a number of anomalies indicating that we failed to build in an *active* degree of freedom into the formalism leading to Eq (3).

The importance of the Nernst Principle to *low temperature physics* comes about from the fact that a number of interesting systems belong to category (a). In these "macroscopic quantum systems" (London, 1954) the zero-point motion destabilizes the potential barriers favoring disorder. It is a corollary that the entropy function at the temperature already reached should be pressure-independent. This is demonstrated dramatically in the entropy-free crystallization of both helium isotopes as purely mechanical compressions.

The applications to solid state physics stem from the analysis provided by category (b), which gives us a handle to disentangle the elusive state of order and disorder with respect to a variety of *active internal parameters.*

Inactive parameters which undergo no observable ordering at the working temperature, T_0, are conveniently ignored. Although isotopic distribution falls usually in this category, it is apparent from systems involving hydrogen and helium that this is not a general rule.

ABSOLUTE ENTROPY IN MOLECULAR CRYSTALS

Numerous investigations on molecular crystals were carried out in the 1930's. The concept of absolute entropy was used to extract information on the quantitative degree of ordering at given temperatures. (The method is explained in Frank and Clusius, (1937) and an extensive experimental material is reviewed in Clusius et al. (1937).) A more recent work Colwell et al. (1963) constitutes a considerable improvement along with detailed references to the older literature. It relates specifically to the light and heavy methane which are of current interest in connection with neutron diffraction experiments (see below).

In these treatments the statistical entropy is computed exactly as in the above section, but the caloric (or "thermal") entropy is defined as

$$S_{\mathrm{cal}}(P,T_r) = \int_0^{T_r} \frac{C_P}{T}\, dT\,. \tag{2a}$$

The motivation for the developments above was to adapt this valuable work to the correct use of absolute entropy. This involves the replacement of eq. 2a by eq. 2. The difference is that the original reference temperature $T = 0$ is replaced by a pair of temperatures: the reference point, T_r, and the actual temperature of the experiment, often the low temperature limit of the investigation, T_0. By this artifice the procedure is justified, because the questionable absolute zero is not explicitly used, but the substance of the argument is unaffected, and the existing experiments can be invoked with some care as demonstrations of the method outlined in the last section.

The left hand side of eq. 3 is the residual entropy or the measure of the disorder of the system. Even the older of the above quoted treatments enables us to conclude among other things that the CO crystal exhibits only about 25% ordering, while the HD crystal is perfectly ordered, obviously due to the zero-point energy. These results are so much the more remarkable, because at that time there was no experimental method whatever to establish the state of order of the hydrogen and deuteron atoms.

The experimental situation has since become unrecognizable because of neutron diffraction and NMR. Accordingly, there is a considerably revived interest in the structure of molecular crystals, particularly those containing H and D atoms (Press and Kolmar (1975); Press (1981)); Sherwood (1979); White and Morrison (1980); Lushington et al. (1981)). The current challenge is to account for the new experimental facts in terms of molecular dynamics.

I suggest that the Nernst principle has an important role in this program, as it helps us to choose the dynamic model that is appropriate for a physical system. Thus the appearance of a negative entropy, as foreseen in case (c) of the last section, can be taken as a sign that the usual omission of the nuclear spin in the entropy calculation is not admissible (see Colwell (1963) in the case of methane). It seems that the information contained in the entropy values is indeed used as advocated by the theories just quoted, but we find no explicit indication that the relevance of the entropy parameter is recognized.

DISCUSSION

We have considered the well-known and unresolved controversy surrounding the Nernst Heat Theorem.

We have to credit Nernst with a number of major insights which have stood the test of time.

(i) The direction of chemical gas reactions and the value of the equilibrium constant is determined by the physical properties of the participating substances.

(ii) Classical thermodynamics is incomplete, as its arbitrary integration constant in the entropy is inconsistent with the above proposition.

(iii) There is some dimly felt connection between gas reactions and processes within the same system near absolute zero.

Nernst developed these ideas into a quantitative theory by merely joining his Third Law to phenomenological classical thermodynamics. This turned out to be a very successful device, but only as a temporary expediency because it covered up the fact that quantum statistics is needed.

Thus technically, Fowler was right in his critique, but he was unaware of the bigger issue. If thermodynamics is completed by quantum statistics, then the *Nernst Principle* of the present paper provides the framework for Nernst's real contribution: the propositions (i)-(iii) which still remain to be fully appreciated.

It seems to be a widely shared belief that evolution of physics consists of "normal times" in which conclusions are drawn from accepted principles, occasionally interrupted by "revolutions" in the course of which old principles are rejected in favor of new ones, thus opening up new areas for speculation.

The present paper suggests that development of the Nernst Principle did not fit into this pattern. It was rather a "consolidation of principles," in which the inevitable discrepancies surrounding the great discoveries were sorted out.

I believe that the appeal of classical physics stems largely from the fact that the formulators were also concerned with consolidation, at least *within* the three great systems of classical physics: mechanics, electrodynamics and thermodynamics. This healthy evolution came to a halt at the turn of the century because of the failure to establish harmonious *intersystem relations.*

A manifestation of this attitude was the contention between a *purely phenomenological thermodynamics*, and a *purely statistical mechanics.* The rough edges of this confrontation were removed with the emergence of quantum statistics. Yet a new look at absolute entropy came about by eliminating a discrepancy along the interface of thermodynamics and quantum statistics. We see in eq. 3 a careful dovetailing of phenomenology and statistics which can be in no way described as a "reduction" of one theory to the other. It is apparent that this is the essential condition for the extraction of microscopic information from macroscopic measurement. Accordingly, the present paper could be reasonably designated as a new approach to the *theory of measurement.*

In the usual interpretation this term refers to the limitation of classical type measurements on the atomic level. By contrast, here we examine some of the features of quantum physics which opened up microscopic phenomena to quantitative observation. Of course, the present paper deals only with relatively simple quantum statistical problems. However, the method of "consolidation" here displayed works also for the more fundamental aspects of quantum mechanics.

ACKNOWLEDGEMENTS

I am pleased to thank Herbert Pohl for commenting on the manuscript. Clifford Shull and Sow-Hsin Chen offered helpful guidance to the current literature on molecular crystals.

REFERENCES

K. Clusius, L. Popp and A. Frank, Physica, 4 (1937) 1105.

G.H. Colwell, E.K. Gill and J.A. Morrison, J. Chem. Phys. 19 (1963) 635.

R.H. Fowler and E.T. Sterne, Rev. Mod. Phys., 4 (1932) 635.

A. Frank, and K. Clusius, Zf. Phys. Chem. (B) 36 (1937) 291.

G.E. Gibson and W. Heitler, Z. f. Physik. 49 (1928) 465.

E. Hiebert, *Dictionary of Scientific Biography, XV, Suppl. I* (1978) 432.

G.N. Lewis and M. Randall, *Thermodynamics*, McGraw-Hill, New York, 1923.

G.N. Lewis and M. Randall, *Thermodynamics*, 2nd ed. revised by K.S. Pitzer and L. Brewer McGraw-Hill, New York, 1961, Chap. 12.

F. London, *Superfluids*, two volumes, Wiley, New York, 1954. (also Dover reprint).

K.J. Lushington, Kazuo Maki and J.A. Morrison, J. Chem. Phys., 75 (1980) 4010.

K. Mendelssohn, Cryogenics, 4 (1964) 129.

K. Mendelssohn, *The World of Walter Nernst*, Macmillan 1973.

W. Nernst, Nachrichten d. Ges. d. Wiss. Göttingen, (1906) 1.

W. Nernst, *Theoretische Chemie*, Stuttgart, 8-10. Aufl. 1921, IV.5.

V.P. Peshkov and K.N. Zinovieva, JETP, 5 (1957), 1024.

R. Planck, *Thermodynamik*, 3. ed. 1911. p. 268.

R. Planck, *Theory of Heat*, Transl. H.L. Brose, Macmillan, 1932.

W. Press and A. Kolmar, Solid State Commun., 17 (1975) 405.

W. Press, *Single-Particle Rotation in Molecular Crystals*, Springer Tracts in Modern Physics, 92, 1981.

J.N. Sherwood, (ed), *The Plastically Crystalline State (Orientationally Disordered Crystals)*, Wiley, Chichester, 1979.

F. Simon, Ergebn, d. Exakten Naturwiss., 9. J. Springer 1930, 222.

F. Simon, Physica, 4, (1937) 1089.

L. Tisza, *Generalized Thermodynamics*, M.I.T. Press, 1966, 92-96.

M.A. White and J.A. Morrison, J. Chem. Phys. 72 (1980) 5927.

14. From Theoretical Physics to Biology: The Forward Path of Theory with Herbert Fröhlich

G. J. Hyland

"The visible world is neither matter nor spirit, but the invisible organization of energy" (Pagels, 1982).

Fröhlich's forward path from theoretical physics to biology is a fascinating story of cross-fertilization initiated and sustained by his constant, characteristic awareness of the possible relevance of concepts to areas in which they did not at first arise. In retracing this forward path, we shall encounter not only those areas with which Fröhlich's name is now inseparably associated—notably, dielectric theory and superconductivity—but also an adventurous one of more recent origin, namely that of nonlocal field theory, in the context of which the crucial, unifying concept of coherence finds natural expression, and via which Fröhlich has, during the past few years, been led back (Fröhlich, 1986) to relativistic particle physics.

It was, of course, his early acquaintance (ca. 1937), in collaboration with Heitler and Kemmer (Fröhlich et al., 1938), with meson theory and with the techniques and concepts of relativistic field theory that was ultimately instrumental to his success in breaking (Fröhlich, 1950a) the deadlock that had prevailed in understanding superconductivity. This progress was, of course, contingent upon the prior identification of the field mediating the interaction between the particles (electrons) within the *non*relativistic, solid-state context under consideration, and it was here that his earlier experience and work in the field of dielectrics proved indispensable. For already, in connection with his early work on dielectric breakdown, he had successfully treated (Fröhlich, 1937) the interaction of an electron with the lattice of an ionic crystal—by considering the ions not individually, but rather *collectively*, in terms of their associated polarization *field*. Somewhat later, around 1948, it occurred to Fröhlich that this field could be treated in a way similar to the relativistic case. This immediately implied (in addition to the emission and absorption of vibrational quanta by the electron—a process already inherent in Bloch's theory of electrical conductivity) the existence of a nondivergent self-energy and in the case of more than one electron an *interaction* between them that under certain conditions could be *attractive*. The rest is well known: the former effect—the first to be investigated, in collaboration with Pelzer and Zienau (Fröhlich et al., 1950), led to the development of what is now known as polaron theory, whilst the second formed the basis (Fröhlich, 1950a) of the modern theory of superconductivity, via which Fröhlich (1950b) explained the isotope effect.

Fröhlich's fundamental contribution to the theory of superconductivity, and the application of field theoretical methods to solid state physics that then* fol-

*It is of interest to note that as early as 1931 (in only his *second* publication) he had already applied (Fröhlich, 1931) to the free electrons in a metal the relativistic field theoretic technique now known as "second quantization."

lowed, now ranks among the most outstanding examples of fruitful cross-fertilization between different branches of physics—an advance that stemmed from his characteristic broad outlook and from his constant awareness of the possible relevance of concepts and techniques to branches of physics in which they did not at first arise. Such an advance which stands as a " . . strong indictment against fragmentation and overspecialization in theoretical physics" (Fröhlich, 1966)—a situation so common today.

Cross-fertilization on an even more spectacular scale, however, is what characterizes Fröhlich's forward path from theoretical physics to biology. For to recognize the possible significance of physical concepts to biological systems, which are in many ways quite different from those usually considered in physics, requires an even broader and more unprejudiced insight, coupled with an ability to identify within the morass of existing biological facts and details just those particular aspects that might be relevant to an approach from physics. But how did such an approach come about in the first place, and why could it even be considered to be appropriate or worthwhile? It actually all started in a rather light-hearted way following a chance meeting with the French biologist Marois, of the Institut de la Vie, whilst Fröhlich was on holiday, sometime during the summer of 1965.

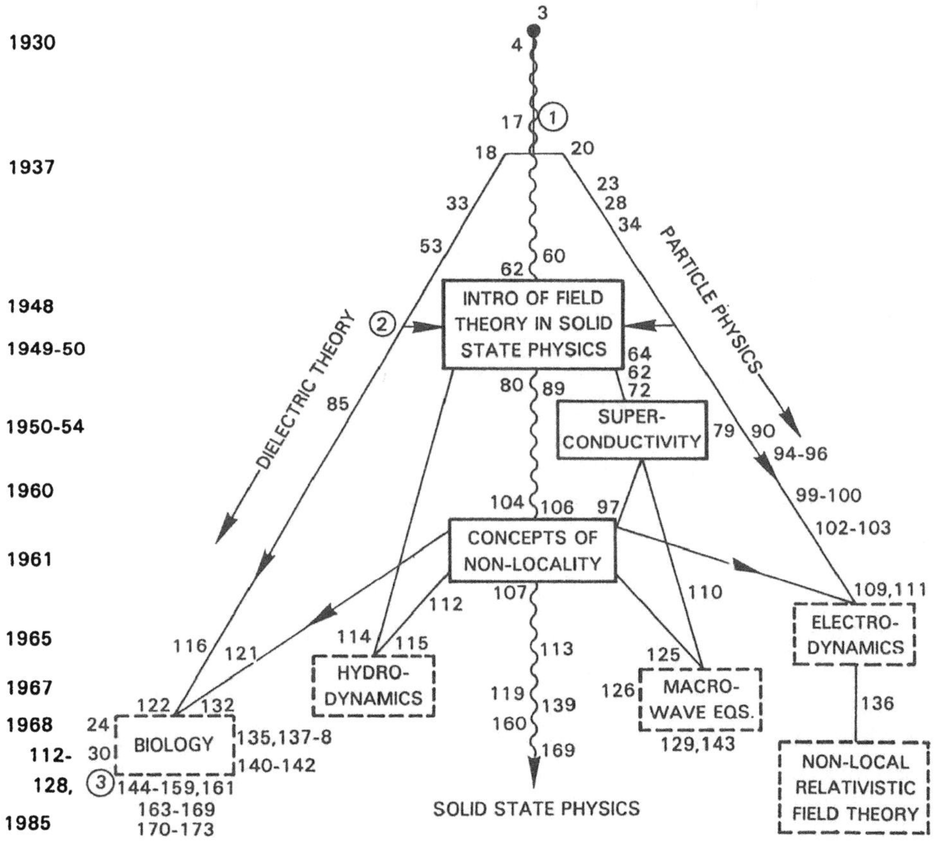

But why, one might ask, should theoretical physics have anything to offer biology? For surely, it was characteristically argued, given the great progress recently made in molecular biology, it would be only a matter of time before the activity and control of macrobiological systems were themselves understandable in microscopic terms. From his experience in physics, however, Fröhlich suspected that this would *not* be the case. For the macroscopic properties in question refer to systems in their *entirety*—and, as such, cannot be formulated in purely microscopic atomic or molecular terms; indeed, the microscopic components of a biological system are the *same*, whether or not the system is active. Thus, for example, the phenomenon of cell division is surely a "*cooperative*" phenomenon, in which the relevant, highly complex, microscopic (molecular) processes cooperate in a highly correlated (coherent) way to produce—at a certain stage in the development of the system, a macroscopic result, which then, in turn, influences the future development of the system. In other words, there must be a high degree of organization, or what in physics is often called "*order*"—an order which, however, from recent advances in structural microbiology, was apparently not of a spatial kind; but Fröhlich knew that other, more subtle, kinds of order can occur in physics. Indeed, the empirical facts that the biological "cooperation" in question evidently becomes effective only at a certain stage of cellular evolution and can, moreover, be completely suppressed by withholding metabolic energy indicated that the relevant order must not only be of a *dynamical* nature, but must also involve what in physics are called "*nonlinear*" processes.

Now, by a fortuitous conjunction of circumstances Fröhlich, at the time of his meeting with Marois, was involved with macroscopic quantum systems—notably superfluid helium (He^4) and superconductors—and, in particular, with the derivation of their characteristic macroscopic properties from microphysics, the most notable of which is, of course, the highly efficient, dissipationless mode of *flow* exhibited by both systems. The stability of these flows is indicative of a high degree of *correlation* between the motion of individual particles, whereby they all act *coherently* in concert to ensure an overall highly organized behavior of the whole system. This is a clear example, then, of a nonspatial "order" in physics—He^4 remaining *liquid* down to absolute zero. As is well known, the order here actually has to do with motion—in particular, with the *momentum* of the helium atoms; for momentum is the atomic counterpart, so to speak, of the hydrodynamic current that is the macroscopic quantity exhibiting the ordered behavior. This order is best described (Fröhlich, 1969) in terms of so-called "macroscopic wave-functions" which express—in certain nonlocal quantum mechanical fields, known as the reduced density matrices—the existence of long-range phase correlations (i.e. coherence), called by Yang (1962), "off-diagonal long-range order" (ODLRO). It was in terms of these fields that the connection of micro- and macrophysics was to be established (Fröhlich, 1973). It is, of course, the "off-diagonality" of the order that reflects its *motional* character, spatial order manifesting itself diagonally.

But what could possibly be the relevance of this type of order—which occurs only at very low temperatures and in systems near thermal equilibrium—to the kind that apparently prevails near room temperature in biological systems which, in their active state, are relatively stable, yet necessarily, in some respects, quite *far* from thermal equilibrium? Furthermore, what, in biological systems, plays the role of the electrons in a superconductor or the atoms in superfluid helium, which are the entities upon which coherence is imposed? For biological systems are, at an atomic level, very much more complicated than are the systems normally dealt with in physics.

Fröhlich, however, believing that the ODLRO concept was not restricted to systems in thermal equilibrium, had some ideas on how to proceed. At the international conference on theoretical physics and biology of the Académie Francaise held in Versailles during the summer of 1967, Fröhlich gave the opening paper (1) entitled, "Quantum Mechanical Concepts in Biology." In this short paper, which (apart from the appendices) does not contain a single equation, Fröhlich laid, in a qualitative and necessarily speculative way, the foundations of most of his subsequent elaborations. In particular, potential carriers of the conjectured coherence were identified and suggestions made as to how the coherence would be established and stabilized. What is not generally known, however, is that the origins of these proposals concerning the identity of the carriers of the coherence actually went back over almost 30 years, to when Fröhlich was at the University of Bristol, in Professor Mott's laboratory. It was thus in Bristol—possibly in the Kardomah Cafe in Piccadilly—that the forward path of theory from theoretical physics to biology really began, in a conversation with Max Reiss, a well-known endocrinologist.

"FROM THEORETICAL PHYSICS TO BIOLOGY"

Key dates and events -1.

1937 **Theory of Electrical Breakdown in Ionic Crystals (interaction of electron with polarization field)**
Discusses dielectric properties of cell membranes with Max Reiss, Willis Jackson and Victor Rothschild.

c.1965 **Chance meeting with the biologist M.MAROIS (Institut de la Vie)**
Connection of micro & macro physics.
Theory of macroscopic quantum systems.
(Macroscopic wavefunctions in superconductors and superfluids)
Appreciation of general significance of COHERENCE.

1967 **1st International Conference on Theoretical Physics and Biology-Versailles.**
Opening paper by H.F. entitled "Quantum Mechanical Concepts in Biology." (contains conceptual basis of future developments- qualitative and speculative).

From Reiss Fröhlich learned that the membranes of biological cells had recently been found to support an electric double layer, across which a potential difference of about 100 mV exists.* Fröhlich, who at that time was working on the theory of electrical breakdown, immediately saw that this indicated that biological materials must possess quite extraordinary dielectrical properties. For the associated enormous electric field $\sim 10^5$V/cm (if the layer is about 100 Å thick) would, in *non*biological materials, actually lead to dielectric breakdown, unless special precautions were taken!

Now in his work (1937) on breakdown in ionic crystals, already mentioned, Fröhlich had formulated the interaction of an electron with the lattice in terms of the polarization field of the latter, particular importance being attached to the long polarization waves corresponding to anti-phase vibrations of the positive and negative ions at the Reststrahlen frequency. The possibility of similar anti-phase elastic

DIPOLAR PROPERTIES OF CELL MEMBRANES (circa 1937)

1. Cell membranes support on electric double layer

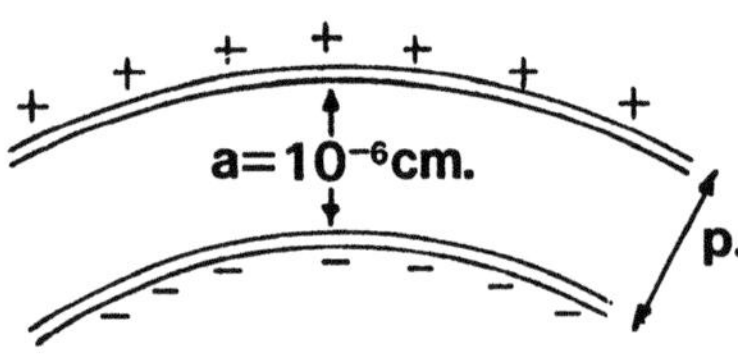

p.d~100mV→10⁵V/cm field

↓

enormous

(→diel. breakdown in other situations!)

2. VIBRATIONAL PROPERTIES:

Vibration of inner and outer layers against one another results in vibration of electric polarization field

$$\text{frequency } \nu = \frac{\text{sound speed}}{2a} \simeq 0.5 \times 10^{11}\text{Hz}$$

$$10^{10} \lesssim \nu \lesssim 10^{12} (10^3\text{GHz})$$

$$\updownarrow$$

$$3\text{cm} \leq \lambda \leq 0.3\text{mm}$$

i.e. the electric field arising from membrane oscillations is in the mm μ wave region

Expt. inaccessible in 1937

*The associated energy difference of unit electric charge across the membrane amounts to about 3 kT at room temperature, suggesting that the transport of single charges obtains significance against thermal noise.

vibrations also existed, of course, in the case of the oppositely charged membrane layers; the frequency of these vibrations, and hence that of the associated electric polarization field could easily be estimated by making reasonable assumptions about the elastic properties of the membrane. This Fröhlich did. The frequency turned out to be about 10^{11} Hz, corresponding to radiation in the millimeter microwave region—a region which was, at the time, experimentally inaccessible, however. "How typical," Fröhlich must have thought, "for Nature to take advantage of our experimental incapabilities!" Encouraged, however, by his experimentalist friend, Willis Jackson, who maintained that such frequencies would indeed be available before too long (perhaps he was aware of the then secret development of RADAR, which itself uses microwaves, but of somewhat longer wavelengths), a sympathetic biologist, in the person of Victor Rothschild, was approached for suggestions of appropriate systems to investigate once the experimental facilities became available; blood cells were suggested. But it was now 1939 and no further progress was made.

These old ideas were not forgotten, however, and were duly resurrected 30 years later, following the meeting with Marois, when Fröhlich realized that therein lay the possibility of a carrier of coherence that had the attractive feature of *not* involving microbiological entities directly, since such are of forbidding complexity. Instead, only a very general macroscopic property of the *whole* biological system was required—namely, the *collective* longitudinal electric oscillations associated with whatever type of vibrating dipolar units characterized a particular system. For in addition to the regions of cell membranes contemplated originally, numerous other entities could now be identified (conceivably having somewhat different vibrational frequencies)—such as, for example, the hydrogen-bonds which occur along giant molecules. These collective modes of vibration, the existence of which follows from the Maxwell equations, must thus be considered to be a quite general feature of diverse biological systems, and a clearly attractive one!

But how could *coherence* be imposed on these collective modes? This was the crucial question to which Fröhlich addressed himself at the Versailles conference. It will be recalled that the order in superfluid He^4, for example, is established by a phase transition to a state whose entropy is not significantly lower than it is in the normal (nonsuperfluid) state—indicating that the ordering somehow involves only *very few* degrees of freedom; nevertheless those few degrees of freedom dominate the remainder and dictate the macroscopic properties of the system. As is well known, the ordering here arises from the impossibility, once a certain temperature (=2°K) has been reached, of uniformly sharing out among all the modes of excitation of the system the *entire* energy decrease associated with a given reduction in temperature—one particular mode getting *more* than its fair share, whereby a *subset* of the system becomes completely de-thermalized and attains an order that would otherwise not obtain until absolute zero. This subset constitutes a macroscopic quantum system, consisting of all those atoms N_0 whose momentum has been reduced to the *same* (zero) value; since all the atoms are identical we have, so to speak, an "N_0-fold replication" of the state of a single atom, which extends throughout the system as a (coherent) matter field: this field is the macroscopic wave function.

Could it be—Fröhlich perhaps thought—that a similar situation prevails in the case of biological systems? In this case, however, any attempt to induce order by energy *extraction* would not seem appropriate since the activity of such systems clearly relies on continual *input* of energy! There exists in physics, however, a well known example of a nonequilibrium system in which *order* can be induced by actually *supplying* energy: the example is, of course, the (optically pumped) laser. Thus, argued Fröhlich, perhaps the behavior of biological systems could, like superfluid He^4, be dominated by just a few degrees of freedom if, by *supplying* energy, one (or a few) of the collective modes of vibration received, so to speak, *more* than its fair share, thereby becoming much more strongly excited than the others, and thus lifted far from thermal equilibrium. In other words, the required behavior could follow if the input energy was *not* completely thermalized, but instead partly channelled into a single mode wherein it is stored *non*thermally. Provided this mode can be stabilized, the biological system will then exhibit long-range phase correlations—the individual dipolar units contributing coherently to a single *giant* dipole oscillation of the *whole* system.

It thus remained to specify a possible mechanism of stabilization so that an enormous energy supply would not be necessary in order to sustain the coherent mode. In this connection, Fröhlich turned once again to dielectric theory, and in particular, to one of the most important points made in his now famous book on dielectrics (Fröhlich, 1949), namely, the dependence of the energy associated with an electric polarization field on the *shape* of the system that is polarized, whereby, if the shape can be changed, a lowering of this energy can be achieved. Since typical components in biological systems, such as cells, for example, are indeed rather easily deformable elastically, such a stabilizing mechanism appeared, in principle, to be viable. This was particularly so when coherent vibrations were excited; for in this case distortions of the participating units no longer intercancel (as they do, to a first approximation, in the case of incoherent (thermal) vibrations) but could conceivably be *constructively* synthesized into a giant (coherent) distortion of the *whole* system. Fröhlich's Versailles paper did not end there, however; he was quick to appreciate the possible biological significance of these (electrostrictive) deformations: in particular, the possibility of non-chemical forces that depend both on the strength of biological activity (i.e., on the metabolic energy supply) and on the state of biological evolution. This is because the elastic restoring forces in a cell membrane, for example, weaken as the cell expands with growth; older cells are thus more easily deformable and hence can then be more strongly coupled to the coherent vibrational mode. The associated increased stress at the cell surface could thus act as a stimulus for cell division!

This, then, was the state of Fröhlich's ideas at the time of the First International Conference on Theoretical Physics and Biology in 1967; necessarily somewhat speculative and qualitative, but sustained by his characteristic intuition that he was on the right track. His view was not, however, universally shared by the participants—the dissenting voices not always being those of biologists—as is evident upon reading the reports (1) of the various discussions at the Conference!

But this was not the first time that Fröhlich had encountered incredulity in connection with his (seemingly) heretical ideas. For what could have been considered more incredible in 1950 than his claim that the same interaction responsible for electrical resistivity—namely, that between the electrons and the ions—was actually responsible for the converse phenomenon, superconductivity! In fact the level of prejudice was, at first, such that one of the discoverers of the isotope effect had to work secretly by night (Fröhlich, 1983) in order to escape the censure of his superiors! But experimental support for Fröhlich's biological speculations took somewhat longer to appear than it had in the case of superconductivity; in fact six years were to elapse before, quite by accident, Frohlich himself happened upon a report of a meeting in the Soviet Union devoted to the effects of coherent millimeter microwave radiation on diverse biological systems: but more of this later. In the meanwhile, undeterred, he persisted with his biological investigations; a second international conference was planned, and even before the Proceedings of the first had appeared in print (1), he had succeeded, using a simple model of a biological system involving a *non*linear coupling between the vibrating units and their surroundings (treated as a heat bath), in quantitatively confirming some of his ideas—in particular, those concerning the establishment of coherent vibrations. The essential point to emerge was the necessity of a minimum *rate* of energy supply. These results appeared first (2) as a *Physics Letter* and then more extensively developed in an article (3) published in the same year, 1968.

Again, before the second paper appeared, the idea of coherent vibrations was extended from animal to vegetable life, to cover not only metabolically fed systems but also those externally fed by sunlight; contact was thus made with photosynthesis. This development was published (4) in *Nature* and reported (6) at the second Versailles Conference in 1969, which also was opened by Fröhlich.

The next step was to consider, in greater quantitative detail, the crucial (electrostrictive) stabilizing mechanism based on the deformability of typical components of biological systems. From this emerged a remarkable, and *quite general* (again nonlinear), property—namely, the existence of a metastable state characterized by a nonvanishing mean value of electric polarization, i.e., by a permanent static displacement, reminiscent of that in ferro-electrics. This result was published as a letter (8) in 1970; its main concern was to point out the possible relevance of this new excitation to resolving ". . .the essential mystery of the enormous catalytic power of enzymes" (Koshland and Neet, 1968). Again these developments were actively discussed at the third Versailles Conference (9) in 1971, after Fröhlich's general introduction in which, for the first time, what is *probably the most important consequence* of the existence of coherent electric vibrations was qualitatively noted. This is the possibility of a long-range (selective) interaction between systems of almost equal vibrational frequencies which is much *stronger* ($\propto R^{-3}$) than the familiar van der Waals ($\propto R^{-6}$) force and which, moreover, can be switched on and off, according to whether or not the rate of metabolic energy supply is sufficient to excite coherent vibrations. In turn, the actual value of the critical energy supply rate depends, via the electrostrictive stabilizing mechanism, on the evolutionary state of the system; as already noted, *older* cells, for example, are *more* easily deformed than young ones. Exciting possibilities of long-range recognition and attraction thus arise, which depend on the stage of biological evolution, and which could never occur electro*statically*. Clearly such possibilities could be of crucial importance to numerous biological processes, such as, for example, the pairing of homologous chromosomes in meiosis (pointed out by Holland (1972), in which was probably the first application of Fröhlich's ideas), enzyme-substrate attraction (18), and the control of cell division, important in the cancer problem (29).

Also noted there for the first time was the possibility of understanding the extreme sensitivity of certain biological systems to the weakest external stimulus in terms of a "triggering" action, whereby such a stimulus simply supplies the *difference* between the available, *sub*critical metabolic energy and the minimum critical value necessary for the establishment of coherent vibrations in the system. An application to the stimulation of nerve impulses immediately presents itself—an idea that was to be pursued in much greater detail somewhat later in 1977, in what represents the pinnacle of Fröhlich's biological speculations: essentially a synthesis of the three ideas already introduced, namely, the coherent electric vibrations, the

QUANTUM MECHANICAL CONCEPTS IN BIOLOGY

H.F.(Versailles) 1967

1. **The components of biological materials have remarkable dielectric properties and are easily deformed elastically.**

2. **Biological systems are relatively stable, yet in some respects far from thermal equilibrium.**

3. **Biological systems probably exhibit some kind of "order" of a non-spacial nature.**

THE FRÖHLICH CYCLE (1967)

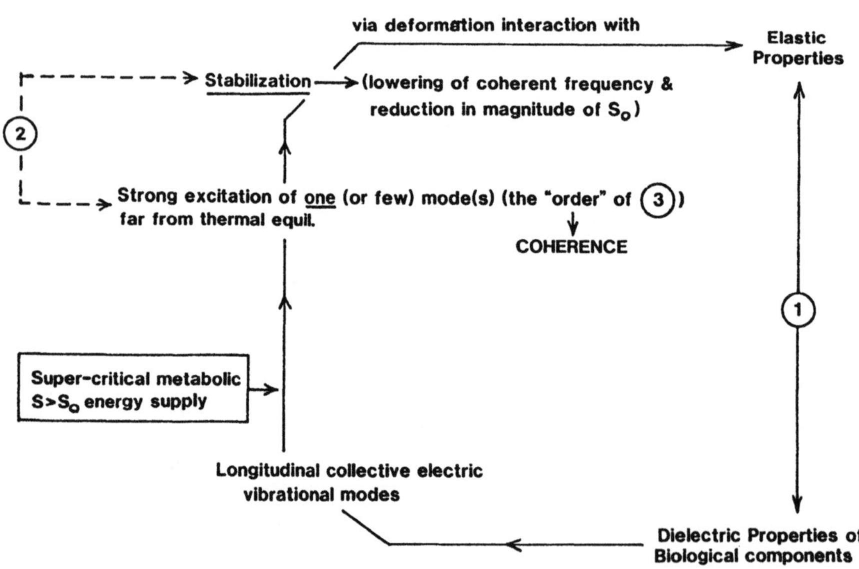

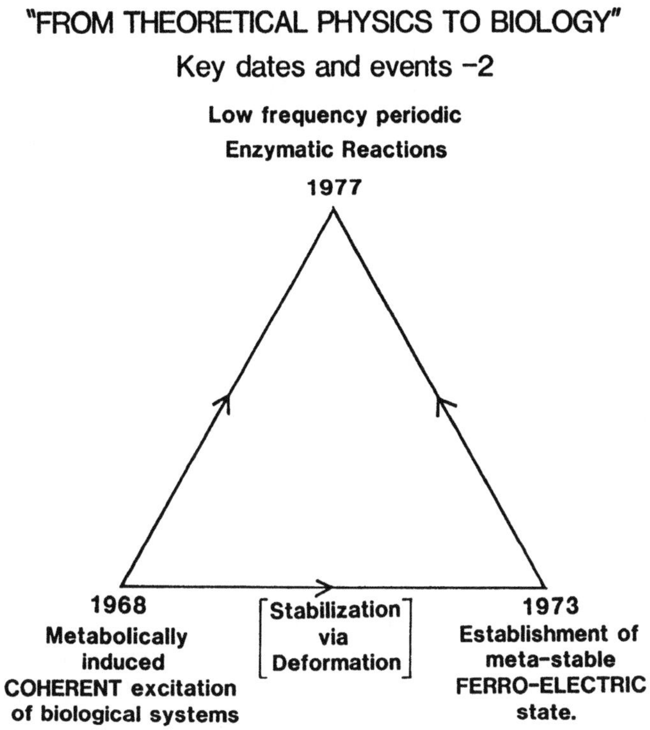

associated long-range selective interaction, and the highly polar metastable states. The result is the so-called "brain-wave" model (22), in terms of which the establishment of low-frequency electric vibrations over large areas of the brain might be understood, together with its extreme sensitivity to external fields that are far too weak to be able to generate a nerve signal directly.

In the meanwhile, however, the first indications of experimental support for Fröhlich's basic idea of coherent electric vibrations in biological systems were starting to appear. The first encouraging evidence was actually found by Fröhlich himself, quite by accident one Saturday morning, on a visit to the departmental library in an attempt to keep warm during a period of heating restrictions, for which the winter of 1974-5 will be remembered in the UK! There he happened upon a report (Devyatkov, 1974) in the Soviet literature of a meeting of the USSR Academy of Sciences held in January 1973; the report contained nine contributions reporting the influence of coherent millimeter microwave radiation* on a variety of biological systems ranging from yeast to the bone marrow of mice! The observed effects were found to be characterized by *two* general features: (a) they depend very strongly on frequency, exhibiting a typical resonance behavior, *precisely* in the theoretically predicted frequency regime near 10^{11} Hz, (b) there exists a critical power level below which no effects are observed and above which the observed effects are

*The earlier work of S. Webb (Webb and Dodd, 1968; Webb and Booth, 1969) was, at this time, unknown to Fröhlich.

nearly independent of the power supply; this was *precisely* the behavior predicted by Fröhlich's model calculations some six years earlier! Furthermore, the relevant power supplies were found to be very *low*—of the order of microwatts per centimeter^2—much too small to produce any heating; i.e., the observed effects were essentially *nonthermal.*† Indeed, subsequent experiments in France (Berteaud et al., 1975) revealed that irradiation of *E. coli.* bacteria by low-power microwaves actually *decreases* their rate of growth, whilst heating increases it!

Since there is no evolutionary need for biological systems to be sensitive to external microwave radiation, one must conclude that the systems *themselves* make use of vibrations in this frequency region. Fröhlich's metabolically generated coherent electric vibrations thus represent a novel possibility in this connection—but clearly, *others* could equally well be devised. An alternative experimental approach is to attempt actually to detect the existence of a metabolically excited coherent biological vibration *spectroscopically*—by laser Raman effect, for example. Here too the results are encouraging but not, of course, conclusive, nor even always reproducible; what *is* definitely established, however, is (i) a strong dependence of the Raman spectrum on the activity of the bacteria investigated (9), certain lines appearing only when the system is metabolically fed, and (ii) the *non*thermal nature of the excitation, the ratio of the anti-Stokes/Stokes lines being close to the classical value of *unity* (28), rather than to the thermal equilibrium (Boltzmann) value of about 0.5. Again, most interesting from the point of view of biological control are the recent comparative Raman investigations (Webb et al., 1977) on normal and cancerous tissues, which have been interpreted (Cooper, 1978) within the framework of Fröhlich's theory; but again, other interpretations are possible.

Since blood was the first system to be suggested for investigation in connection with Fröhlich's pre-war ideas concerning the possibility of electrical oscillations in cell membranes, mention should also be made of the fascinating interpretation of rouleaux formation in blood, in terms of Fröhlich's selective, frequency-dependent, long-range interaction, which has recently been presented. (Rowlands, 1983; Sewchand and Rowlands, 1983).

In attempting to assess the present status of Fröhlich's remarkably novel and attractive approach, caution should be exercised against too pessimistic a verdict solely on the grounds that encouraging as the available experimental evidence undeniably is, it is not as compelling as is the inherent intellectual appeal of his thesis. From a positive point of view, the present somewhat nondefinitive state of affairs should be accepted more as a biological "fact of life" than as an indictment. For in biology, the prevailing experimental conditions are often necessarily *quite different* from those encountered in physics, owing to such problems as the lack of long-range homogeneity and the difficulty of obtaining samples of a size and shape suitable for experimental investigation, *without* interfering with their biological activity. In addition, biological effects are often multicausal, whilst the systems themselves are of much greater complexity than are usually encountered in physics,

†It is of interest to note that the permitted leakage limit on microwave ovens in Western Europe and the USA is also of this order, this being about one thousandth of their operational power; in the USSR, however, the limit is a thousand times lower.

and have, moreover, evolved over large periods of time to perform certain specific tasks. Thus, for example, the sensitivity of the eye to a single light quantum could never have been predicted solely from a knowledge of the properties of its molecular constituents—constituents ". . . which no electronics engineer would dream of using in the construction of an image converter!" (26)

In consequence of this "evolved specificity" it is permissible to ask what is the purpose (i.e., role) of a particular excitation in a given biological system—a question which in physics is strictly forbidden except when dealing with machines or similar constructions. Indeed, not only is it permissible to ask such questions but, before complete understanding can be claimed, positive answers to them must be found: for structural knowledge alone is insufficient, *function*, also, must be established! In fact, consideration of *machines* and the complicated tasks which they can perform lends some support to the premise that the establishment of the physical bases of macroscopic phenomena of the natural world is a much less straightforward task than it is often claimed to be, *notwithstanding* the fact that there is apparently no difference between the atomic and molecular constituents of animate and inanimate matter. For a macroscopic system—be it physical, biological or whatever—is *more* than simply the sum of its parts, and even in the case of *machines* an understanding of their macroscopic properties and behavior in physical

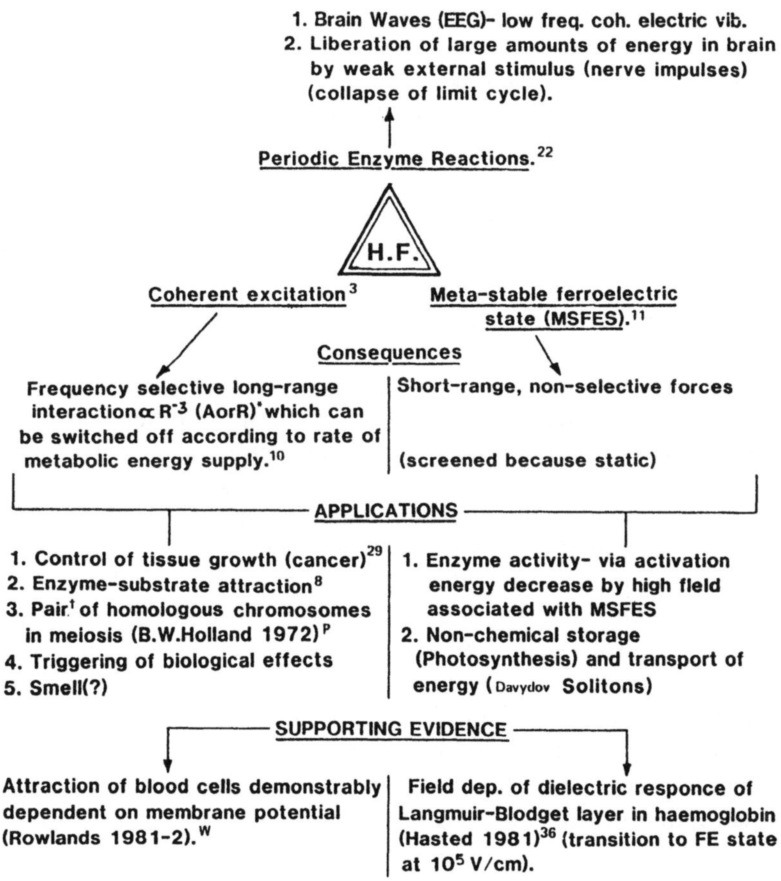

*A = Attractive, R = Repulsive †Pair = Pairing

terms can be far from trivial, often requiring great ingenuity of approach. Given the much greater *micro*scopic complexity of even the simplest evolved biological system, compared with that of the most sophisticated machine, it is then surely not unreasonable to anticipate a corresponding increase in the complexity and subtlety of its associated collective *macroscopic* properties, an understanding of which could well require the introduction of quite new physical concepts. However, as we have, I hope, seen in retracing Fröhlich's forward path from theoretical physics to biology, certain collective concepts that are *already* familiar in physics have found some surprising applications to biological systems, entailing, for the first time, the possibility of a nonchemical control of the whole system, and one which can be switched on and off according to the level of metabolic activity. It is, of course, precisely this activity that distinguishes animate and inanimate matter.

Thus just as in the case of superconductivity, the genius of Herbert Fröhlich has shown that the introduction of new interactions is *not* necessary in order to understand certain macroscopic properties of biological systems, contemporary physics already containing at least some of the required interactions. That this is, in fact, the case only becomes apparent, however, once the appropriate concepts have been identified. This, of course, requires creativity, particularly in the case of biological systems whose inherently nonequilibrium character precludes an approach *via* linear response. Even in the case of physical systems for which this is no longer necessarily true, it should not be forgotten that the connection between "micro" and "macro" is far from systematic—characteristic macroscopic properties invariably being *collective* ones, whose description requires concepts that do not even exist at the microscopic level of the constituent particles!* Such concepts are still, in fact, being discovered, a recent example is the *macroscopic wave functions* of superconductors from which their characteristic macroscopic properties can be derived.

Whatever, in the case of biological systems, may be the extent of the ultimate acceptance of Fröhlich's highly novel ideas, the potential value of unanticipated perspectives on more general questions that arise from them should not be overlooked and could well prove to be of comparable importance, as Fröhlich himself has stressed on numerous occasions. For progress occurs *via* the (correct) answers to questions; but the questions themselves must first be asked. I conclude by giving two examples, both due to Fröhlich (7). The first concerns whether there is a minimum size for an (active) biological system. From the standpoint of physics the answer is in the affirmative, in consequence of the existence of fluctuations that increase in intensity as the number of micro-biological components decreases. Clearly then, there must be a lower limit to any such decrease (in size), below which the collective electric modes—essential to the existence of Fröhlich's coherent excitations in the system (and which are, of course, basic to the envisaged mechanism of biological control)—can no longer be defined, leaving only incoherent heat vibrations. The second question arises from further consideration of the parallel between machines and biological systems, and concerns the following: in principle, it is possible to build machines which, like biological systems, have the capability of reproducing themselves. The question immediately arises, however, as to the construction of the very first reproducing system. In the case of

*A particularly simple example of this which Fröhlich is fond of citing is that of *friction.*

the machines this can, of course, be attributed to the intelligence of its inventor (which is something that perhaps cannot be expressed in purely physical terms), whilst in the case of a biological system one is immediately confronted with questions pertaining to the very origin of life itself! But that is another story!

Thus the forward path of theory from theoretical physics to biology must necessarily be of a somewhat different nature from that which characterizes the link between theoretical and experimental physics, a much *greater* degree of interplay between experiment and theory now being required. But the journey along this path has only just begun, indeed, the path itself is hardly yet well defined, and much work remains to be done, despite the growing numbers of physicists and biologists, both theoretical (Various authors, 1980) and experimental (Various authors and dates), who have taken up this path. A novel and unique start has, however, been made possible through the original insight and persistence of one man—Herbert Fröhlich—who has, over the past 20 years convincingly demonstrated that theoretical physics does have a fundamental role to play in biology. Perhaps the most important aspect of that is the apparent possibility of nonchemical biological control based on quantum mechanical coherence; eventual confirmation of this would then, of course, be equivalent to the demonstration that macroscopic quantum effects are not confined to the inanimate world of traditional physics but are actually involved in a fundamental way in the very processes of life itself!

"We have found a strange foot-print on the shores of the unknown. We have devised profound theories, one after another, to account for its origin. At last, we have succeeded in reconstructing the creature that made the foot-print. And Lo! it is our own" (Eddington, 1953).

ACKNOWLEDGEMENT

It is a pleasure and a privilege to acknowledge the long friendship and guidance received from H. Fröhlich during the past 20 years, and not least in connection with the preparation of this article, which is respectfully and gratefully dedicated to him on the occasion of his 80th birthday:

AD MULTOS ANNOS!

REFERENCES

(In the interest of completeness all the biological publications of H. Fröhlich between 1967 and 1985 are listed below in chronological order; the reference numbers in the text refer to this list. Other references are identified by author(s) and date).

THE BIOLOGICAL PUBLICATIONS OF H. FRÖHLICH (1967-1985)

1. Proc. 1st International Conference on Theoretical Physics and Biology (Versailles, 1967), M. Marois, North Holland, Amsterdam, 1969, pp. 13-22.

2. Phys. Lett. **26A**, 402-3 (1968).

3. Int. J. Quantum Chem. **2**, 641-649 (1968).

4. Nature **219**, 743-744 (1968).

5. Nature **221**, 976 (1969).

6. "Introduction Générale," Proc. 2nd Interntional Conference on Theoretical Physics and Biology (Versailles, 1969), M. Marois, Ed., CNRS, Paris, 1971, pp. 17-19.

7. In *Wohin fuhrt die Biologie?*, M. Lohmann, Ed., Verlag Karl Hasner, München, 1970, pp. 147-173.

8. Nature **228**, 1093 (1970).

9. "General Introduction," Proc. 3rd Int. Conf. *From Theoretical Physics to Biology,* (Versailles, 1971), M. Marois, Ed., S. Karger, Basel, 1973, pp. 2-6.

10. Phys. Lett. **39A**, 153-154 (1972).

11. Coll. Phenom. **1**, 101-109 (1973).

12. In *Synergetics*, H. Haken, Ed., Teubner, Stuttgart, 1973, pp 241-245.

13. Phys. Lett. **44A**, 385 only (1973).

14. In *Progress in Synergetics*, H. Haken, Ed., North Holland, Amsterdam, 1974, pp 263-267.

15. Phys. Lett. **51A**, 21-22 (1975).

16. Phys. Lett. **53A**, 129-130 (1975) (with N.A.G. Ahmed, J.H. Calderwood and C.W. Smith).

17. Proc. 1975 Conference on Dielectric Materials, Measurement and Applications, Cambridge 1975, I.E.E. Conf. Publ. No. 129, pp 44-47 (with N.A.G. Ahmed, J.H. Calderwood and C.W. SMith).

18. Proc. Nat. Acad. Sci. (U.S.A.) **72**, 4211-4215 (1975).

19. Coll. Phenom. **2**, 155-166 (1976) (with N.A.G. Ahmed, J.H. Calderwood and C.W. Smith).

20. Biopolymers **15**, 219-225 (1976) (with L. Genzel, F. Keilmann, M.W. Makinen, T.P. Martin, G. Winterling and Y. Yacoby).

21. In *75 Jahre Quantentheorie*, W. Brauer, H.W. Streitwolf and K. Werner, Eds., Akademie-Verlag, Berlin, 1977, pp 45-51.

22. Neuroscience Res. Prog. Bull. **15**, 67-72 (1977).

23. Biosystems **8**, 193-194 (1977) (dedicated to L. Onsager).

24. Reply to M.A. Livshits, Biophysics (USSR) **22**, 777-778 (1977).

25. Nuovo Cimento **7**, 399-418 (1977).

26. In *Synergetics*, H. Haken, Ed., Springer, Berlin, 1977, pp 241-246.

27. Phys. Lett. **62A**, 463-466 (1977) (with W. Grundler and F. Keilmann).

28. Phys. Lett. **63A**, 407-408 (1977) (with S.J. Webb and M.E. Stoneham).

29. IEEE Trans. Microwave Theory Techniques **26**, 613-617 (1978).

30. Adv. Electron. Electron Phys. **53**, 85-152 (1980).

31. Coll. Phenom. **3**, 139-146 (1981).

32. Z. Naturforsch, **36b**, 208-212 (1981) (with H. Bilz and H. Büttner).

33. Reply to M.Y. Yushina, Phys. Lett. **93A**, 105-106 (1982).

34. Bioelectromagnetics **3**, 45-46 (1982).

35. In *Physics as Natural Philosophy*, A. Shimony and H. Feshbach, Eds., MIT Press, Cambridge, MA, 1982, pp 287-293 (Essays in honor of L. Tisza).

36. In *Coherent Excitation in Biological Systems*, H. Fröhlich and F. Kremer, Eds., Springer, Berlin, 1983, pp. 1-5.

37. Int. J. Quantum Chem. **23**, 1589-1595 (1983).

38. Phys. Lett. **110A**, 480-481 (1985).

39. In *Modern Bioelectrochemistry*, F. Guttmann and H. Keyzer, Eds., Plenum New York, 1985, pp. 241-261.

OTHER REFERENCES

A.J. Berteaud, M. Dardalhon, N. Rebeyrotte and D. Averbeck, Comp. Rend. Acad. Sci. **281D**, 843-846 (1975).

M.S. Cooper, Phys. Lett. **65A**, 71-73 (1978).

N.D. Devyatkov, Sov. Phys. Usp. **16**, 568-579 (1974).

A.S. Eddington, *Space, Time and Gravitation*, Clarendon Press, Oxford, 1953, p 201.

H. Fröhlich, Z. Phys. **71**, 715-719 (1931).

H. Fröhlich, Proc. R. Soc. London **A160**, 230-241 (1937).

H. Fröhlich, *Theory of Dielectrics*, Clarendon Press, Oxford, 1949, 2nd Ed. 1957.

H. Fröhlich, Phys. Rev.**79**, 845-856 (1950).

H. Fröhlich, Proc. Phys. Soc. London **A63**, 778 only (1950).

H. Fröhlich, "Superconductivity and the Many-Body Problem," in *Perspectives in Modern Physics*, R.E. Marshak, Ed., Interscience, New York, 1966, pp 539-552.

H. Fröhlich, J. Phys. Soc. Jpn. **26** (supplement), 189-195 (1969).

H. Fröhlich, Nuovo Cimento **3**, 490-534 (1973).

H. Fröhlich, "History of the Theory of Superconductivity," in *Advances in Superconductivity*, B. Deaver and J. Ruvald, Eds., Plenum, New York, 1983, pp 1-11 (p. 6 in particular).

H. Fröhlich, Proc. Meson—"50"—to be published (1986).

H. Fröhlich, W. Heitler and N. Kremmer, Proc. R. Soc. London **A166**, 154-177 (1938).

H. Fröhlich, H. Pelzer and S. Zienau, Phil. Mag. **41**, 221-242 (1950).

B.W. Holland, J. Theor. Biol. **35**, 395-397 (1972).

D.E. Koshland and K.E. Neet, Ann. Rev. Biochem. **37**, 359 and 380 (1968).

H.R. Pagels, *The Cosmic Code: Quantum Physics as the Language of Nature*, Simon and Schuster, New York, 1982.

S. Rowlands, in *Coherent Excitations in Biological Systems*, H. Fröhlich and F. Kremer, Eds., Springer, Berlin, 1983, pp 145-161.

L.S. Sewchand and S. Rowlands, Phys. Lett. **93A**, 363-364 (1983).

Various authors, Adv. Electron. Electron Phys. **53**, 85-152 (1980).

Various authors and dates, in sources cited for Fröhlich, 30, 31, 36, and 39.

S.J. Webb and D.E. Dodds, Nature **218**, 374-375 (1968), S.J. Webb and A.D. Booth, Nature **222**, 1199-1200 (1969).

S.J. Webb, R. Lee and M.E. Stoneham, Int. J. Quantum Chem., Quantum Biol. Symp. **4**, 277-284 (1977).

S.J. Webb and M.E. Stoneham, Phys. Lett. **60A**, 267-268 (1977).

C.N. Yang, Rev. Mod. Phys. **34**, 694-704 (1962).

Part II

Biophysics

15. On Davydov Solitons at 310 K

A.C. Scott

ABSTRACT

Recent numerical studies of Davydov solitons at physiological temperature are in disagreement with a previous theoretical analysis by Davydov. It is suggested that the disagreement arises because the specific assumption concerning the introduction of temperature in the numerical experiments leads to a mechanism for soliton decay (a soliton-exciton transition) that is forbidden by the Franck-Condon principle. The temperature dependence of the Davydov soliton must be governed by other mechanisms.

INTRODUCTION

A. S. Davydov proposed in 1973 that self-localization of peptide vibrations in proteins might play a role in the storage and transport of biological energy (Davydov and Kislukha, 1973; Davydov, 1973). Since that time, he and colleagues have produced a large body of work, which has been summarized (Davydov, 1973, 1982a-b), to explore ramifications of this germinal idea. Among these studies was a detailed analytical investigation of the effects of physiological temperature (say 310 K) on the formation and propagation of biological solitons which indicated that these "Davydov solitons" should function at 310 K (Davydov, 1980). Recently some numerical studies have appeared that question this conclusion (Lomdahl and Kerr, 1985; Lawrence et al., 1986; see also van Velzen and Tjon (1986)). It is the aim of this paper to examine the discrepancy in results between Davydov and his critics and to suggest a reason for the misunderstanding.

DAVYDOV'S ANALYSIS

Inspection of the α-helix structure of protein reveals "channels" with the atomic sequence

---H–N–C = O---H–N–C = O---H–N–C = O---

which Davydov assumes to be governed by the Fröhlich energy operator (Fröhlich, 1954)

$$\hat{H}_0 = \hat{H}_{CO} + \hat{H}_{ph} + \hat{H}_{int}. \tag{1}$$

$\hat{H}_{CO}$ is an energy operator for the CO stretch (amide-I) vibration including the effects of nearest-neighbor interactions, $\hat{H}_{ph}$ is an energy operator for longitudinal (acoustic) sound waves, and $\hat{H}_{int}$ introduces interactions between phonons and amide-I vibrations.

Davydov's analysis shows that $\hat{H}_0$ has extended (exciton) eigenfunctions with wave number k of the form

$$| \psi_{ex} (k) > = \sum_n a_{n,k}(t) \, \hat{\phi}_T \, \hat{b}_n^\dagger | 0 > \tag{2}$$

and self-localized (soliton) eigenfunctions of the form

$$| \psi_s > = \sum_n a_{n,s}(t) \, \hat{\phi}_0 \hat{\phi}_T \, \hat{b}_n^\dagger | 0 > . \tag{3}$$

Here the index n counts molecules along the chains, $\hat{b}_n^\dagger$ is a creation operator for amide-I vibrational quanta, $a_{n,k}(t)$ and $a_{n,s}(t)$ are probability amplitudes for finding an amide-I quantum at the nth molecule, $| 0 >$ is the ground state for amide-I quanta and phonons, $\hat{\phi}_0$ is an operator that generates the distortion of the longitudinal sound system that leads to self-localization, and $\hat{\phi}_T$ is an operator that thermalizes the phonon states. Thus

$$\hat{\phi}_T = \prod_q (\nu_q!)^{-1/2} \, (\hat{\gamma}^{\dagger q})^{\nu_q} \tag{4}$$

where $\hat{\gamma}^{\dagger q}$ is the creation operator for quanta in a phonon mode of wave number q and

$$\nu_q = [\exp(\hbar \frac{\omega_q}{kT}) - 1]^{-1} \tag{5}$$

is the average number of phonons in the mode of wave number q and frequency ω_q.

NUMERICAL STUDIES

The numerical studies (Lomdahl and Kerr, 1985; Lawrence et al., 1986) that disagree with Davydov's conclusions (Davydov, 1980) start with the average Hamiltonian

$$H_0(T = 0) = < \psi_S \; (T = 0) | \, H_0 \, | \psi_S \; (T = 0) > \tag{6}$$

where $| \psi_S(T = 0) >$ is calculated from eq. 3 with $T = 0$ K, which implies that $\hat{\phi}_T = 1$. Variation of this average Hamiltonian leads to equations for the parameters in $| \psi_S >$ that are identical to classical equations (Scott et al., 1985). These are

$$\left(i\hbar \frac{d}{dt} - E_0 \right) a_{n,s} + J(a_{n+1,s} + a_{n-1,s}) = \chi(\beta_n - \beta_{n+1}) \, a_{n,s} \tag{7a}$$

$$M \frac{d^2\beta_n}{dt^2} - W(\beta_{n+1} - 2\beta_n + \beta_{n-1}) = \chi \, [\, | a_{n+1,s} |^2 - | a_{n,s} |^2] \tag{7b}$$

where E_0 is the fundamental energy of the amide-I vibration, $-J$ is the nearest neighbor interaction energy, M is the mass of an amino acid, W is the spring constant of the hydrogen bond between oxygen and nitrogen, χ is the derivative of the amide-I vibrational energy with respect to the length of the adjacent hydrogen bond, and β_n is the longitudinal displacement of the nth molecule.

In the numerical studies, physiological temperature has been accounted for by introducing fluctuation and dissipation into eq. 7b. Numerical integration of eqs. 7 with physically reasonable parameters then demonstrates that self-localized (soliton) solutions decay into extended (exciton) solutions after a few picoseconds.

To show that their calculation of soliton decay does not depend on the method of introducing temperature, Lomdahl and Kerr (1985) proceed as follows: First they use the fluctuation and dissipation terms to "prepare" the system close to the desired temperature. Then they turn off the fluctuation and dissipation terms, introduce a Davydov soliton at the center of the chain, and continue the evolution deterministically using eqs. 7 exactly. They find that "the evolution of the soliton is the same. . . . . . it disappears within a few picoseconds."

In these numerical studies, transitions from an initial soliton state into exciton states are induced by thermal energy in the acoustic system

$$M \frac{d^2\tilde{\beta}_n}{dt^2} - W(\tilde{\beta}_{n+1} - 2\tilde{\beta}_n + \tilde{\beta}_{n-1}) = 0\,, \tag{8}$$

where $\tilde{\beta}_n$ is the thermal motion of the nth molecule. Inspection of eqs. 7 shows that this effect is equivalent to introducing an extra term of the form $\chi(\tilde{\beta}_n - \tilde{\beta}_{n-1})a_{n,s}$ into the right-hand side of eq. 7a. This, in turn, is equivalent to perturbing the average Hamiltonian by

$$H' = \sum_n f_n\,(t,T)\,|\,a_{n,s}\,|^2 \tag{9}$$

where

$$f_n \equiv \chi(\tilde{\beta}_n - \tilde{\beta}_{n-1}) \tag{10}$$

and has the property that

$$f_n\,(t,0) = 0. \tag{11}$$

Such a perturbation corresponds to replacing the original energy operator $\hat{H}_0$ with $\hat{H}_0 + \hat{H}'$ where

$$\hat{H}' = \sum_n f_n\,\hat{b}^{\dagger n}\hat{b}_n. \tag{12}$$

QUANTUM ANALYSIS AT 310 K

In this section I sketch a quantum calculation of the transition rate from an initial soliton state into exciton states under the perturbation of the energy operator in eq. 12. It is convenient to begin with a mixed wave function of the form

$$|\,\psi > \; = C_0(t)\,|\,\psi_S > + \sum_K C_k\,(t)\,|\,\psi_{ex}\,(k) > . \tag{13}$$

The Schrödinger equation

$$i\hbar\,\frac{\partial}{\partial t}\,|\,\psi > \; = [\hat{H}_0 + \hat{H}']\,|\,\psi \tag{14}$$

then requires that

$$i\hbar \dot{C}_0 | \psi_S > + i\hbar \sum_k \dot{C}_k | \psi_{ex}(k) > \; = \; C_0 \hat{H}' | \psi_S > + \sum_k C_k \hat{H}' | \psi_{ex}(k) > \quad (15)$$

In order to describe the thermal decay of a soliton, it is appropriate to assume that at $t = 0$

$$C_0(0) = 1, \text{and}$$

$$C_k(0) = 0 \text{ for all } k. \quad (16)$$

Next it is necessary to evaluate the squared magnitude of the projection of the soliton wave function onto the exciton wave functions, $| < \psi_S | \psi_{ex} > |^2$. This contains the factor

$$F = | < 0 | \hat{\phi}^{\dagger T} \hat{\phi}_0 \hat{\phi}_T | 0 > |^2 \quad (17)$$

which is the squared magnitude of the projection of the wave function describing distortion of the sound system for a soliton onto that for an exciton. It is appropriate to call this a *Franck-Condon factor* (FCF) since it is similar to a well known quantum mechanical restriction on transitions between electronic states in diatomic molecules (Franck, 1925; Condon, 1928).

It is convenient to write

$$\hat{\phi}_0 = \prod_q \exp\,[\Gamma_q \hat{\gamma}^{\dagger_q} - \Gamma_q^*(t) \hat{\gamma}_q] \quad (18)$$

for then one can use a boson identity from Davydov (1980) to show that

$$< 0 | \hat{\phi}^{\dagger T} \exp\,[\Gamma_q \hat{\gamma}_q^\dagger - \Gamma_q^* \hat{\gamma}_q] \hat{\phi}_T | 0 > \; = \exp[-(\nu_q + 1/2) | \Gamma_q |^2]. \quad (19)$$

Expressing the FCF as a function of temperature, we can see that *it is always less than its zero temperature value.* Thus for T > 0 K,

$$F(T) < F(0) = \exp\,[-\sum_q | \Gamma_q |^2]. \quad (20)$$

For a stationary soliton, the Γ_q*'s* are related to the β_n*'s* in eq. 7b by the discrete Fourier transform

$$\Gamma_q = \sum_n \left(\frac{M_{\omega_q}}{2N\hbar} \right)^{1/2} \beta_n \exp\,(-iqna). \quad (21)$$

Here N is the total number of amino acid units in the channel, a is the one-dimensional lattice spacing, and

$$\omega_q = \omega_a \sin\,(qa/2) \quad (22)$$

where ω_a is the band edge frequency for acoustic modes. Writing the sum in eq. 20 as

$$\sum_q | \Gamma_q |^2 = \left[\frac{M}{2} \hbar \sum_n \beta_n^2 \right] \left[\frac{1}{N} \sum_q \omega_q \right], \quad (23)$$

and noting from eq. 22 that the average value of ω_q is $2\omega_a/\pi$, gives

$$\sum_q |\Gamma_q|^2 = \frac{M_{\omega_a}}{\pi \hbar} \sum_n \beta_n^2 . \tag{24}$$

This is an example of Parseval's theorem. Thus, if the displacements of the amino acid residues from their equilibrium positions by the presence of the soliton (the β_n's) are known, the zero-temperature FCF can be calculated as

$$F(0) = \exp\left[-\frac{4}{\pi} \hbar \frac{W}{\omega_a} \sum_n \beta_n^2 \right] \tag{25}$$

where I have used the fact that the cut-off frequency for acoustic modes is $\omega_a = 2\sqrt{W/M}$ (Brillouin, 1953).

To apply eq. 25 in a simple case, assume a stationary soliton to be initiated at a distance N' molecules from the end of a channel that has a total length of N molecules. The soliton will expand the channel length by χ / W (Davydov, 1979). If the center of mass of the channel does not change, the molecules to one side of the soliton will be displaced by $\beta_n = (N'/N)(\chi/W)$ and those to the other side by $\beta_n = -[(N-N')/N](\chi/W)$. Thus from eq. 25

$$F(0) = \exp\left[-\frac{4}{\pi} \frac{(\chi^2/W)}{\hbar \omega_a} N'(1-\frac{N'}{N}) \right] . \tag{26}$$

For the specific calculation by Lomdahl and Kerr discussed above, $N = 100$ and the soliton is introduced at the center of the chain. Assuming $(\chi^2/W) \approx \omega_a$, we calculate that

$$F(0) \approx 10^{-14} . \tag{27}$$

The implication of eqs. 27 and 20 is thus that *transitions from a soliton state to exciton states along a single channel are quantitatively forbidden by the Franck-Condon principle at both zero and finite temperature.* Davydov sometimes refers to this as the "topological stability" of his soliton. Thus the assumption used by Lomdahl and Kerr (1985) and by Lawrence et al. (1986) to introduce temperature dependence into eq. 7b is invalid.

To see this more precisely, we can return to eq. 15 armed with the facts that

$$< \psi_{ex}(k) \mid \psi_S > \doteq 0 \tag{28}$$

and

$$< \psi_{ex}(k) \mid \hat{H}' \mid \psi_S > \doteq 0 . \tag{29}$$

Then from the initial conditions expressed in eq. 16:

$$\dot{C}_k \doteq 0 , \tag{30}$$

which in turn, implies:

$$| C_0(t) |^2 \doteq 1 . \tag{31}$$

Thus $\hat{H}'$ does not induce transitions from solitons to excitons.

CONCLUDING COMMENTS

The above remarks should not be construed to imply that physiological temperature has no effect upon the behavior of a Davydov soliton. The introduction of temperature dependence must, however, be correctly formulated. Davydov (1980) describes two such effects: (i) a decrease in the nearest neighbor interaction energy $(-J)$ produced by violation of the phase coherence of the intramolecular excitations, and (ii) a decrease in the nonlinearity parameter (χ^2/W) caused by a decreased coupling of the exciton with lattice deformations. As the temperature is increased, soliton wave functions look more and more like exciton wave functions. Eventually the soliton states become identical to exciton states and effectively disappear. This behavior is contained in the many-body calculation by Alexander and Krumhansl (1986); see also Krumhansl (1986).

Finally it is noted that the present considerations apply also to the "molecular dynamics" calculations pioneered by Karplus and his colleagues (McCammon, et al. 1977). Since these calculations are also classical and at physiological temperature, they cannot be expected to reveal the soliton proposed by Davydov.

ACKNOWLEDGEMENT

It is a pleasure to thank T. W. Barrett and J. A. Krumhansl for helpful comments during the preparation of this manuscript.

REFERENCES

D. Alexander and J.A. Krumhansl, Phys. Rev. B: Solid State, in press, (1986).

L. Brillouin, *Wave Propagation in Periodic Structures*, Dover, New York, 1953.

E.U. Condon, Phys. Rev. **32**, 858 (1928).

A.S. Davydov and N.I. Kislukha, Phys. Stat. Sol. (b) **57**, 465 (1973).

A.S. Davydov, J. Theor. Biol. **38**, 557 (1973).

A.S. Davydov, Phys. Scripta **20**, 387 (1979).

A.S. Davydov, Sov. Phys. JETP **51**, 397 (1980).

A.S. Davydov, *Biology and Quantum Mechanics*, Pergamon, New York, 1982a.

A.S Davydov, Sov. Phys. Usp. **25**, 898, (1982b).

J. Franck, Trans. Faraday Soc. **21**, 536 (1925).

H. Fröhlich, Adv. Phys. **3**, 325 (1954).

J.A. Krumhansl, this volume (1986).

A.F. Lawrence, J.C. McDaniel, D.B. Chang, B.M. Pierce, and R.R. Birge, Phys. Rev. A: Gen. Phys. **33**, 1188 (1986).

P.S. Lomdahl and W.C. Kerr, Phys. Rev. Lett. **55**, 1235 (1985).

J.A. McCammon, B.R. Gelin, and M. Karplus, Nature **267**, 585 (1977).

A.C. Scott, P.S. Lomdahl, and J.C. Eilbeck, Chem. Phys. Lett. **113**, 20 (1985).

G.A. van Velzen and J.A. Tjon, "Numerical studies on the stability of Davydov solitons," (preprint) (1986).

16. Comparisons of Optical Absorption by Impurity Center and by Polarons in Crystalline Acetanilide

J. A. Krumhansl

With 1 Figure

ABSTRACT

Acetanilide (ACN) is of interest because of its similarity to α-helical protein. Nonlinear interactions between molecular (e.g., C=O) excitons and lower-frequency phonons in ACN have been shown by Scott and others to produce a highly localized (soliton) polaron. The optical signature is a narrow line detached from the general amide-I vibrational excitations; this line has a strong temperature dependence. The purpose of this report is to compare the theory of that temperature dependence with those developed earlier for the "zero-phonon" and side band absorption by localized modes in crystals. We show how there are additional contributions to the polaron absorption beyond those proposed by Scott; the complete expression leads to the same form for the low-temperature dependence as that traditionally found for zero-phonon lines.

This paper is dedicated to the memory of Denise M. Hobbins Alexander, who was tragically killed October 23, 1985; it is based on many discussions with her on this topic.

INTRODUCTION

Careri et al. (1984) have performed significant experiments on a hydrogen-bonded polypeptide analog, crystalline acetanilide (ACN). The theoretical model developed by Scott (1985, Careri et al., 1984) for the Davydov (1982) soliton in polypeptides is based on a Hamiltonian of the form

$$H = H_{\text{amide-I}} + H_{ph} + H_{\text{int}} \tag{1}$$

referring to the C=O vibrational excitons, the phonons, and the interaction respectively. Specifically, in model form the H_{ph} is described in a basis of $(\hat{p}_n , \hat{q}_n)$ the momenta and displacements of phonons, and the $\hat{b}_n, \hat{b}_n^+$:

$$H_{ph} = \sum_n \chi_n \hat{q}_n \hat{b}_n^+ \hat{b}_n \tag{2}$$

which are C=O vibration operators. This is traditionally called the linear coupling model, with χ_n defined as the nth phonon mode coupling constant.

Analysis of the coupled amide-phonon system following Davydov's method leads to the result that with ACN parameters the soliton effectively degenerates into a small polaron trapped at one site. The lattice phonons that interact with the C=O vibrations to form the polaron are thereby statically displaced from equilibrium, so that optical excitation of the polaron requires transitions between the

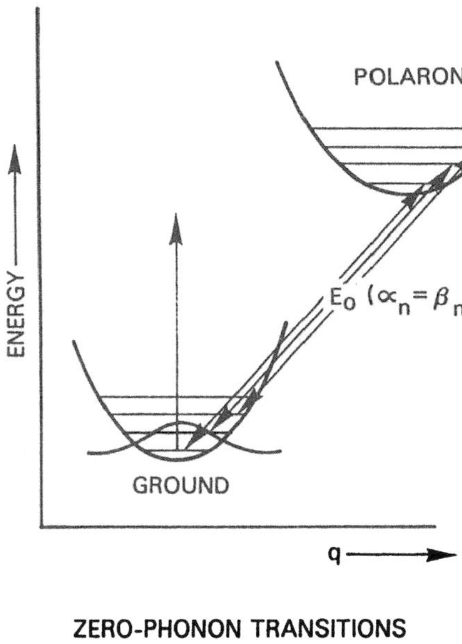

Fig. 1 — Configuration diagram of ground and polaron states for one of the interacting phonons.

ground state phonons and displaced phonons. The calculation of the temperature dependence of this process is the subject of the present paper, as well as the previous work of Scott (1984) and Alexander (1985) for ACN, and a host of others in connection with phonons and localized impurity state interactions (Fitchen, 1968). An extensive examination of this problem was first developed by Holstein (1958), and the finite-temperature many-body treatment is summarized by Mahan (1981). The formation of the small (i.e., localized) polaron state itself is not the issue here; there is general agreement on that matter. The signature of optical production of the polaron state is a narrow line appearing in ACN at aproximately 1650 cm^{-1}, approximately 17 cm^{-1} below the C=O band at 1667 cm^{-1}. The 1650 cm^{-1} shows a strong temperature dependence, decreasing rapidly form a maximum at T = 0 by an order of magnitude at room temperature. This temperature dependence is our main concern. We compare the theory developed for "zero-phonon" lines of impurity states with those for polarons in ACN.

BACKGROUND

First, we quote the main results of Scott's theory, with some slight changes in notation. A configuration coordinate representation of the excitation of a C=O polaron is shown in Fig. 1; the coordinate of the nth phonon is q_n.

The polaron state is that with nominally one quantum of C=O energy, $\hbar\omega_{C=O}$, shifted by the coupled phonon displacement energy. Respectively, for the ground (taken as zero energy) and polaron states (for the nth phonon contribution):

$$E_{\text{ground}} = \frac{1}{2} M \omega_n^2 q_n^2 \tag{3a}$$

$$E_{\text{polaron}} = \hbar\omega_{C=O} + \frac{1}{2} M\omega_n^2 q_n^2 - M\omega_n^2 q_n d_n \tag{3b}$$

$$= \left[\hbar\omega_{C=O} - \frac{M\omega_n^2 d_n^2}{2}\right] + \frac{M\omega_n^2}{2}(q_n - d_n)^2 \tag{3c}$$

where d_n is the static displacement of the nth phonon in the polaron state, given by $(\chi_n/2W_n)$, and $W_n = M\omega_n^2$. In eq. 3c the first term shows the C=O energy shift due to the phonon (trapping) interaction. Attention should be called to the fact that all models presently under discussion ignore the phonon renormalization due to the C=O excitation. Thus, in eqs. 3 the curvature of both ground and excited curves is the same; ordinarily this is a very good approximation and is assumed to hold here. In quantized form, the energies are

$$E_{\text{ground}} = \left(\alpha_n + \frac{1}{2}\right)\hbar\omega_n$$

$$E_{\text{polaron}} = E_{on} + \left(\beta_n + \frac{1}{2}\right)\hbar\omega_n \tag{4}$$

where $E_{on} = \hbar\omega_{C=O} - (\chi_n^2/2W_n)$, and (α_n, β_n) are the ground- and excited-state phonon vibrational quantum numbers.

Actually, the above has been schematically simplified to one oscillator; a real polaron is formed by interaction with many phonons and more properly the shifted C=O vibrational term should be

$$E_0 = \hbar\omega_{C=O} - \sum_n (\chi_n^2/2W_n). \tag{5}$$

The complete excitation spectrum is given by a sum over terms

$$E_{\text{exc}}^{n,\alpha,\beta} = E_0 + (\beta_n - \alpha_n)\hbar\omega_n \tag{6}$$

and in principle n can range over acoustic and optical phonons. The χ_n are large for optical modes, moderate for mid-range acoustic phonons, and vanish at $\omega_n = 0$.

At this point we encounter a fundamental point in Scott's analysis. He considers the 1650 cm^{-1} transition to take place between $\alpha_n = 0, \beta_n = 0$ only, for which it is clear that the excitation energy is E_0, the polaron energy. At zero temperature these will indeed be the only processes. However at finite temperature additional transitions ($\alpha_n = 1, \beta_n = 1$; $\alpha_n = 2, \beta_n = 2$;...) *also have excitation energy E_0 and lead to C=0 polaron production; in all other theories they are included.* This follows because in the linear approximation (which is very good) the ground state phonon spectrum and displaced state have the same spacings between levels (Fig. 1). Excitation processes in which $\alpha_n = \beta_n$ result in no emitted phonons, thus the designation "zero-phonon."

POLARON OPTICAL ABSORPTION

As just noted the optical excitation of a C=O polaron can take place by many processes, as long as there is no net emission of phonons, i.e. ground-state phonon quantum number α_n and excited state β_n are the same. However, the excited-state phonon coordinate is shifted by the amount d_n. The matrix element between ground and excited phonon states is required in order to compute the transition probability. The overlap between displaced harmonic oscillator wave functions is discussed by Fitchen (1968) in connection with the zero-phonon absorption by impurities. With obvious changes in notation:

$$|<\alpha_n|\beta_n>|^2 = e^{-S_n}\frac{\alpha_n!}{\beta_n!}S_n^{(\beta_n-\alpha_n)}[L_{\alpha_n}^{(\beta_n-\alpha_n)}(S_n)]^2 \tag{7}$$

where $S_n = (m\omega_n d_n^2/2\hbar) = (\chi_n^2/2\hbar\omega_n W_n)$, and L_s^t is the Laguerre polynomial. This result is obtained in the small polaron limit (i.e. $|a|^2 = 1$ in Scott's notation).

At T = 0 the only processes contributing to the polaron (zero-phonon) excitation are $\alpha_n = 0$, $\beta_n = 0$ whence, since $L_0^0 = 1$, the transition probability is proportional to

$$W_{00}^n(0) = e^{-Sn} \tag{8}$$

sometimes called the Franck-Condon factor.

At finite temperature the probability of initial or final states being just $\alpha_n = 0$, $\beta_n = 0$ respectively is $[1-\exp(-\hbar\omega_n/kT)]$, so that the temperature dependence of this *special* process (only) is

$$W_{00}^n(T) = e^{-S_n}[1-\exp(-\hbar\omega_n/kT)]^2 \tag{9}$$

which is the temperature dependence of the polaron optical absorption proposed by Scott, in particular with a *single* optical phonon having a temperature-dependent frequency in order to fit the data.

However, if all $\alpha_n = \beta_n \neq 0$ contributions for the polaron absorption are taken into account, the temperature-dependent transition strength involving the nth phonon is given by [see (5-8b), in Fitchen (1968)]:

$$W_0^n(T) = \exp\left[-S_n\coth\left(\frac{\hbar\omega_n}{2kT}\right)\right] I_0\left[S_n\operatorname{csch}\left(\frac{\hbar\omega_n}{kT}\right)\right] \tag{10}$$

where I_0 is the zeroth-order imaginary argument Bessel function, and this is the contribution of one (the nth) phonon.

In addition it has been shown that within the linear coupling-harmonic phonon model the contributions to the *total* "zero-phonon" absorption are multiplicative:

$$W_0(T) = \prod_n W_0^n(T). \tag{11}$$

For a general set of phonons interacting with the C=O excitation and over a wide temperature range one would simply have to evaluate eq. 1 numerically to determine the polaron optical absorption strength.

However, at low temperature it is possible to make useful interpretive approximations. Note that if $\hbar\omega << kT$, $\operatorname{csch}(\hbar\omega/kT)\rightarrow 0$, and $I_0(0+) \simeq 1$. In addition, the thermal average quantum number $\bar{\alpha}_n$ can be related to eq. 10 by

$$\coth\left(\frac{\hbar\omega_n}{kT}\right) = (2\bar{\alpha}_n + 1). \tag{12}$$

Then, approximately [(5-22) in Fitchen (1968)]:

$$W_0(T) \simeq \exp\left\{-\sum_n{}' S_n\,(2\bar{\alpha}_n+1)\right\}, \tag{13}$$

where the prime on the sum implies that $\hbar\omega_n \lesssim kT$.

Proceeding farther in approximation and following the theoretical developments for impurity zero-phonon lines, at very low temperature we expect the acoustic phonons to dominate in the condition $\hbar\omega_n << kT$. In the Debye approximation for those modes, and using Maradudin's [Fitchen (1968)] result that $S_n \propto \omega_n^{-1}$ one finds that eq. 13 reduces to

$$W_0(T) = \exp\left[-\bar{S} - \beta T^{D-1}\right] \tag{14}$$

where $\bar{S} = \sum_n S_n$ and D is the dimensionality of the material. For an isolated linear polymer, $D = 1$, but in crystalline ACN, $D = 3$. One expects this simplified approximation to hold only in the very lowest temperature range; many linear and sheet polymers show crossover "effective dimensionality" at intermediate temperature, particularly if highly anisotropic. In the case of crystalline ACN the low-temperature intensity of the polaron absorption is very well fit by $\log(I/I_0) \simeq -\beta T^2$. This is the result obtained by Alexander (1985). However, as Fitchen notes, the assumption of a single coupled phonon does not deviate much from the above except at the high and low temperature extremes. This statement also applies to the eq. 9 used by Scott. More refined low-temperature data is necessary to see the "pure" $\exp[-\beta T^2]$ behavior. However, at higher temperature one must remember to include the Bessel function term, I_0, from eq. 10; it has the effect of causing the absorption to fall off less rapidly with increasing temperature than the exponential alone, eq. 14, would predict.

DISCUSSION

The main purpose of this report has been to examine the nature of the low-temperature optical absorption in the 1650 cm^{-1} region of ACN. This is believed to be the signature of a C=O vibrational excitation trapped as a highly localized soliton, i.e., "small polaron."

There are several distinct quantities involved in the overall description when many phonon modes are taken into account in the polaron formation.

First, the polaron energy shift from the free C=O vibrational frequency (1667 cm^{-1}); this is given by

$$\Delta E = -\sum \frac{\chi_n^2}{2W_n}. \tag{15}$$

Second, the zero-temperature Franck-Condon factor in the optical absorption given by

$$W_0(0) = \exp\left[-\sum_n S_n\right] = \exp\left[-\sum \frac{\chi_n^2}{2\hbar\omega_n W_n}\right]. \tag{16}$$

Finally, the finite temperature correction given in general by eqs. 10 and 11, and in the low-temperature region by

$$W_0\,(T) \simeq W_0(0)\exp\left[-2\sum_n S_n\,\bar{\alpha}_n\right] \tag{17}$$

where $\bar{\alpha}_n$ is the thermal average quantum number $\bar{\alpha}_n \simeq (kT/\hbar\omega_n)$ if $(\hbar\omega < kT)$.

It is seen that these three quantities involve distinctly different moments (i.e. $\omega_n^0, \omega_n^{-1}, \omega_n^{-2}$) of quantities (x_n^2/W_n). As a result one can say that in a general way all modes contribute equally to the polaron trapping energy eq. 15, while the low-frequency modes dominate the finite temperature diminution of the absorption strength eq. 17 at intermediate temperatures. Beyond this general statement more elaborate models must be examined. In a more detailed analysis and discussion (Alexander and Krumhansl, 1986) we found approximately equal contributions to ΔE from an acoustic band with upper limit 33 cm^{-1} and from an optical mode at $\approx$ 70 cm^{-1}. At the same time the acoustic modes dominated the temperature dependence of the absorption. These specific results agree with the general arguments just given.

It is reasonably accurate to say that, generally, optical modes dominate the polaron formation, while acoustic modes are more important at low temperature for decreasing the optical absorption.

We hope that we have shown how the complete expression for the temperature dependence of optical absorption by the polaron involves more processes than Scott has considered (i.e. all $\alpha_n = \beta_n$, not just $\alpha_n = \beta_n = 0$). *This point is the major source of conceptual and quantitative difference between Alexander (1985) and Scott ((1985), Careri et al (1985)).*

In a larger sense these differences are secondary, and we emphasize that we subscribe to the main features proposed by Scott and by Careri in explaining the 1650 cm^{-1} absorption at low temperature in crystalline ACN as due to excitation of a small polaron, the limiting case of a highly compact soliton.

At the same time there are practical questions that arise when one asks about the biological relevance of the polaron excitation at room temperature. Lomdahl and Kerr (1985) have carried out quantitative computer simulations of the model under discussion in the presence of a thermal reservoir and find that with the parameters generally agreed upon from optical data the polarons are not stable at room temperature. This also appears formally in a complete many-body theory (Lomdahl and Kerr, 1985) though as yet that has not been quantified and the assumption concerning temperature dependence used by Lomdahl and Kerr (1985) is quite different from that proposed here (cf Scott, 1986). And beyond these considerations one must remember that ACN is a crystalline surrogate of a real protein, and that recent experiments (Layne et al., 1986) have not found evidence for polaron excitation in an in vivo situation.

In spite of these caveats we believe that the experiments on ACN have been important in their own right in the context of excitations in hydrogen-bonded materials, but more needs to be done. Inconsistencies still exist between infrared and Raman measurements, and the possibility of Fermi resonance explanations is not considered here. These will be discussed elsewhere.

We acknowledge continued constructive criticism by A.C. Scott, and useful discussions with I.J. Bigio and B.I. Swanson. This work was supported by the Naval Air Systems Command H.Q. as part of a general study of energy transport in biomolecular materials.

REFERENCES

D.M. Alexander, Phys. Rev. Lett. **54**, 138 (1985).

D.M. Alexander and J.A. Krumhansl, Phys. Rev., (1986), in press.

G. Careri, U. Buontempo, F. Galuzzi, A.C. Scott, E. Gratton, and E. Shyamsunder, Phys. Rev. B: Solid State **30**, 4689 (1984).

A.S. Davydov, *Biology and Quantum Mechanics*, Pergamon, New York, 1982.

D.B. Fitchen, in *Physics of Color Centers* W.B. Fowler Ed., Academic, New York, 1968.

T. Holstein, Ann. Phys. **8**, 325 (1959).

S.P. Layne, I.J. Bigio, A.C. Scott, and P.S. Lomdahl, Proc. Natl. Acad. Sci. U.S.A., **82**, 7599 (1986).

P.S. Lomdahl and W.C. Kerr, Phys. Rev. Lett. **55**, 1235 (1985).

G.D. Mahan, *Many Particle Physics*, Plenum, New York, 1981, pp. 282-93, 523-33.

A.C. Scott, Phil. Trans. R. Soc. London **A315**, 423 (1985), and references therein.

Scott, A.C., 1986, this volume.

17. The Vibrational Soliton: an Experimental Overview

I. J. Bigio

With 4 Figures

ABSTRACT

To date the most convincing evidence of vibrational solitons in biopolymers has been found in two very disparate systems: Davydov-like excitations in hydrogen-bonded linear chains (acetanilide and N-methylacetamide) which are not biopolymers but plausible structural paradigms for biopolymers, and longitudinal acoustic modes of possibly nonlinear character in biologically viable DNA.

INTRODUCTION

The first biophysical theory to address the possibility of collective phenomena in biological systems was developed by Fröhlich (1968). In this generic theory a collection of oscillators interact nonlinearly as they exchange quanta with a heat bath. When pumped metabolically, a coherent state of the oscillators evolves that is reminiscent of a Bose-Einstein condensation. Later in 1973, Davydov and coworkers (Davydov, 1973; Davydov and Kislukha, 1973) more specifically proposed that energy might be transported efficiently in alpha-helical protein molecules by means of a vibrational soliton. The excitation that they suggested invokes the coupling, through a nonlinear interaction, of amide-I vibrations (mainly C=0 stretch) of the peptide groups with low-frequency phonons along the alpha-helix chain (mainly involving stretch of the hydrogen bonds between peptide groups). More recently Scott (Scott, 1982) extended the theory to the continuous limit (the excitation is distributed over a large number of peptide groups) resulting in the nonlinear Schrödinger equation for which the soliton is a traveling-wave solution. To date, to our knowledge, no definitive evidence of Davydov solitons in alpha-helix molecules has been published.

However, Scott (in collaboration with Careri and coworkers) has also done a similar analysis (Careri, 1983, 1984) for acetanilide (ACN), which is a linear hydrogen-bonded polypeptide, in order to explain the anomalous spectral feature which appears at low temperatures at 1650 cm^{-1}. In this case the amide-I vibrations couple with an out-of-plane motion of the hydrogen-bonding proton, producing a new resonance reduced in energy by about 15 cm^{-1} from the fundamental at 1665 cm^{-1}. The self-trapped state is presumed to be stationary, and numerical modeling (Eilbeck et al., 1984) shows it to be quite localized. Figure 1a shows a simplified schematic of the molecular structure of ACN. Since many amides form hydrogen-bonded linear chains, structures like that of N-methylacetamide (NMA) pictured in Figure 1b would be expected to exhibit the same interaction.

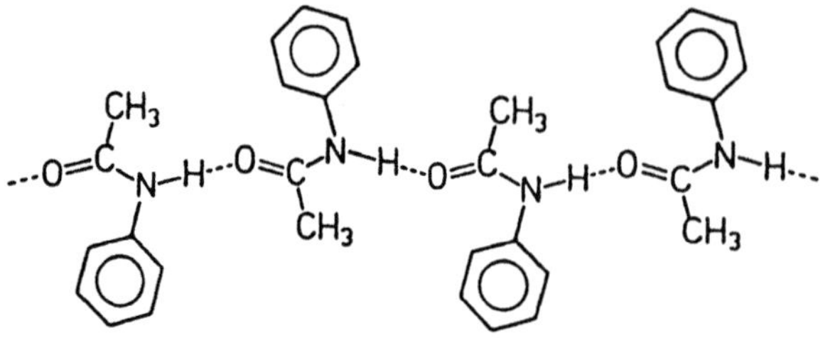

Fig. 1a — Simplified structure of ACN.

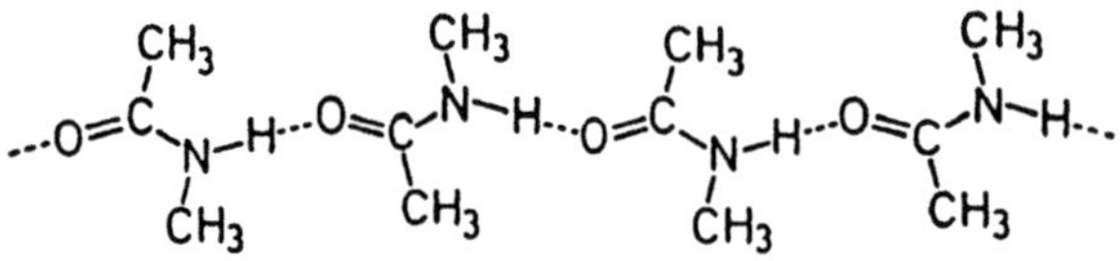

Fig. 1b — Simplified structure of NMA.

More recently, Scott has written a different nonlinear-dynamics theory (Scott, 1985) to describe the experimental results of Edwards et al. (1984, 1985) in which microwave resonances in DNA were demonstrated for the first time.

TECHNIQUES

Among the various laser and optical techniques, Raman and infrared spectroscopies are certainly the most common. Brillouin spectroscopy, IR-laser stimulation, nonlinear spectroscopies and hybrid techniques can also be employed.

Raman spectroscopy can address a large variety of transition symmetries since (in biopolymers) most dipole-allowed transitions are also Raman active. There is also the advantage of using visible wavelengths in a scattering mode to study infrared transitions, whereas IR-absorption spectroscopy (which works directly at the infrared wavelengths) often suffers from the media being opaque at those frequencies, and from other sample-presentation problems; but infrared spectroscopy is generally much more sensitive than Raman for those transitions which are weak.

Brillouin spectroscopy employs difficult techniques that invoke multi-pass Fabry-Perot interferometers (Lindsay et al., 1984), but phonons and other excitations can be observed in the frequency range of <50 MHz to 50 GHz which is generally not available to other techniques (except, under certain circumstances, for direct microwave resonance absorption).

Nonlinear spectroscopies, especially coherent anti-Stokes Raman spectroscopy (CARS), can also be useful, particularly with troublesome highly fluorescent samples. When picosecond lasers are used for CARS, transition lifetimes can be measured with resolutions to less than 10^{-12} s.

Infrared lasers can be built which can often be tuned to the vibrational transition of interest. Then an interesting hybrid technique is to observe the optical phonon region with Raman spectroscopy while stimulating a specific vibrational transition with an infrared laser. This would be of interest, for example, if it were suspected that the vibration couples with a phonon via a nonlinear interaction. Such a scheme is represented schematically in Figure 2.

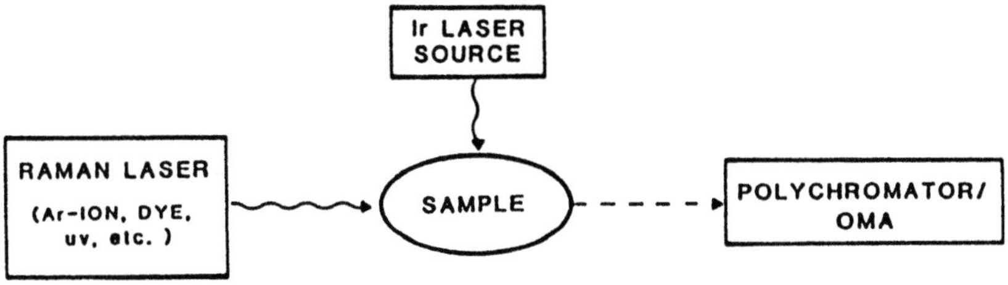

Fig. 2 — A hybrid technique to study nonlinear dynamical interactions.

Most recently microwave spectroscopy has proven to be a useful technique for the frequency range of 100 MHz to ~40 GHz. Appropriate sample preparation and presentation are essential for success, and Edwards et al. (1985) have succeeded in using this method to study resonances in DNA which have generated much interest.

EXPERIMENTAL RESULTS

Reports of unusual Raman spectroscopic features in growth-synchronized cultures of metabolically active bacteria (Webb, 1980) have been difficult to duplicate. Other authors (Layne, 1985, and references therein) using different experimental techniques have reported that experimental artifacts and transient bursts of fluorescence in these cultures may be responsible for the previously reported spectra.

At this time, the most plausible experimental evidence for solitons in biopolymers is the Raman and IR spectroscopic data on ACN. While Careri and coworkers (Careri, 1973; Careri et al., 1983) have considered and discarded structural changes, phase transitions, etc. or Davydov splitting as possible explanations for the temperature dependence of the 1650 cm^{-1} line, Johnston and Swanson (1985) have suggested a Fermi resonance between the amide-I fundamental and a combination mode (the sum of the amide-II at ~1550 cm^{-1} and an optical phonon at ~100 cm^{-1}). In their scheme the temperature dependences of the frequencies of the components of the combination mode are such that the Fermi coupling becomes stronger at lower temperatures, thus explaining the temperature variation of the 1650 cm^{-1} line intensity. Their publication also shows some isotopic-substitution data which is claimed to be confirmatory of the Fermi resonance explanation.

However, careful examination of the isotopic data and the addition of more recent data (overtone spectra) reopen the question to examination. The relative considerations are given below.

(1) Both the IR spectra (Careri et al., 1983) and the Raman spectra (Johnston and Swanson, 1985) show that the 1650 cm^{-1} line becomes *more* intense than the 1665 cm^{-1} line for temperatures below ~100 K. Since there are no other appropriate lines for intensity to be "borrowed" from, this indicates, in a Fermi interaction, a crossing of the original frequencies. Thus it is straightforward to determine the sign of the anharmonic defect of the combination

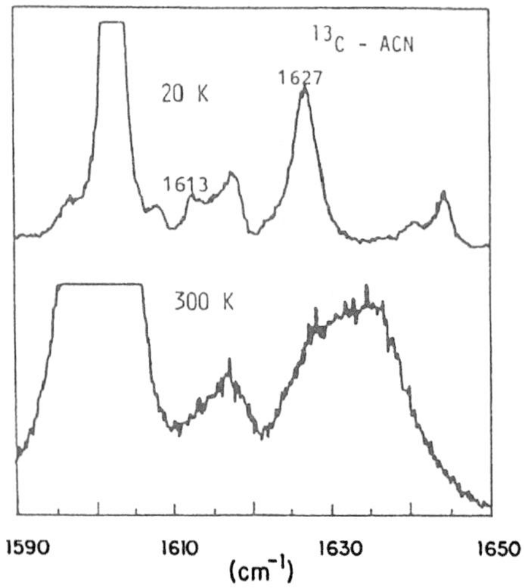

Fig. 3 — Raman spectra of ^{13}C-substituted ACN.

mode and to estimate its magnitude. The result is that for both the ^{12}C and the ^{13}C data to fit a Fermi resonance explanation, the anharmonicities for the two isotopes would have to be opposite in sign and greatly different in magnitude. However, there is no physical reason for the sign of the anharmonicity to change, or for its magnitude to change significantly, under isotopic substitution.

(2) Johnston and Swanson also mention an expected frequency shift of 30 cm^{-1} for the amide-I mode under ^{13}C substitution. However, a simple reduced-mass estimate gives a shift of 37 cm^{-1}. Figure 3 shows the ^{13}C data on an expanded scale for the amide-I region. The new location for the amide-I is down-shifted by the expected 37 cm^{-1} to $\sim$1627 cm^{-1}, and it should be noted that in the cold spectrum there is also a smaller feature at $\sim$1613 cm^{-1}, which is just 37 cm^{-1} down from the 1650 cm^{-1} line of the ^{12}C data. Since this line finds itself in the midst of strong features assigned to the benzene ring, it is to be expected that interference from these resonances would dampen the nonlinear interaction thus reducing the intensity of the "soliton" line.

(3) Under ^{15}N substitution or deuteration of the peptide group, the effects are precisely as predicted in Scott's theory, but they do not disagree with a Fermi resonance interruption.

(4) The hallmark of a theory is a predictive capability. Scott and coworkers (Scott et al., 1986) have recently demonstrated that the self-trapped-state theory can predict the frequencies of the overtones of the vibrational soliton in ACN with reasonable accuracy. In this paper they derive the anharmonicity from the self-trapped-state equations and use it to directly calculate the overtone frequencies. Experimentally in each case a temperature dependent feature was found (with IR-absorption spectroscopy) at a frequency quite close to that of the predicted overtone. While there are numerous other temperature-dependent spectral features at various wavelengths, the requirements for positive identification of the identified features were several and quite rigorous.

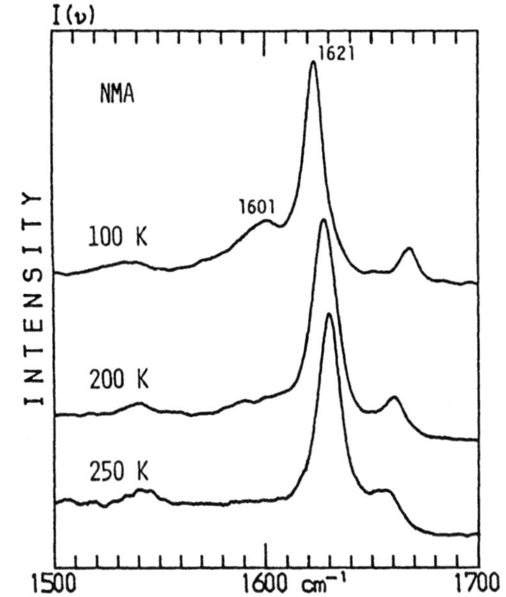

Fig. 4 — Raman spectra of polycrystalline NMA at three temperatures.

As mentioned earlier, NMA (and other hydrogen-bonded polypeptides) would be expected to exhibit nonlinear-dynamical effects similar to that in ACN. In an attempt to verify this, in collaboration with 0. Faurskov Nielsen we have looked at the Raman spectrum of NMA at a few temperatures (Bigio et al., 1986). (NMA is a liquid at room temperature and presents experimental problems of its own). Very preliminary data are shown in Figure 4. Here again there is a feature which appears only at reduced temperatures, in this case at about 1601 cm^{-1} (at 100 K) which is approximately 20 cm^{-1} below the amide-I line at 1621 cm^{-1}. The 20 cm^{-1} frequency separation is approximately what would be expected for a self-trapped state. Obviously more work is required on NMA to confirm the identification. This work is in progress (including the hybrid technique depicted in Fig. 2) and it is hoped that further results will be published in the near future.

Edwards and coworkers (Edwards et al., 1984, 1985) have recently demonstrated the existence of surprisingly narrow resonances in the microwave-absorption spectrum of DNA. The DNA was prepared so that every molecule in the sample was of the same length and sequence (from cloned plasmids with site-specific cuts). The resonances correspond to longitudinal acoustic modes of the molecules, but the frequencies measured lead to some potential confusion about the sound velocity. As mentioned above, Scott has offered a theory which incorporates nonlinearity (soliton-like character) in the molecular dynamics to explain the results (Scott, 1985). Both experiment and theory are currently benefiting from continued work and further refinement.

CONCLUSIONS

The best evidence to date for solitons in biopolymers has been found in molecular structures which are not really biological in nature but which are reasonable structural paradigms that can be studied rigorously. However, microwave techniques have been used with plasmid DNA and have resulted in data which may also have an exciting interpretation based on a soliton analysis (albeit different from the Davydov soliton). In general this field will benefit from enhanced interactions between theoreticians and experimentalists.

ACKNOWLEDGEMENTS

This work was supported in part by the U.S. Naval Air Systems Command HQ, the U.S. Department of Energy, and the Danish Natural Science Research Council. The author would especially like to thank the Danish Research Council for travel funds for collaboration with Prof. O. Faurskov Nielsen at the H.C. Orsted Institute of the University of Copenhagen.

REFERENCES

I.J Bigio, O.F. Nielsen, and C.T. Johnston, Proc. Hayashibara Forum on Computer Analysis for Life Science, Ohmsha Press, (in press, 1986).

G. Careri, in: Cooperative Phenomena, eds. H. Haken and M. Wagner, (Springer-Verlag, 1973) 391.

G. Careri, U. Buontempo, F. Carta, E. Gratton, and A.C. Scott, Phys. Rev. Lett. **51** (1983) 304.

G. Careri, U. Buontempo, F. Galluzzi, A.C. Scott, E. Gratton and E. Shyamsunder, Phys. Rev. **B30** (1984) 4689.

A.S. Davydov and N.I. Kislukha, Phys. Stat. Sol. (b), **59**, (1973) 465.

A.S. Davydov, J. Theor. Biol. **38**, (1973) 559.

G.S. Edwards, C.C. Davis, J.D. Saffer, and M.L. Swicord, Phys. Rev. Lett. **53**, (1984) 1284.

G.S. Edwards, C.C. Davis, J.D. Saffer, and M.L. Swicord, Biophysical Journal **47**, (1985) 799.

J.C. Eilbeck, P.S. Lomdahl, and A.C. Scott, Phys. Rev. B **30**, (1984) 4703.

H. Fröhlich, Int. J. Quantum Chem. **2**, (1968) 641.

C.T. Johnston and B.I. Swanson, Chem. Phys. Lett. **114**, (1985) 547.

S.P. Layne, I.J. Bigio, A.C. Scott, and P.S. Lomdahl, Proc. Nat. Acad. Sci., U.S.A. **82** 7599 (1985).

S.M. Lindsay, J.W. Powell, and A. Rupprecht, Phys. Rev. Lett. **53**, (1984) 1853.

A.C. Scott, Physica Scripta **25**, (1982) 651.

A.C. Scott, Physica Scripta **32**, (1985) 617.

A.C. Scott, E. Gratton, E. Shyamsunder, and G. Careri, (in press, 1986).

S.J. Webb, Phys. Rep. **60**, (1980) 201.

18. Long-lived and Large-coherence Length Excitations in the DNA Double Helix

E. W. Prohofsky

With 1 Figure

ABSTRACT

We discuss the lifetime and mean free path or coherence length of vibrational modes of the DNA double helix in the light of recent calculations and experimental observations. The helix is essentially a long polyion and we consider the range of forces along the helix that are important in determining the vibrational mode behavior. The factors that are likely responsible for the observed long lifetimes and large coherence lengths are reviewed.

INTRODUCTION

By far the majority of people currently carrying out theoretical calculations on the dynamics of biomolecules use methods such as molecular dynamics simulations which for technical reasons examine small regions and short-time behavior of biomolecules. Fröhlich very early realized that the large macromolecular assemblies found in biological structures could support very low-frequency modes that could be coherent excitations over large distances and live for long times. Recently experimental evidence has accumulated which indicates that at least in the macromolecule, the DNA double helix, long-lived modes with large coherence lengths do exist. These modes are present in all frequency regimes and likely dominate the dynamics of the helix. That this fact has taken so long to be recognized after the suggestions of Fröhlich is a result of the popularity of those theoretical methods that are inherently incapable of detecting modes of this type.

When theoretical methods are employed that are compatible with detecting long-lived long-coherence vibrational modes, one finds very good agreement between theory and experiment. In fact we have made a large number of predictions, many prior to experimental observation, that have been in agreement with subsequent observations. In this paper I will list some of these predictions and their relations to observations and then discuss some possible reasons for the unusual success of this theoretical approach.

THE PHYSICAL SYSTEM

The biologically active double helix is always surrounded by water, ions, and proteins. Extracted DNA that is not denatured is always surrounded by at least ions and water of hydration. If these factors are removed, the DNA denatures to a random coil configuration. Even in fibers and films there is considerable, $\approx$50% by weight, water of hydration. It is commonly believed that a macromolecule such as the DNA helix in such intimate contact with water would have its low-frequency modes damped out by that water. This is because the helix is relatively thin, i.e., its radius is 10-15 Å depending on how one defines the boundary, so that all helix atoms are relatively close to the helix surface and therefore the water. The one calculation (Dorfman and Van Zaudt, 1983) that probed the probable damping did indicate that the acoustic modes of a helix-fully dissolved in water would be heavily overdamped. This calculation, however, was based on the application of standard hydrodynamic behavior for the water and the helix-water interface. It is clear that this is not an appropriate approximation for the helix.

On the other hand, there are some reasons to expect that the coupling between the helix and its surroundings would not be very strong. Many processes occur on DNA in which the energy is supplied by the splitting of an ATP molecule. Processes such as the breaking of the hydrogen bonds holding a base pair together require energy that is of the same order of magnitude as the energy of the ATP. This implies that the operation is relatively efficient, i.e., that much of the initial energy is maintained in the DNA-enzyme system and that only some fraction of it is radiated away into the surrounding water due to relaxed motion of the helix and enzymes. One way of achieving this form of thermodynamic efficiency would be for internal motion of the helix to be in some way dynamically decoupled from the surrounding medium. This decoupling would be detected as long lifetimes of some vibrational modes. It may have been an evolutionary requirement for such an isolation of internal dynamics to exist to bring about the necessary thermodynamic efficiencies of the helix-enzyme system.

EVIDENCE FOR RESONANT MODES IN THE DOUBLE HELIX

The observation of resonant microwave absorption (Edwards et al., 1984) by DNA in dilute solution is the most straightforward indicator that excitations exist in the single macromolecule of the double helix and that they have surprisingly long lifetimes. The simplest modes in matter are those of compressional acoustic waves. In a long thin system of finite length such as the double helix, standing acoustic waves can occur. These waves are familiar as the standing waves of an organ pipe. Such waves occur when an integer number of half wavelengths fit the distance between ends and allows for either node or antinode boundary conditions. Of these standing waves only those with odd half wavelengths would have net coupling to a microwave field (Kohli et al., 1981). The even half wavelength modes would cancel the net absorption. The wavelengths for a strand of length L would be given by

$$\left(n + \frac{1}{2}\right)\lambda = L \tag{1}$$

and the frequency ν by

$$\nu = \frac{C}{L}\left(n + \frac{1}{2}\right) \tag{2}$$

where C is the velocity of sound on the helix. The frequencies have the ratios of 1:3:5:7- - -. This pattern is observed when care is taken to have DNA strands of identical length. The frequencies do depend on the length of the DNA strand and have an acoustic velocity for the linear pieces the same as that observed by Brillouin scattering (Maret et al., 1979; Hakim et al., 1984). In the experiment described, the DNA is in very dilute solution and the individual DNA molecules are separated. The sound wave must be associated with the single DNA strands. Similar resonances are observed in closed natural plasmid DNA but in that case the standing waves are integral whole wavelength modes as needed for that topology:

The intact plasmids are supercoiled. The observation of the standing waves predicted for these entities implies that the acoustic wave is able to propagate around all the bends and twists resulting from the supercoiling. The velocity is greater than that in the linear DNA but that is consistent with observations in fibers that the velocity increases as the packing of the fibers is tighter.

The fundamental mode, i.e., the one with $n = 0$ in eq. 1, is seen for a segment of DNA $\approx$1000 base pairs (b.p.) in length. This necessarily implies a coherence length of that magnitude as the wave must be coherent over the length of the sample to express the standing wave resonance. The authors report lifetimes $>$ 300 ps and coherence lengths $>$ 1000 b.p. for DNA in dilute solution.

The discussion till now has centered on acoustic modes. Most of the vibrational modes of a complex system like DNA are optical modes, that is, they are not characterized by a simple inverse relationship between frequency and wavelength. One might expect that the long lifetime and large coherence lengths are more likely for acoustic modes than optical modes. This assumption is likely wrong, as the acoustic modes should couple more strongly to the aqueous surroundings as acoustic vibrations tend to drag water along in their motions. The optical modes are such that the center of mass of each unit cell remains fixed in space. There is no gross motion for the surrounding water to couple to. In any case, there is experimental data which allows us to estimate coherence length for the low-lying optical modes as well.

It is more difficult to determine lifetime and coherence length for the optical modes because the experimental material is different. In the microwave experiments referred to, the DNA was identical single-molecule strands made from identical plasmids. The data from Raman and Brillouin measurements comes from DNA fibers and films. The individual molecules are of different lengths and base sequences. In addition the fibers are considerably disordered in their relative positions in the macroscopic state of aggregation. The observed lines are heterogeneously broadened and one is unlikely to be measuring intrinsic linewidths. Nonetheless reasonable lines are detected. It is observed that the lines do show k-conservation scattering (Urabe et al., 1983; Lindsay and Powell, 1983). Because of the polarization selection (Lindsay and Powell, 1984) of the observed lines one knows that these lines correspond in the DNA helix to modes of wavelength of 10-11 base pairs. For the k-conservation to be expressed, the coherence length must be many wavelengths or at least of the order of hundreds of b.p. This lower limit does not depend on any detailed theoretical model.

By using a second observation one can get a better estimate of the coherence length but it does depend on a particular theoretical model. Our calculations

Table 1. Predicted Frequencies and Experimental Observations

A-DNA		B-DNA	
Theory (cm^{-1})	Experiment (cm^{-1})	Theory (cm^{-1})	Experiment, (cm^{-1})
11.5[a]	12.5[a]	11.5	12.
14.1	15	14.8	16
20[a]	22[a]	83[b]	85[b]
34	33		
64	65		
100	100		

[a]Signifies predicted soft modes that are seen to soften experimentally.

[b]Signifies hydrogen-bond modes that are seen experimentally to behave anomalously at melting.

(Devi-Prasad and Prohovsky, 1984; Lindsay et al., 1984) of the spectrum of A poly (dG) · poly (dC) show a number of modes that satisfy the selection rules for Brillouin and Raman scattering (see Table 1). One mode, the 12-cm^{-1} mode is a zone center mode which corresponds to infinite wavelength. That it is of such wavelength is verified by its being seen in (z,z) polarization scattering. We predict that it is the conformational soft mode of the A-to-B conformation change. Another mode or group of modes at $\approx$22 cm^{-1} are at wavelength of eleven base pairs. This wavelength is verified as this mode is seen in (x,z) or (y,z) scattering polarization. This mode is predicted to be closely related in character to the 12-cm^{-1} mode. This relationship is borne out by the experimental observation that it also softens as predicted at the A-to-B conformation change. For the shift in frequency from 22 cm^{-1} at eleven base pairs to 12-cm^{-1} to be expressed the coherence length must be of the order of thousands of base pairs. It appears that this low-lying optical mode has a coherence length as large or larger than the acoustic modes above 2.5 GHz. As already suggested, the optical modes are expected to have less coupling to water around the helix and therefore longer lifetimes and coherence lengths.

In addition to the particular modes discussed above, the lattice calculations predict a number of modes that could be observed by Raman scattering. The comparison of these predictions with results of observation is shown in Table 1. An overall pattern of quantitative agreement exists in the cases where particular characteristics can be assigned, such as soft-mode behavior, or hydrogen-bond character (85 cm^{-1}). In addition to the low-frequency spectrum discussed here we have used lattice methods to calculate vibrational modes at much higher frequencies and relatively accurate refinements compared to Raman observations have been achieved.

The existence of large coherence lengths does dictate the kinds of theoretical analysis that can be applied. The lattice approach is the only feasible approach for such large coherent excitations. For example, a molecular dynamics simulation of these modes would require a calculation cell of thousands to tens of thousands of base pairs. It would also have to simulate for times of nanoseconds after equilibrium. The use of smaller simulation cells would miss completely the dispersion, i.e. change in frequency with wavelength, that is so important at the lower frequencies. One would also not be able to calculate the correct selection rules.

FACTORS IN DNA WHICH FAVOR EXCITATIONS OF LARGE COHERENCE LENGTH

The principal cause of the large coherence lengths in DNA is the stiffness of the helix caused by the long-range electrostatic-like forces between elements of the helix. Again the importance of this factor was pointed out some time ago by Fröhlich. In DNA there is a large negative charge present as each phosphate group is the negative ion of an acid. There are two such groups per unit cell and the double helix is roughly a long polyion with charge $2e^-$ per unit cell. The polyion is neutralized by counterions for stationary equilibrium. Relative motion between the bound negative charges and the counterions is demonstrated by the large dielectric constant of long-chain DNA in aqueous solution. Takashima (1966) has measured this dielectric response as a function of frequency and found that the additional dielectric response has fallen off by frequencies above tens of kilohertz for DNA samples of several lengths. The counterions seem to be unable to respond and shift along the helix at higher frequencies. The negative charge attached to the stiff helix can be resonant at higher frequencies. Moving charge can couple to the electric field at high frequencies even though the material is neutralized by counterions at zero frequency. The counterions likely complicate the interaction with fields up to kilohertz frequencies but are likely immobile for frequencies beyond tens of kilohertz. For frequencies above hundreds of wavenumbers the nonbonded forces are insignificant compared to strong valence forces.

The initial indication of the importance of long-range forces arose in an attempt to fit the observed acoustic (Maret et al., 1979; Hakim et al., 1984) velocity to a theoretical lattice model. It was noted that the velocity was larger than might be expected for an open structure such as the double helix. If one included no nonbonded forces the velocity was orders of magnitude low, as compression could then take place by simple rotations of bonds opposed by weak torsional force constants (Syster and Prohovsky, 1974). We calculated the acoustic velocity for several models having forces of different range (Mei et al., 1981). We found that the observed acoustic velocity could be fit for forces having the range expected for van der Waals forces but that it would involve force constants of the size of those for valence bond stretch. This seemed to be unreasonable. The same would hold for ring current forces, etc., which also are of short range. When, on the other hand, we fit the observed acoustic velocity with electrostatic-like force constants we found that only reasonable net charges were needed. By reasonable we mean charges of the order of, but less than, those net charges predicted by ab initio electron orbital calculations. This is not to say that short-range nonbonded forces do not exist, but only that the integrated effect of longer-range ones overpower the short-range forces for compressional acoustic waves and that one cannot therefore refine values for short-range forces from acoustic data.

Water of hydration complicates calculations of acoustic velocities. Dorfman and Van Zandt (1984) have carried out calculations in which the loading and viscosity of water on the helix modes have been taken into account. They find that in fibers and films the compressional velocity is reduced by a factor of two from what it would be for a unloaded double helix. The argument about the need for long-range forces was an order-of-magnitude effect and not affected by factors of two. In fact the water loading is in the direction of strengthening the need for long-range rather than short-range forces.

By using a model incorporating long-range nonbonded interactions with effective dielectric constant determined by fitting the acoustic velocity data (Devi-Prasad and Prohovsky, 1984), a number of other low-lying frequencies were predicted. For A-DNA the model predicted a zone center (i.e., $\theta = 0$, where θ is the phase shift from one base pair to the next) mode at 12 cm^{-1} and $\theta = \psi$ (i.e., phase shift equal to the helix pitch angle) modes at 14 cm^{-1} and $\approx$22 cm^{-1}. The model predicted wavevector conservation and polarization selection rules and all predictions are in agreement with experimental observation. In addition the model predicted a mode at 600 MHz which is also observed (Lindsay and Powell, 1983). The initial prediction was at 900 MHz (Pulnam et al., 1981) but with parameters from more recent acoustic velocity measurements, the agreement between theory and experiment is excellent. The long-range-force model has successfully predicted the frequency for most modes prior to observation. It is possible that these frequencies could be fit with a model which does not involve long-range forces but such a model would almost certainly involve anomalously large nonbonded interactions.

The second class of arguments for long-range forces has to do with the phenomenological behavior of certain modes rather than the qualitative agreement of predictions and observations. The 22 cm^{-1} mode (Urabe and Tominaga, 1982; Lindsay et al., 1983) is observed to soften on approaching the A-to-B conformation change. The theory predicts that this mode will be on the same branch as the conformational soft mode and (Devi-Prasad and Prohovsky, 1986) that it should soften at the conformation change. Analysis of structural transitions making use of renormalization group theories indicate that soft-mode transitions are impossible for one-dimensional systems unless there are long-range interactions. The fact that mode softening occurs for the one-dimensional DNA helix is therefore an indication of long-range interactions. The actual range will determine the degree of softening that can occur before the system breaks up into an order-disorder transition. The experimental observation of the true soft mode can be used to determine the actual range of interaction but at this point the data is not good enough to carry out such an analysis. However, the degree of softening already observed is indicative of relatively long-range interaction.

The 600-MHz mode is another case where the existence of a mode at such a frequency is a stronger test of the long-range nature of the nonbonded interactions. We predict that this mode is a defect mode associated with the terminus of the helix. No plausible alternative explanation for a mode of this linewidth at this frequency has been proposed. In the theory the existence of this mode and its character result from the termination of the long-range forces. Terminating the valence bonds along the backbone do not cause a terminus mode anywhere near this frequency. It seems to be sensitive to the particular nonbonded interaction model. The existence of local modes and soft modes therefore depends phenomenologically on long-range interactions.

The third class of argument for long-range forces is the observation of resonant microwave absorption (Edwards et al., 1984). When we introduced the long-range force model we realized that its existence indicated that resonant absorption should also occur. Both require effective charge able to couple to electric fields associated with internal vibrational motion. It was this connection that led to the predictions of resonant absorption (Kohli et al., 1981). The observation of resonant absorption certainly enhances the probability that long-range electrostatic interactions are present in the double helix for the frequency range above the dielectric relaxation frequency. The upper limit of importance of nonbonded forces is at ≈ 100 cm^{-1} where the stronger valence forces determine the frequencies of vibration.

Another factor which helps maintain coherent excitations is the similarity in size and mass of the DNA base pairs. The two pairs AT and GC are similar in size so that both can fit into the helix without distortion. The masses of the two pairs differ by one Dalton. This is out of a total mass of ≈ 600 Daltons per base pair unit cell. Clearly the mass differences for the chain because of differing base composition are of the order of those caused by fluctuation in isotopic composition. The position of atoms in the two base pairs do differ, and there is still the major difference in orientation of the base pairs, i.e., whether, for example, the G or C is attached to a particular backbone. These differences, which do break the symmetry of the molecule, certainly affect the spectrum and since sequence and orientation vary in an almost random way one could expect lines to be at least heterogeneously broadened. Certainly our calculations in the middle infrared frequency range indicate that base pair heterogeneity can cause large changes in the spectrum at these higher frequencies.

We have carried out calculations for all the repeating polymers of DNA that can occur if the repeat size is limited to two base pairs. We find that each of these six helices have different spectra but there are similarities which can nonetheless give rise to a band in a completely random mix of base pairs. This broad band can be seen as a resonant band of natural DNA. We concentrated on the interbase hydrogen-bond stretch motion. This motion is one that tends to separate the two bases making up a base pair. For a GC pair there are three hydrogen-bonds whereas an AT base pair has only two. In spite of this substantial difference we find that for all base pair combinations there are modes of this type motion between 65 and 110 cm^{-1}. The situation is displayed in Fig. 1. We plot the average of the eigenvector projections onto the hydrogen bond stretch for zone center modes at the frequencies shown. The height of the bars correlate roughly with the degree to which that mode is a pure hydrogen-bond stretch mode. Each polymer shows many such modes, which can be considered to be due to splittings of the basic mode caused by different behavior in other parts of the unit cell. However, the most striking fact is that almost all the modes for all the polymers that represent hydrogen-bond stretch cluster in the frequency region shown. In a natural DNA helix, where all combinations of base pairs occur, one expects a pass band roughly made up of many broadened localized modes in this frequency region. Urabe et al. (1982) has observed a mode in DNA centered about 85 cm^{-1} and with a breadth which spans very closely the spread of frequencies shown in Fig. 1. Because this mode softens and broadens and ultimately disappears at the helix strand separation melting temperature, they assign this mode to hydrogen-bond motions. This softening and eventual disappearance of the hydrogen-bond mode on approach to the strand separation melting temperature is what we predict on the basis of a modified self-consistent phonon calculation (Gao et al., 1984).

Because of the selection rules for Raman scattering, all the observations of these modes are at or near the infinite wavelength limit. The pass band associated with the 85 cm^{-1} band can be seen because of the long-wave nature of this mode and the averaging of the contributions from various regions that results from the

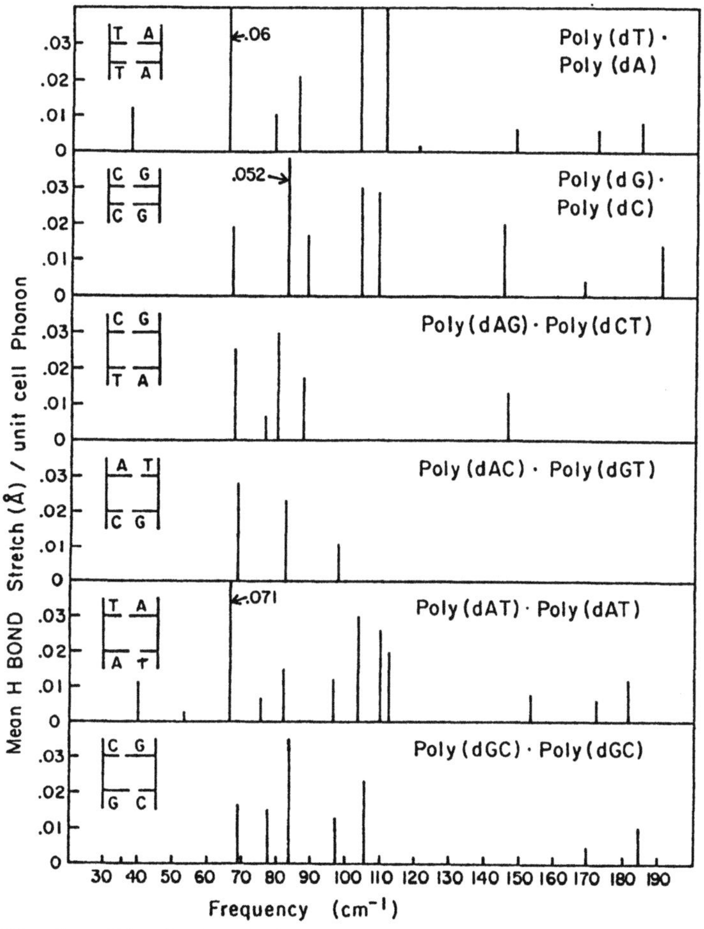

Fig. 1 — The hydrogen-bond coherent stretch modes. The figure shows the amplitude of the average hydrogen-bond stretch per phonon at the indicated frequency for the indicated DNA polymer. Only those modes that had in-phase stretch for all hydrogen-bonds of a given base pair are shown. The normalization of the amplitudes is one quantum per unit cell. The extended phonon amplitudes should be multiplied by $N_{-1/2}$ where N is the number of base pairs.

long-range interactions. If one were to observe short-wavelength modes, the dispersion between the various regions of differing base sequence would destroy the coherent pass band.

The lower-frequency optical modes discussed in earlier sections no doubt have large coherence lengths because they are long-wavelength modes in which the long-range forces are the dominant forces. These forces extend over regions so large that the base-sequence variations are averaged out so that any one region looks similar to any other and we are dealing with bulk vibrational properties. Although the modes below 35 cm^{-1} are technically optical modes, they do involve the coherent motion of large sections of helix rather than motions where nearby atoms move in opposition to each other. For example, whole base sections move relatively coherently. This is a situation in which the averaging process can occur giving rise to bulk effective force constants.

It appears that the mass density uniformity coupled with the average bulk forces make the DNA seem uniform over distances and motions associated with low-frequency long-wavelength modes. The result is a number of long-life, large coherence-length modes. These modes are not overdamped by the surrounding water. The result is a very rich spectrum which is likely to be of importance in the biological functioning of the DNA helix.

ACKNOWLEDGMENT

This work was supported in part by National Institute of Health Grant GM 24443 and the Naval Air Systems Command HQ.

REFERENCES

K.V. Devi-Prasad and E.W. Prohofsky, Biopolymers **23**, 1795 (1984).

K.V. Devi-Prasad and E.W. Prohofsky, Biopolymers, (1986), in press.

B.H. Dorfman and L.L. Van Zandt, Biopolymers **22**, 2639 (1983).

B. Dorfman and L.L. Van Zandt, Biopolymers **23**, 913 (1984).

G.S. Edwards, C.C. Davis, J.D. Saffer, and M.L. Swicord, Phys. Rev. Lett. **53**, 1284 (1984).

Y. Gao, K.V. Devi-Prasad, and E.W. Prohofsky, J. Chem. Phys. **80**, 6291 (1984).

M.B. Hakim, S.M. Lindsay, and J. Powell, Biopolymers **23**, 1185 (1984).

M. Kohli, W.N. Mei, E.W. Prohofsky, and L.L. Van Zandt, Biopolymers **20**, 853 (1981).

S.M. Lindsay and J. Powell in *Structure and Dynamics: Nucleic Acids and Proteins*, E. Clementi and R.H. Sarma, Eds., Adenine, New York 1983, p. 241.

S.M. Lindsay and J. Powell, Biopolymers **22**, 2045 (1983).

S.M. Lindsay and J.W. Powell, Phys. Rev. Lett. **53**, 1853 (1984).

S.M. Lindsay, J.W. Powell, E.W. Prohofsky, and K.V. Devi-Prasad in *Structure and Dynamics of Nucleic Acids, Proteins and Membranes*, E. Clementi and R. Sarma, Eds., Adenine, New York, 1981.

G. Maret, R. Oldenbourg, G. Winterling, K. Dransfeld, and A. Rupprecht, Colloid Polym. Sci. **257**, 1017 (1979).

W.N. Mei, M. Kohli, E.W. Prohofsky, and L.L. Van Zandt, Biopolymers **20**, 833 (1981).

B.F. Putnam, L.L. Van Zandt, E.W. Prohofsky, and W.N. Mei, Biophys. J. **35**, 271 (1981).

J.M. Syster and E.W. Prohofsky, Biopolymers, **13**, 2505 (1974).

S. Takashima, J. Phys. Chem. **70**, 1372 (1966).

H. Urabe and Y. Tominaga, Biopolymers **21**, 2477 (1982).

H. Urabe, Y. Tominaga, and K. Kubota, J. Chem. Phys. **78**, 5937 (1983).

19. Far Infrared Spectra of Biomolecules

J. B. Hasted

With 5 Figures

Modern pharmacology currently is developed in terms of structural theory, based, in popular terms, on the idea of molecular keys fitting physiological locks, and thereby allowing or preventing metabolisms from taking place. But it might be plausible that these same molecules transmit or receive certain specific vibration frequencies that could bring about similar metabolic action. There is no direct evidence for such a hypothesis, but there is at least ample experimental proof of the richness of biomolecular vibrational structure.

There is also the exciting possibility that Fröhlich modes (Fröhlich, 1968a, 1968b) might play an active role in metabolic processes, so that it becomes necessary to explore the spectral structure of the important molecules experimentally. Although the coherence proposed by Fröhlich applies primarily to longitudinal modes, these may give rise to electromagnetic activity precisely because biomolecules abound with polar groups and show strong spectral absorptions.

Especially rich in absorptions are the far infrared and submillimeter bands, which have now been made accessible by Fourier transform spectroscopy (FTS) (Chantry et al., 1969; Chamberlain and Gebbie, 1971); this complements the Raman techniques whose development was stimulated by the availability of rare gas-ion-lasers. Using the FTS facilities developed extensively at the National Physical Laboratory, U.K., we have been able to study systematically spectra of amino acids, peptides, and chemically related molecules (Husain et al., 1984a), polypeptides, and proteins (Husain et al., 1986), as well as saccharides and polysaccharides (Husain et al., 1984b). Assignment of the vibrational modes is assisted by the existing normal-mode analyses, which have been made for the amino acids glycine (Machida et al., 1977), alanine (Machida et al., 1978), and serine (Machida et al., 1979), and for polyglycine and polyalanine (Moore and Krimm, 1976). Although there are many complications, a proportion of the absorption bands can be grouped together, in the manner traditional in infrared spectroscopy, and interpreted in the light of the normal-mode analyses. The results of this exercise (Husain et al., 1984a) are summarized below:

Infrared absorption bands of amino acids and related molecules with assignments

Group	Frequency, cm^{-1}	Assignment[a]
1	< 100	predominantly lattice modes
2	100-120	
3	118-145	
4	145-165	lattice modes, and possibly CH_2Xt.
5	165-245	CO_2t.
6	245-310	possibly CCt.
7	285-315	possibly OHt.
8	225-360	CH_3t.
9	315-440	CCNd.
10	425-485	molecules possessing a benzene ring.
11	425-510	NH_3t.
12	490-560	CO_2r.
13	520-540	CCOd assignment in serine.
14	560-620	CO_2w.
15	$\gtrsim$ 620	CO_2b.

Oligopeptide spectra can be classified within these groupings, but polypeptide spectra (Husain et al, 1986) fall more naturally into seven groupings, which are based on the observed spectrum and normal-mode analysis of poly-L-alanine by Moore and Krimm (1976):

Infrared absorption bands of oligopeptides

Group	Frequency, cm^{-1}	Assignment
Poly 1	100-150	$NC^{\alpha}Cd$.
Poly 2	220-240	$C^{\alpha}C^{\beta}t$.
Poly 3	Broad background absorption in proteins.	
Poly 4	250-350	$NC^{\alpha}Cd$.
Poly 5	375-420	α helical structure.
Poly 6	445-495	$C^{\beta}b$.
Poly 7	500-600	α helical structure.

Protein spectra (Husain et al., 1986), whilst very much less rich in structure, can also to some extent be analyzed in terms of these groupings.

Saccharide and polysaccharide spectra (Husain et al., 1984) are not easily interpreted in the absence of normal mode analyses, but similar division into ten different groupings has been made.

Our spectra were all taken at two temperatures, 100 K and 300 K, so that some idea could be obtained about any dependence on temperature that each characteristic frequency might show. These temperature shifts have now been studied and show features of interest. Attention has also been paid to the linewidths and intensities. All these parameters are of assistance in assigning vibration modes, and may indicate the participation of Fröhlich modes. Spectra reported by Genzel et al. (1983) are available for comparison and in general support the present conclusions; some important differences will be discussed later in this article.

Temperature shifts of line frequencies are towards lower frequencies as the temperature is raised; the very few "negative shifts" can usually be explained in terms of unsuspected additional structure, obscured at higher temperatures, but showing up at lower temperatures, sometimes at those as low as 6 K. In general, unresolved structure is a serious complicating factor in the investigation of temperature shifts, and care has been taken in the present studies to avoid inclusion of any doubtful cases. For line frequencies at 100 K and 300 K the computerized data have been examined and the difference between the maxima given to the nearest wavenumber. This procedure is not applicable to shoulders, which have been avoided as far as possible, but estimated graphically in some clearcut cases.

It is immediately clear that although differences exist even between temperature shifts of lines with identical interpretations, these are in general much smaller than the differences between different groups. The mean differences, with standard deviation, have been calculated for each group, and are listed for amino acids and related molecules in Table 1. They vary between the largest mean temperature shift 8.9 ± 3.0 and the smallest 2.1 ± 1.3. This variation assists in the interpretation of individual lines or multiplets, for example in distinguishing CO_2t, a line with large shift, from CCt, whose shift is small.

It is encouraging to find that the spectra of some chemically related molecules (Husain et al, 1984a) show very similar behavior; these molecules are: deuterated glycine, benzoic acid, dinitrobenzoic acid, urea, and sodium acetate. In Table 1 the line intensities, relative to the strongest line in the spectrum, have been treated similarly, and again, fairly large variations, with relatively small standard deviations, have been found. Relative intensities are also of assistance in the interpretation of lines in overlapping groups. However, the correlation between temperature shift and intensity is weak, as might be expected.

The large temperature shifts all fall between 130 and 350 cm^{-1}, but within this range there are some small shifts as well; no smooth variation with wavenumber is apparent. In general, we would expect small shifts at large frequencies where quantum transitions are only weakly affected by interactions with neighbors, and small shifts at sufficiently low frequencies, where the classical Arrhenius behavior produces large relative shifts that are small in absolute terms. The classical-quantum boundary at about kT (200 cm^{-1} at 300 K) gives rise to large relative shifts. A comparison of amino acid, polypeptide, and protein shifts and characteristic 100 K frequencies, with standard deviations, is shown in Fig. 1. The data for related molecules are omitted for clarity, but they correlate well with amino acid data.

From this figure it will be seen that the polypeptide and protein temperature shifts correlate to some extent with the amino acid values. The data are summar-

Table 1. Mean and standard deviation frequencies, temperature shifts, and intensities for absorption bands of amino acids: glycine, alanine, valine, leucine, isoleucine, serine, threonine, cysteine, cystine, methionine, proline, hydroxyproline, phenylalanine, tyrosine, tryptophane, aspartic acid, glutamic acid, lysine, arginine, histidine, asparagine, and glutamine, and for related molecules: deuterated glycine, benzoic acid, dinitrobenzoic acid, urea, and sodium acetate. All data from Husain et al. (1984a).

		Amino acids			Related molecules		
Group	Assignment	Frequency cm^{-1}	Shift cm^{-1}	Intensity	Frequency cm^{-1}	Shift cm^{-1}	Intensity
1	*l*	83 ± 7	2.5 ± 1.4	0.23 ± 0.10	84 ± 4	1.5 ± 1	0.22 ± 0.06
2	*l*	107 ± 7.8	3.3 ± 1.9	0.34 ± 0.4	106 ± 3	3 ± 1.4	0.34 ± 0.16
3	*l*	131 ± 7	5.7 ± 2.3	0.46 ± 0.25	133 ± 6	5.9 ± 1.9	0.45 ± 0.20
4	CH_2Xt	153 ± 8	5.5 ± 2.9	0.43 ± 0.20	155 ± 3	6.3 ± 2.8	0.45 ± 0.21
		177 ± 5	6.4 ± 2.5	0.63 ± 0.25	184 ± 3	7.1 ± 1.7	0.93 ± 0.13
5	CO_2t	203 ± 25	8.6 ± 2.2	0.73 ± 0.21	215 ± 36	8 ± 5.5	0.92 ± 0.11
6		289 ± 8	2.2 ± 2.1	0.47 ± 0.21	289 ± 4	2.9 ± 2.1	0.54 ± 0.18
		303 ± 6			305 ± 2		
7	OHt	292 ± 5	1.8 ± 2.2	0.39 ± 0.23	–	–	–
		306 ± 5					
8	CH_3t	244 ± 29	8 ± 4	0.72 ± 0.25	249 ± 14	5 ± 5	0.99 ± 0.02
		346 ± 4	8.9 ± 3		335		
9	CCNd	354 ± 20	3.6 ± 1.4	0.81 ± 0.15	359 ± 27	2.2 ± 2	0.75 ± 0.14
9′		392 ± 11	2.3 ± 2.2	0.62 ± 0.27	395 ± 8	1.9 ± 1.7	0.49 ± 0.22
		406 ± 10					
11	NH_3t	455 ± 23	2.1 ± 1.3	0.50 ± 0.18	457 ± 22	2.5 ± 1	0.55 ± 0.08
12	CO_2r	534 ± 9	2.3 ± 1.8	0.63 ± 0.17	542 ± 6	1.6 ± 2.1	0.62 ± 0.24
		548 ± 7					

Abbreviations used:

l, lattice vibration; t, torsion; d, deformation; r, rocking; w, wagging; b, bending.

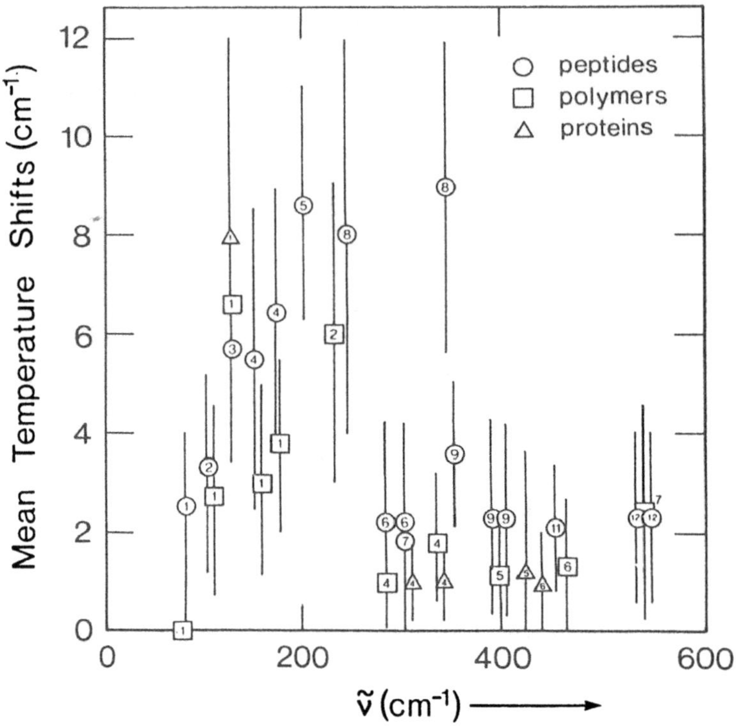

Fig. 1 — Mean temperature shifts of characteristic vibration frequencies of amino acids ○, polypeptides □, and proteins Δ, as a function of mean line frequency. Numbers indicate the groups 1-15 and poly 1-7 mentioned in the text.

ized in Table 2. Polymerization appears to affect the temperature shifts less significantly than one would expect from the differences between interpretations of amino acid and polymer lines. For amino acids, the large shift modes include torsion of the terminal groups CH_3 and CO_2, as well as some lattice modes. One of the amino acid backbone modes, CCNd, displays a moderately large shift. After polymerization, there is no CO_2 group to contribute, but for poly-L-alanine the $C^{\alpha}C^{\beta}$t mode is of course a torsion of the CH_3 group, and it is gratifying to find that this correlates well both in frequency and in temperature shift. The polypeptide backbone (NC^{α}Cd) and α helix modes display only weak temperature shifts, with the exception of the NC^{α}Cd line at 129 cm^{-1} mean frequency.

A similar analysis has been made of oligosaccharide and polysaccharide spectra (Husain et al., 1984b), the results being displayed in Table 2 and Fig. 2. As in the case of peptides and polypeptides, there is an exceptionally large temperature shift for the lines around 250 cm^{-1}; this supports their proposed connection (Husain et al., 1984b) with CCt. There is also a large shift around 150 cm^{-1}, as with the CH_2OHt line in serine and other peptides. The lines around 300 cm^{-1} have reasonably small shifts, which supports their proposed connection (Husain et al., 1984b) with OHt.

Table 2. Mean and standard deviation frequencies, temperature shifts, and intensities for absorption bands of polypeptides: polyglycine, poly-L-alanine, poly-L-leucine, poly-L-phenylalanine, and for proteins: hemoglobin, serum albumin, lysozyme, insulin, salmine, and collagen. All data from Husain et al. (1986).

		Polypeptides			Proteins	
Group	Assignment	Frequency cm^{-1}	Shift cm^{-1}	Intensity	Frequency cm^{-1}	Shift cm^{-1}
1	l	79 ± 4	0 ± 2	0.27 ± 0.15		
	l	112 ± 5.5	2.7 ± 1.9	0.25 ± 0.10		
	$NC^{\alpha}Cd$	129 ± 14	6.6 ± 1.7	0.56 ± 0.32	128 ± 5	8 ± 4
1′		159 ± 7	2.9 ± 2.1	0.42 ± 0.20		
		180 ± 9	3.7 ± 1.6			
2	$C^{\alpha}C^{\beta}t$	232 ± 6	6 ± 3	0.51 ± 0.3	–	–
3	OHO	–	–	–	165 ± 12	7 ± 5
4	$NC^{\alpha}Cd$	287 ± 20	1 ± 0.9	0.45 ± 0.16	312 ± 5	0.9 ± 0.6
		339 ± 8	1.8 ± 1.8	0.36 ± 0.08	342 ± 5	
5	α helix	399 ± 13	1.1 ± 1.2	0.72 ± 0.22	423 ± 45	1.2 ± 2.4
6	$C^{\beta}b$	466 ± 15	1.3 ± 1.3	0.72 ± 0.28	440 ± 25	0.9 ± 1.2
7	α helix	543 ± 31	2.4 ± 2.3	0.83 ± 0.14	573	
8	OHO	–	–	–	577 ± 23	

It might be expected that only intense lines with minimal temperature shift would couple to Fröhlich modes, or would exhibit any coherence. The backbone modes in polypeptides and proteins (Groups 4-7, Table 2) to some extent meet these criteria. Extensive further investigations must be undertaken.

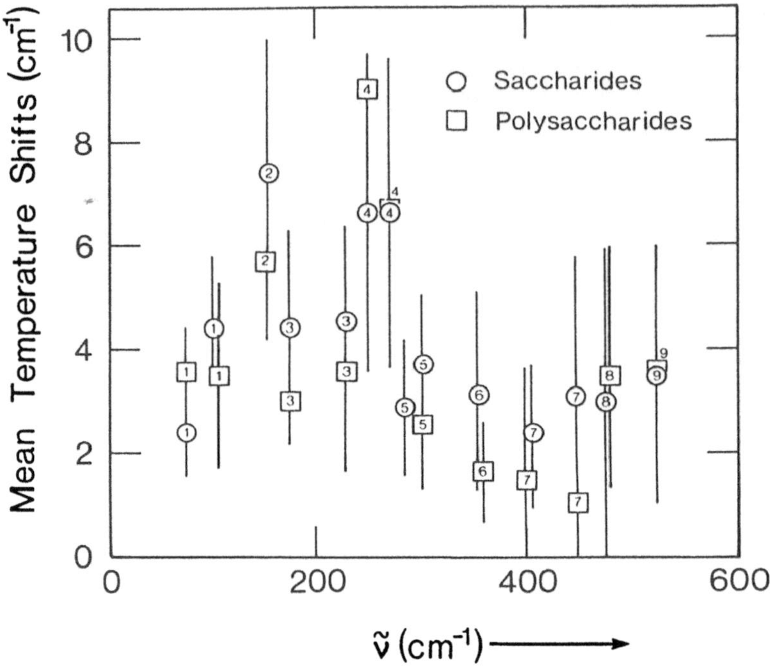

Fig. 2 – Mean temperature shifts of characteristic vibration frequencies of saccharides ○ and polysaccharides □, as a function of mean line frequency. Numbers indicate the group classifications of the original paper (Husain et al., 1984).

Protein spectra include, although only to a small extent, the information given by the spectra of their twenty peptide constituents. It is of interest to compare, as in Fig. 3, our spectrum of the protein salmine sulfate (Husain et al., 1986) with that of the amino acid arginine (Husain et al., 1984a), which is its principal (70%) constituent. When a normal-mode analysis is conducted on a polyatomic molecule, a damping of the modes is added to the computation, and the lines are broadened and also shifted to lower frequencies. One might interpret the effect of polymerization upon the arginine spectrum in these terms, even though salmine is not an artificial polymer.

It would be premature to imagine that the far infrared spectra of these biomolecules are well understood. We have noted that there are a few serious discrepancies between our own amino acid spectra and those of the Stuttgart group (Genzel, et al., 1983). The following lines, among the strongest in our 300 K and 100 K spectra, do not appear even in the five published Stuttgart (T = 6 K) spectra (Gly, Ala, Tyr, Val, Leu), let alone at higher temperatures.

Alanine	177 cm^{-1}	(CO_2t)
	225 cm^{-1}	(CH_3t)
Tyrosine	216 cm^{-1}	(CO_2t)
Glycine	366.5 cm^{-1}	(CCNd)

Figures 4a, b, c illustrate 100 K transmission functions (inverse spectra) comparatively.

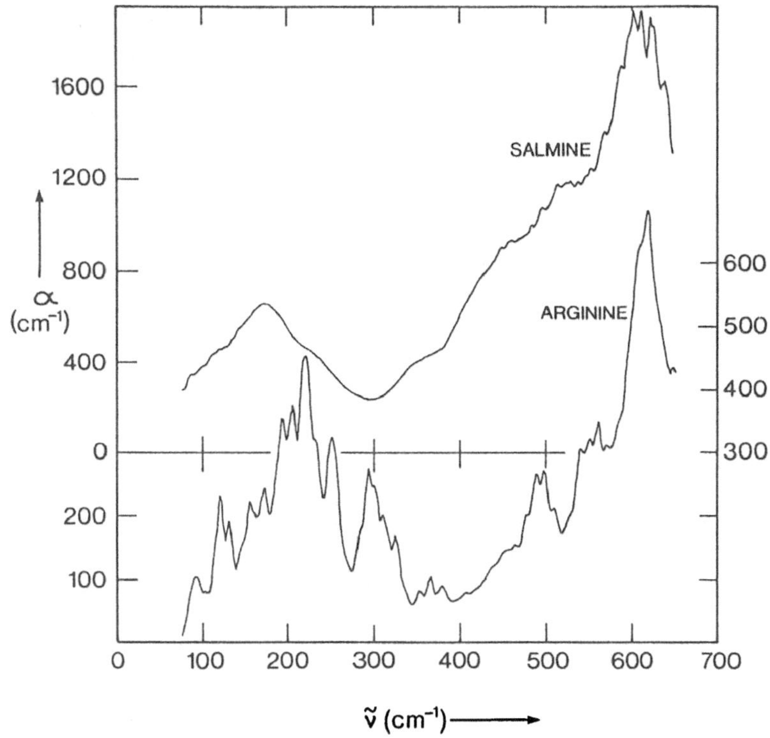

Fig. 3 — Comparison of far infrared spectra of arginine (Husain et al., 1984) and salmine sulfate (Husain et al., 1986).

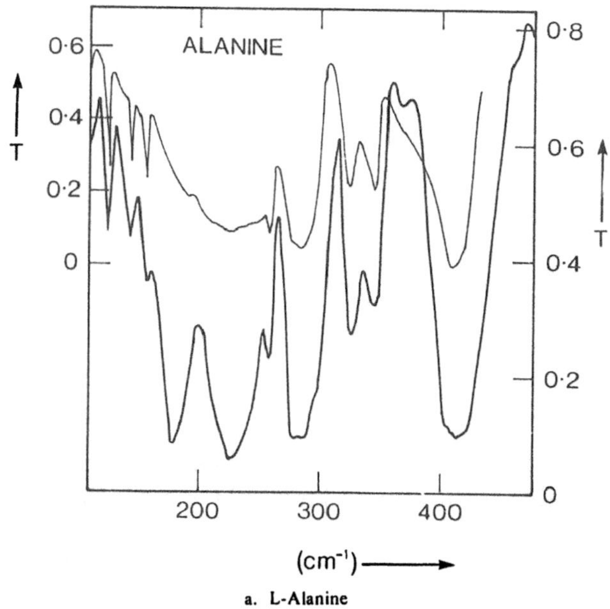

a. L-Alanine

Fig. 4 — Comparison of amino acid transmission functions of London (Husain et al., 1984a) and Stuttgart (Genzel et al., 1983) groups.

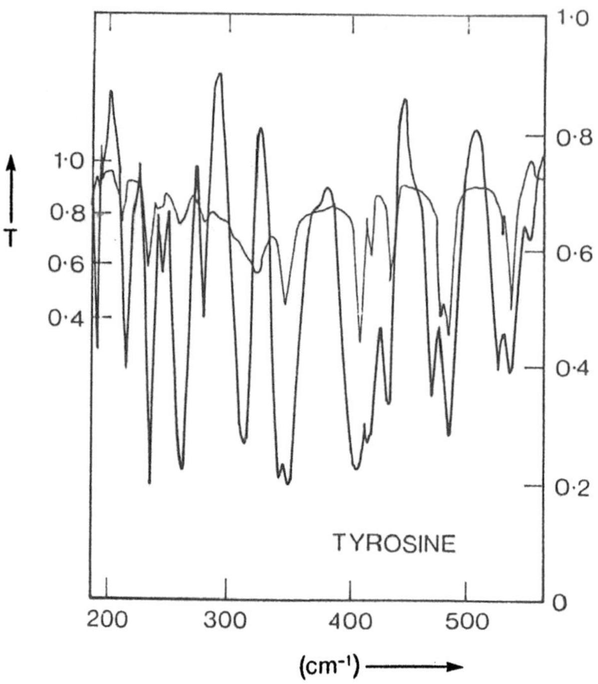

b. Tyrosine

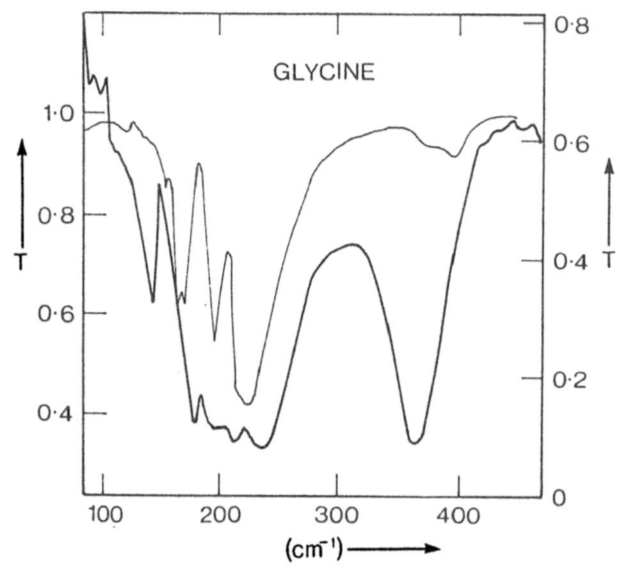

c. Glycine

Fig. 4 (Continued) – Comparison of amino acid transmission functions of London (Husain et al., 1984a) and Stuttgart (Genzel et al., 1983) groups.

The clue to these discrepancies may lie in the obtrusion of the very broad 200 cm^{-1} intermolecular OHO stretch bend, which is well known in the spectrum of liquid water (Hasted et al., 1985), and which also obscures any sharp bands which might otherwise appear in the spectra of (hydrated) proteins. This band appears not to sharpen even at temperatures as low as 6 K, as can be clearly seen in the

Stuttgart glycine thick-sample transmission function (Fig. 5). In samples that show this feature, other features can be obscured. It is more than possible that hydration even of amino acid crystals is capable of changing the characteristic features of the spectra, and this must be investigated with systematic variation of humidity and temperature. Structural change induced by hydration is surely an important chemical phenomenon with possible far-reaching biological consequences.

Table 3. Mean and standard deviation frequencies, temperature shifts, and intensities for some absorption bands of saccharides and analogs: glucose, galactose, inositol, fructose, ascorbic acid, maltose, cellobiose, sucrose, and lactose. Also for some polysaccharides: cellulose, starch, pectin, agar, agarose. All data from Husain et al. (1984b).

Oligosaccharides					Polysaccharides			
Group	Assignment	Frequency, cm^{-1}	Shift, cm^{-1}	Intensity		Frequency, cm^{-1}	Shift, cm^{-1}	Intensity
		75 ± 4	2.4 ± 1.0	0.28 ± 0.16	Cell.	75.5	3.5	
					Agse..	79	—	
1	*l*	102 ± 2	4.4 ± 1.6	0.28 ± 0.09	Cell.	107	5	s
					Agar.	112	2	
2	CH_2OH	157 ± 6	7.4 ± 2.6	0.55 ± 0.1	Cell.	151	6.5	
					Pect.	158	5	
3	COC	174 ± 3	4.4 ± 1.9	0.51 ± 0.26	Cell.	174	3	s
		230 ± 4	4.5 ± 1.9	0.57 ± 0.24	Cell.	236	3	s
					Mean	231 ± 4	3.6 ± 0.6	
4	CCt	250 ± 6 271 ± 4	6.6 ± 3.0	0.45 ± 0.20	Cell.	252	9	
					Cell.	264	—	
					Mean	253 ± 1 269 ± 3	6.7 ± 2.0	
5	OHt	287 ± 4	2.9 ± 1.3	0.73 ± 0.24				
		301 ± 4	3.7 ± 1.2	0.60 ± 0.20	Cell.	306	2.5	s
					Mean	304 ± 3	2.3 ± 1.3	
6		355 ± 9.5	3.1 ± 2.0	0.44 ± 0.18	Cell.	352	2	s
					Cell.	363	1.5	
					Mean	357 ± 8	1.8 ± 0.3	
7		408 ± 3.5	2.4 ± 1.2	0.5 ± 0.22	Cell.	401	0	s
					Mean	404 ± 5	1.5 ± 2	
		447 ± 8	3.1 ± 2.7	0.45 ± 0.22	Cell.	450	1	s
						460	3.5	s
					Starch	440	0	
8		477 ± 9	3.0 ± 3.0		Cell.	510	—	
					Mean	480 ± 17	3.5	
9	CCOd	523 ± 4 534 ± 3	3.5 ± 2.5	—	Cell.	522	3.7 ± 2.9	s
					Cell.	533		s
					Starch	522		
					Pect.	531		

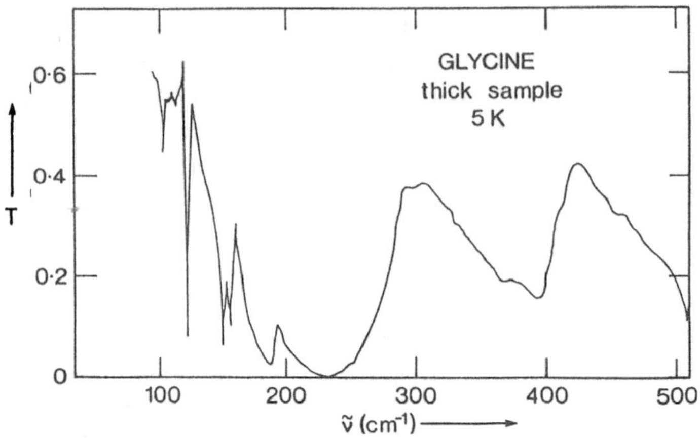

Fig. 5 — Thick-sample glycine transmission function obtained by Stuttgart group (Genzel et al., 1983).

REFERENCES

J. Bandekar, L. Genzel, F. Kremer, and L. Santo, Spectrochim. Acta **39A**, 357 (1983).

J. Chamberlain and H.A. Gebbie, Appl. Opt. **10**, 1184 (1971).

G.W. J. Chantry H.M. Evans, Chamberlain, and H.A. Gebbie, Infrared Phys. **9**, 85 (1969).

H. Fröhlich, Phys. Lett. **26A**, 402 (1968a).

H. Fröhlich, Int. J. Quatum Chem. **2**, 641 (1968b).

L. Genzel, F. Kremer, A. Poglitsch, and G. Bechtold, in *Coherent Excitations in Biological Systems*, H. Fröhlich and F. Kremer, Springer, Berlin/Heidelberg/New York/Tokyo, 1983, p. 58.

J.B. Hasted, S.K. Husain, F.A.M. Frescura, and J.R. Birch, Chem. Phys. Lett. 118, 622 (1985).

S.K. Husain, J.B. Hasted, D. Rosen, E. Nicol, and J.R. Birch, Infrared Phys. **24** 201, (1984a).

S.K. Husain, J.B. Hasted, D. Rosen, E. Nicol, and J.R. Birch, Infrared Phys. **24** 209 (1984b).

S.K. Husain, J.B. Hasted, D. Rosen, E. Nicol, and J.R. Birch, "FIR spectra of polypeptides and proteins" (To be submitted to Infrared Phys., 1986).

K. Machida, M. Izumi, and K. Kagayama, Spectrochim. Acta **35A**, 1333 (1979).

K. Machida, A. Kagayama, Y. Saito, T. Uno, and Y. Kuroda, Spectrochim. Acta **33A**, 569 (1977).

K. Machida, A. Kagayama, Y. Saito, and T. Uno, Spectrochim. Acta **34A**, 909 (1978).

W.H. Moore and S. Krimm, Biopolymers **15**, 2439, 2465 (1976).

J.F. Rabolt, W.H. Moore, and S. Krimm, Spectrochim. Acta **10**, 1065 (1977).

20. Spectral Dimensions of Paramagnetic Proteins

H. J. Stapleton

With 11 Figures

ABSTRACT

The temperature dependence of the Raman electron spin-lattice relaxation rate is a sensitive probe of the low-frequency vibrational spectrum of the structure surrounding a paramagnetic ion. If $\rho(\nu)$, the density of these low-frequency vibrational states, varies with frequency as a simple ν^{m-1} power law, then the Raman relaxation rate varies with temperature as T^{3+2m} at low temperatures, where m is the spectral dimension. In proteins m is found to be much reduced from the value of 3 that is normally observed in three-dimensional crystalline solids. Various theoretical models are discussed. Data are presented which contrast electron spin relaxation in proteins with that observed in amorphous materials (where two-level tunneling states are involved in the relaxation mechanism) and glasses (where the temporal profile of the spin recovery follows a stretched exponential).

The temperature dependence of electron spin-lattice relaxation rates can be a sensitive probe of the low-frequency vibrational spectrum of the surrounding structure. After a saturating pulse, a spin system can return to thermal equilibrium by several vibrationally induced mechanisms, or additionally, in the case of amorphous materials, by interacting with localized two-level states, or through spectral diffusion. Each of these mechanisms is characterized by a unique temperature variation and/or a characteristic recovery profile. The three commonly observed, vibrationally induced relaxation mechanisms are the direct, the Raman, and the Orbach (resonant Raman) processes. Only the Raman process involves a summation over all thermally activated vibrational modes and is therefore sensitive to the density of vibrational states, $\rho(\nu)$. If the low-frequency behavior of $\rho(\nu)$ is a simple power law of the form

$$\rho(\nu) \propto \nu^{m-1} \qquad (0 \leqslant \nu \leqslant \nu_{\max}) \tag{1}$$

then the temperature dependence of the Raman relaxation rate for an ion with an odd number of unpaired electrons (a Kramers' ion),

$$1/T_1 = \int_0^{\nu_{\max}} \frac{\rho^2(\nu)\, \nu^4 \exp(h\nu/kT)\, d\nu}{[\exp(h\nu/kT)-1]^2}, \tag{2}$$

becomes

$$1/T_1 \propto T^{3+2m}\, J_{2+2m}(\Theta/T) \tag{3}$$

where $\Theta \equiv h\nu_{\max}/k$ is the Debye temperature, and $J_n(x)$ is a transport integral of the form

$$J_n(x) \equiv \int_0^x \frac{z_n\, e_z\, d_z}{[e^z - 1]^2}\,. \tag{4}$$

At low temperatures ($T << \Theta$), the upper limits on the integrals in eqs. 2 and 4 can be approximated as infinity; the transport integral of eq. 3 is temperature independent; and the Raman relaxation rate becomes a simple power law in temperature,

$$1/T_1 \propto T^{3+2m} . \tag{5}$$

The exponent, m, in eqs. 1, 3, and 5 is called the spectral dimension, in analogy with its value for periodic lattices in one, two, or three dimensions. The T^9 temperature variation of the Raman relaxation rate in ordinary crystalline solids at low temperature ($T << \Theta$) has been well established experimentally for Kramers' ions (Orbach and Stapleton, 1972; Stapleton, 1972) and for low-spin ferric ions in particular (Bray et al., 1962).

At higher and higher temperatures, the transport integral of eq. 3 has the effect of reducing the temperature of eq. 3 from T^{3+2m} to T^2, in agreement with that expected for a two-quantum process in the classical limit.

In 1980 we published a letter (Stapleton et al., 1980) noting that new data, as well as previously published Raman relaxation rates (Mailer and Taylor, 1973; Herrick and Stapleton, 1976), on low-spin ferric hemoproteins, implied spectral dimensions well below 3.

Figure 1 shows these data for ferricytochrome c (cyt c), myoglobin.OH (Mb), myoglobin azide (MbN_3), and cytochrome P-450. Each set of data is separately fit to the sum of a direct relaxation process (varying with temperature as T or T^2) and a simple T^n power law. The best fitting values of n are indicated and the standard errors are typically 1%. Although it is not shown in Fig. 1, we have extended the relaxation measurements on MbN_3 to a rate of 3×10^5 s^{-1} (corresponding to a temperature of 17 K) using the technique of continuous wave saturation (Allen, 1982). The resulting fit of the Raman rate is a $T^{6.30}$ temperature dependence.

The straight line in Fig. 1 corresponds to a T^9 temperature dependence, associated with a spectral dimension of 3. A spectral dimension of 5/3 would yield a $T^{6.33}$ Raman rate, in close agreement with the data of Fig. 1. It was also noted that 5/3 is the fractal dimension of a self-avoiding random walk in three-dimensional space.

Figure 2 shows a fit to eq. 4 of the same data for all relaxation rates under 10^4 s^{-1}. Two features are apparent; the $T^{6.34}$ temperature variation of the cytochrome c data above 11 K is not reproduced, and unusually low values of the Debye temperature are required to reproduce the $T^{6.34}$ temperature variation of the relaxation rates.

Figures 3 and 4 provide similar relaxation data on 2-iron-2-sulfur ferredoxins that exhibit spectral dimensions of approximately 1.34, corresponding to a $T^{5.68}$ temperature dependence. Figure 3 represents a reanalysis of the data of Gayda et al. (1979), while Fig. 4 is from our work (Wagner et al., 1985). Figure 4 also indicates that the observed spectral dimension may depend upon solvent conditions, which in this case represented differences in the ionic strength, I, of the solvent. Figure 5 shows that a similar variation of spectral dimension with ionic strength occurs in the relaxation data of cytochrome c-551 (Wagner et al., 1985).

We have studied these solvent effects in greater detail (Colvin and Stapleton, 1985) with measurements on myoglobin azide under 11 different solvent conditions

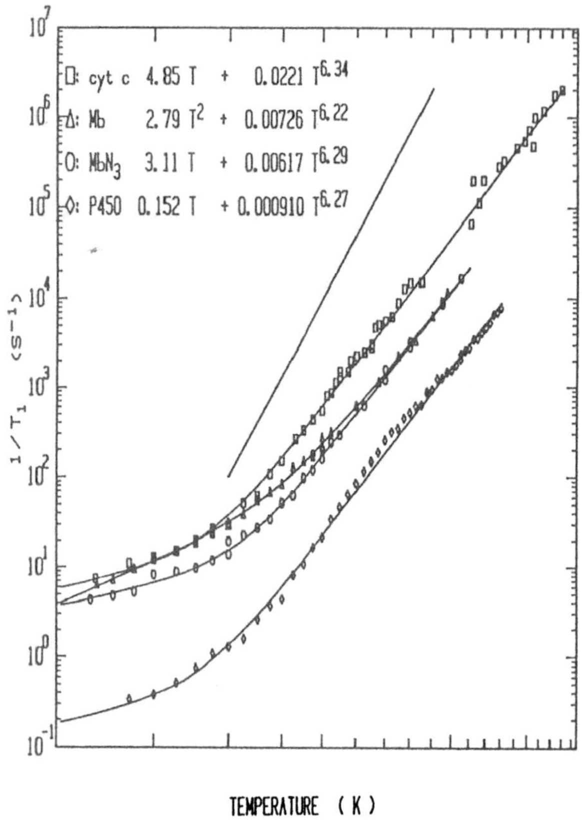

Fig. 1 — Relaxation rates (Allen et al., 1982) for four low-spin hemoproteins: ferricytochrome c (cyt c), myoglobin.OH (Mb), myoglobin azide (MbN_3), and cytochrome P-450 from *Pseudomonas putida* (Herrick and Stapleton, 1976). Each data set is separately fit to the sum of a direct process (varying with temperature as T or T^2) and a simple T^n power law. The best fitting values of n are indicated. Standard errors in n are typically 1%. The straight line represents a T^9 power law. For Mb and MbN_3 only every third datum is displayed between 5 and 11 K to improve clarity. An n value of 6.33 implies a spectral dimension of 5/3. All data corresponding to relaxation rates below $10^4 sec^{-1}$ were measured at X-band microwave frequencies ($\approx$ 9.5 GHz) on frozen protein solutions. The cytochrome c data above $T = 10$ K were taken on a single crystal using a microwave frequency of 24 GHz (Mailer and Taylor, 1973) and have been reduced by a factor of 0.106.

and found two general results. Solvent conditions that suppress the strength of the direct relaxation process, which under typical experimental conditions involves acoustic wavelengths of the order of 1000 Å, also raise the observed spectral dimension. Examples of this are shown in Fig. 6, where the protein solution for one sample was subjected to a partial vacuum in order to remove a larger fraction of the water. Also seen in the results from that same sample is a break in the log-log slope of the relaxation rate vs temperature at about 6 K. This phenomenon can be seen more clearly in Fig. 7, which shows data from a sample of 1.5 mM myoglobin azide in a 50% glycerol-50% water solvent. If a spectral dimension of 5/3 and a temperature of 6 K is assigned, then the integrand of the Raman integral (eq. 2) peaks for phonons with a frequency of 660 GHz (wavelength $\approx$ 38 Å if a sound velocity of 2.5×10^5 cm/s is assumed). This characteristic length is in approximate agreement with myoglobin's molecular size (44 × 44 × 25 Å).

Earlier measurements on cytochrome P-450 (Herrick and Stapleton, 1976) indicated that the log-log slope of the Raman relaxation rate vs temperature was independent of microwave frequency and the g-value chosen for the measurements.

Scholes et al. (1984) have published electron spin relaxation data on both iron and the copper ions in cytochrome c oxidase that yield spectral dimensions of 1.77 ± 0.08 and 1.78 ± 0.07, respectively.

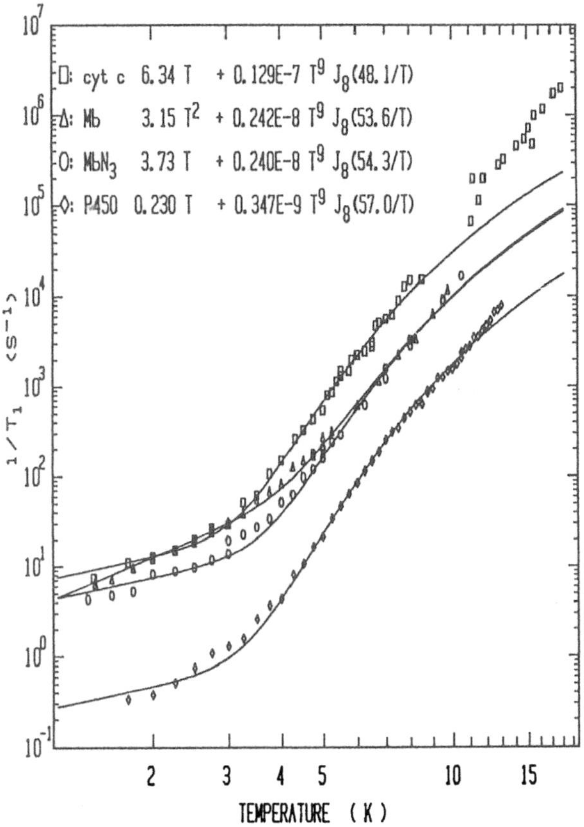

Fig. 2 — A conventional (spectral dimension equal 3) fit of the Raman relaxation data of Fig. 1 for rates under 10^4 s^{-1} (Allen et al., 1982). The fitting curves are extrapolated to higher temperatures. Debye temperatures near 53 K are required to reproduce the weak ($\approx T^{6.3}$) temperature dependence of these rates. The extrapolated curves fall below the data at the higher temperatures.

We have proposed (Stapleton et al., 1980; Allen et al., 1982) that these data are evidence for a density of vibrational states, $\rho(\nu)$, which is determined by the chain fractal dimension of the biopolymer, rather than the three-dimensional space in which it resides. The chain fractal dimension, $\bar{\mathbf{d}}$, of a polymer relates its root mean square (RMS) end-to-end length, R, to N, the number of monomers included in the chain. It is equal to the inverse of the Flory constant of polymer theory. The relationship is

$$R(N)/r = N^{(1/\bar{\mathbf{d}})} \tag{6}$$

where r is the average monomer length and $\bar{\mathbf{d}}$ lies between 1, a linear chain, and 2, an unrestricted random walk. We have computed chain fractal dimensions for about 70 proteins (Wagner et al., 1985) by considering the X-ray coordinates of the alpha carbons, available from the protein data bank (Bernstein et al., 1977). For each polypeptide chain of M residues, we compute r, the RMS alpha carbon-alpha carbon separation, and R $(2 \leqslant N \leqslant M)$, the RMS through-space distance between N consecutive residues of the polypeptide chain. Some consideration is also given for possible geometric self-similarity. Full details are published elsewhere (Wagner et al., 1985). Typical results of log-log plots of $R(N)$ vs N for six proteins of interest are shown in Fig. 8.

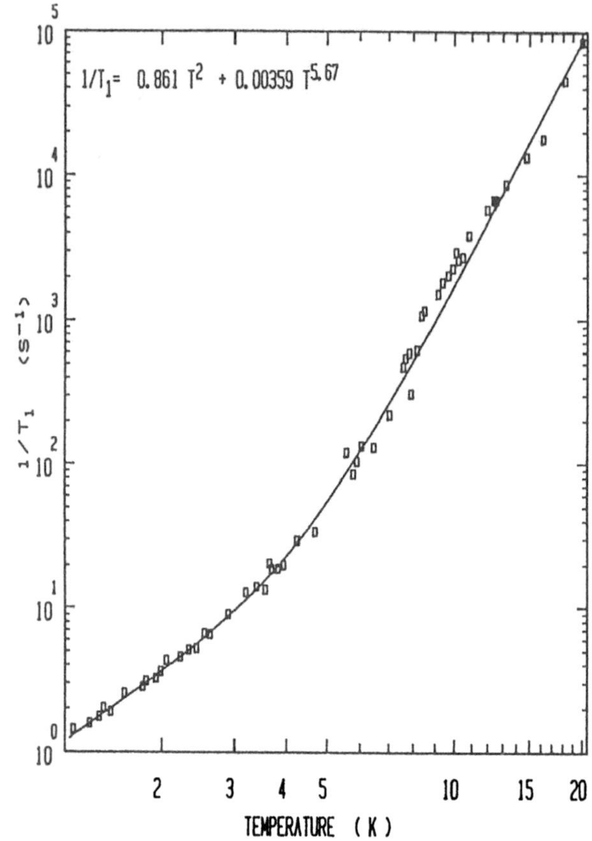

Fig. 3 — Relaxation data on a 2-Fe/2-S ferredoxin (from blue-green algae, *Spirulina maxima*) (Gayda et al., 1979) up to a temperature of 20 K. The curve is our reanalysis (Allen et al., 1982) of these data using a simple power T^n power law for the Raman process. The data imply a spectral dimension of $m = 1.34 \pm 0.06$.

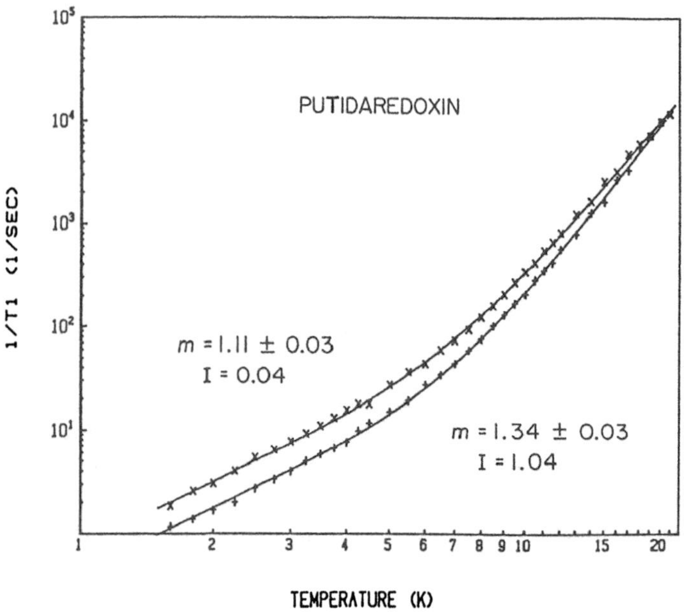

Fig. 4 — Relaxation data on two frozen solutions of *putidaredoxin* (Wagner et al., 1985). The temperature dependence of the Raman relaxation rates implies a spectral dimension $m = 1.11 \pm 0.03$ for the sample in a solution of low ionic strength (I = 0.04) and $m = 1.34 \pm 0.03$ for the sample in a 1.0 M NaCl solution (I = 1.04). Rates were measured on the EPR absorption maximum at g = 1.94.

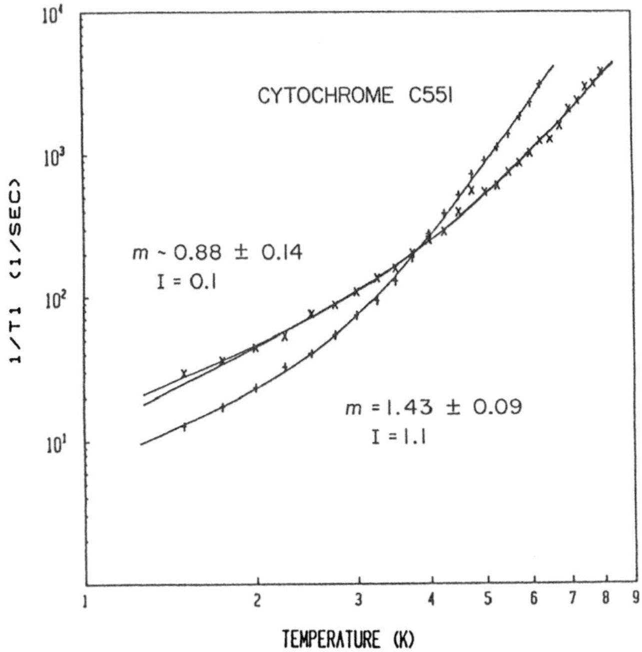

Fig. 5 — Relaxation data on two frozen solutions of cytochrome c-551 (Wagner et al., 1985). The temperature dependence of the Raman relaxation rates implies a spectral dimension near unity for the sample in a solution of low ionic strength (I = 0.1). Data from the sample in a 1.0 M NaCl solution (I = 1.1) indicate $m = 1.43 \pm 0.09$. Relaxation rates were measured at g = 1.04 (I = 0.1) and g = 2.16 (I = 1.1).

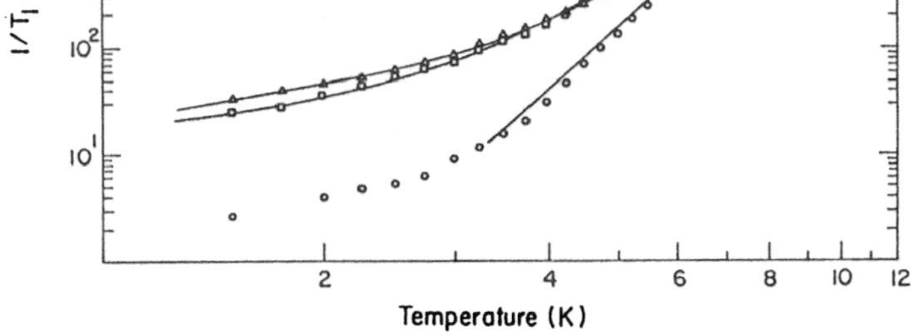

Fig. 6 — Relaxation data from three samples of myoglobin azide under different solvent conditions (Colvin and Stapleton, 1985).

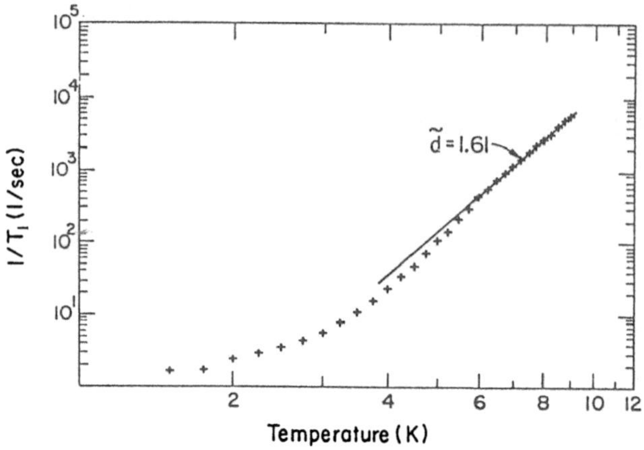

Fig. 7 — Relaxation data on 1.5 mM myoglobin azide in a solvent of 50% glycerol and 50% water (Colvin and Stapleton, 1985).

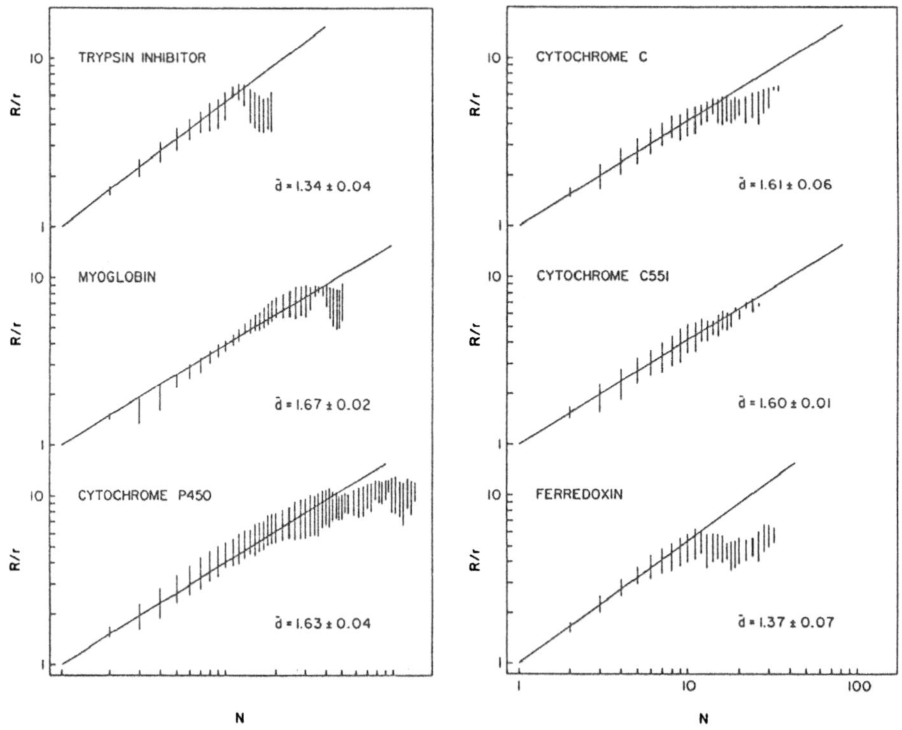

Fig. 8 — Graphic estimates of the chain fractal dimension of six selected protein structures (Wagner et al., 1985) using alpha carbon coordinates from the protein data bank (Bernstein et al., 1977). Vertical lines are twice the standard deviation in log (R/r). For clarity, some data are symmetrically deleted for all plots.

The arguments, which lead to a modified vibrational density of states for a fractal structure, can be briefly summarized as follows. Consider two geometrically or statistically self-similar objects with overall sizes of R and R/s, where s is a scaling factor. If the density of states of the smaller object is $\rho_{R/s}(\nu)$, then the density of states of the larger object in the neighborhood of a given frequency is

$$\rho_R(\nu) = S^{\bar{\mathbf{d}}} \rho_{R/s}(\nu) \tag{7}$$

if perturbations due to the connections can be neglected, i.e., the density of states of the large object is simply the sum of that of its parts. The conglomerate object is expected to display the same normal modes as its scaled counterparts, at correspondingly lower frequencies. The structure of its density of states in the neighborhood of a given normal mode should be the same, differing in the multiplicative factor given by eq. 7 and the frequency scale upon which the modes appear, so

$$\rho_R(\nu_R) d\nu_R = R/s(\nu_{R/s}) d_{R/s} \tag{8}$$

where $\nu_{R/s}$ is the frequency of a mode in the smaller object and ν_R is the frequency at which the corresponding mode is excited in the larger fractal. The difference in the frequency scales is in the differential frequency intervals. The scaling exponent of the mode frequencies describes how the frequency of a given mode varies with the radius of the object in which it is excited. One expects the modes in the larger object to be excited at lower frequencies. The scaling exponent describing this behavior (Alexander and Orbach, 1982; Rammal and Toulouse, 1983) satisfies

$$\nu_{R/s} = S^{\alpha} \nu_R \, . \tag{9}$$

The relationship between the spectral dimension m, the fractal dimension $\bar{\mathbf{d}}$, and the mode frequency scaling exponent A, follows from eqs. 7-9 with the result

$$m = \bar{\mathbf{d}}/a \, . \tag{10}$$

Alexander and Orbach (1982) refer to $m\_$ as the fraction dimension and employ $(1 + \bar{\delta}/2)$ instead of a in eq. 10, where $\bar{\delta}$ is the diffusion constant scaling exponent. Helman et al. (1984) replace a by $d_w/2$, where d_w is the fractal dimension of an unrestricted random walk on a fractal structure. Here the jump probability for a step on the structure is proportional to the strength of the bond joining the mass points .

The relationship between spectral and fractal dimensions is based upon the connectivity of the fractal structure. If an impulse, applied to one end of a polypeptide chain, can reach the other end only by propagation along the backbone, then such a system has a spectral dimension of unity regardless of its fractal dimension, and the value of a in eq. 10 is $\bar{\mathbf{d}}$. Conversely, a high degree of off-backbone connectivity can result in equal values of the spectral and fractal dimensions. This is evident in the notation of Helman et al. (1984) since d_w approaches 2, the fractal dimension of an unrestricted random walk. In this limit there is some confusion in computing the proper fractal dimension of a finite biopolymer. With a high degree of strongly bonded connectivity, the concept of a backbone vanishes and the mass fractal dimension that measures the scaling properties of the total mass with respect to distance (as opposed to the end-to-end length of the chain with respect to distance) is approximately 30% larger than the chain fractal dimension for a typical protein (Colvin and Stapleton, 1985; Stapleton, 1985). The mass fractal dimension is large because it scales all of the monomers as function of radius R, not just those that are contiguous. The conclusion of Helman et al. (1984) that no inconsistency would exist between experiment and the fractal model of relaxation in biopolymers, if cross-linking bonds were included between non-neighboring monomers, seems to ignore this difference between chain and mass fractal dimensions computed from X-ray data.

Table 1 indicates the correlation between the maximum observed spectral dimension, m, and $\bar{\mathbf{d}}$, the chain fractal dimension computed from X-ray structural data on the same or very similar protein structures.

There exists also some degree of correlation between the computed chain fractal dimension and the fraction of a protein's tetrapeptides in an alpha helical or beta turn conformation. This is shown in Fig. 5 of Wagner et al. (1985) where $\bar{\mathbf{d}}$ is plotted against the fraction, f, of tetrapeptides with an end-to-end length of 7 Å or less. The linear correlation is roughly given by $\bar{\mathbf{d}} = 1.20 + 0.65\, f$.

Gō et al. (1983) have published a histogram of the lowest 201 vibrational modes (in vacuo) of the small (58-residue) protein, bovine pancreatic trypsin inhibitor (BPTI). Their analysis considers bond lengths and angles fixed and treats only dihedral angles of the main chain and the side chains as dynamic variables. Since each alpha carbon position is described by two dihedral angles, the main chain has 116 degrees of freedom that correlate with about 116 delocalized modes of the backbone. Their results indicate that vibrational modes of BPTI below a frequency of about 3600 GHz are predominantly delocalized, i.e., they correspond to con-

Table 1 — Correlations between the largest spectral dimensions (m values) observed in the relaxation data of various proteins and the chain fractal dimension computed from X-ray data on the same or very similar protein structures.

Protein	Source	m_{max}	$\bar{\mathbf{d}}$
HEME PROTEINS			
myoglobin.OH	sperm whale	1.61 ± 0.05	1.67 ± 0.02
myoglobin azide	sperm whale	1.65 ± 0.04	1.67 ± 0.02
cytochrome P-450	*Pseudomonas putida*	1.64 ± 0.03	1.63 ± 0.04
cytochrome c	horse	1.67 ± 0.03	
	albacore		1.61 ± 0.04
cytochrome c-551	*P. aeruginosa*	1.43 ± 0.09	1.60 ± 0.01
IRON SULFUR PROTEINS			
ferredoxin	*Spirulina maxima*	1.34 ± 0.06	
	S. platensis		1.37 ± 0.07
putidaredoxin	*P. putida*	1.34 ± 0.03	
	S. platensis		1.37 ± 0.07

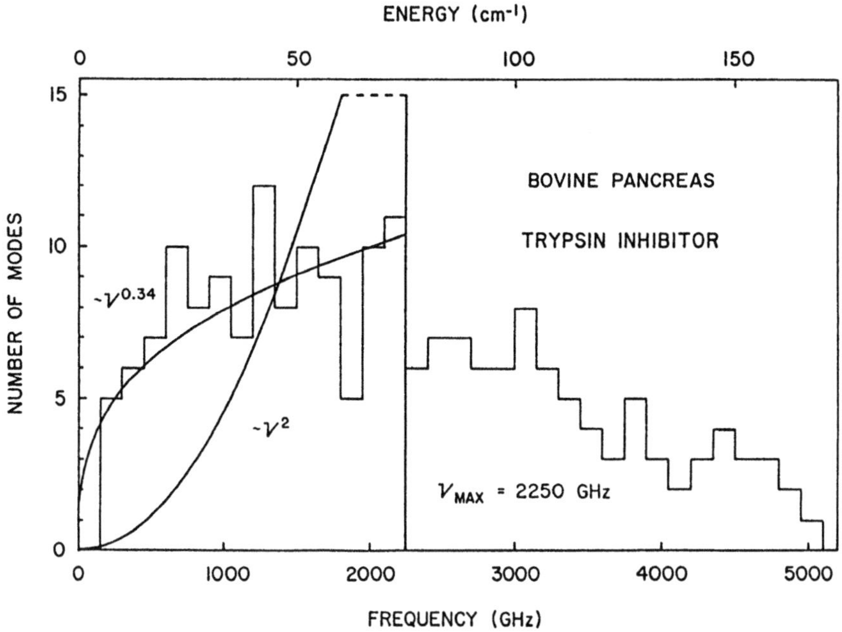

Fig. 9 — Histogram of the lowest 201 vibrational modes in bovine pancreatic trypsin inhibitor (Wagner et al., 1985). Data are reproduced from Gō et al. (1983). Superimposed are normalized densities of vibrational states varying as ν^2 (standard Debye model) and $\nu^{0.34}$ (fractal model). The latter is derived from the estimated chain fractal dimension for BPTI, with the assumption that the spectral exponent is equal to the chain fractral dimension.

certed movements of the entire polypeptide backbone. In Fig. 9 we reproduce their data, estimate a maximum frequency of 2250 GHz, corresponding to a total of 117 vibrational modes, for the main chain vibrations of BPTI, and indicate the normalized density of states predicted by spectral dimensions of 1.34 (the computed chain fractal dimension of BPTI) and 3. The fractal approximation provides a more accurate description of the low-frequency modes than does the classical Debye model.

Recently we have completed relaxation measurements on the copper-containing, 126-residue protein, azurin (Muench et al., 1986). Relaxation data on one sample, taken at temperatures between 1.5 and 9.5 K, reveal a Raman rate that can be equally well fit by a simple power law ($2.53 \times 10^{-4} T^{6.93 \pm 0.12}$) or an expression involving a transport integral ($2.83 \times 10^{-10} T^9 J_8(62/T)$). A log-log plot of relaxation data from another sample, in the temperature range between 4.2 and 22 K, indicate a slowly varying log-log slope. These Raman relaxation rates are better fit by a transport integral ($3.21 \times 10^{-10}\, T^9 J_8(66.75/T)$) than by a simple power law. This fit, to all except the lowest datum, which probably contains a significant contribution from the direct relaxation process, is shown in Fig. 10.

These relaxation data on biopolymers should be compared with results from amorphous or glassy systems. Figure 11 shows relaxation data (Askew et al., 1986) on amorphous silicon; made amorphous by silicon ion implantation into crystalline slabs of silicon. The dangling bonds of amorphous silicon make it intrinsically paramagnetic with a g value of about 2.0055. These relaxation data, taken at 16.5 GHz between temperatures of 0.3 and 4 K, fit a simple $T^{2.36}$ power law rather than

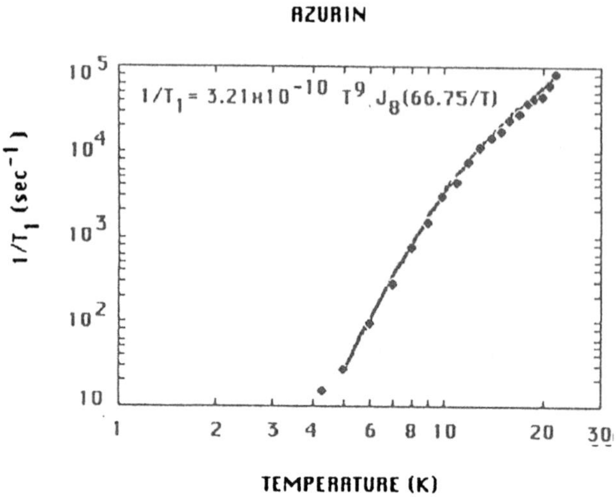

Fig. 10 — Relaxation data on a frozen solution of the copper-containing protein, azurin, between 4.2° and 22° K (Muench et al., 1986). The data can be fit to a conventional Debye model (m = 3) with a $T^9 J_8(66.75/T)$ temperature dependence and a low Debye temperature of 66.75 K.

the coth $(h\nu/2kT)$ dependence expected for a direct (one-phonon) relaxation mechanism, or the $\coth^2$ $(h\nu/2kT)$ dependence expected for a phonon-limited (bottlenecked) direct relaxation mechanism. A relaxation model involving localized two-level tunneling states can explain a $T^{2+\lambda}$ temperature variation when the density of two-level states varies with energy as E^λ. By contrast, the temperature dependence of spin relaxation in a myoglobin sample has been shown to follow a $\coth^2$ $(h\nu/2kT)$ dependence at sufficiently low temperatures (Muench et al., 1984). This was positive identification of a phonon bottleneck and excluded a T^2 process that might be associated with a tunneling state for which λ was zero.

Recent relaxation measurements on a Yb-doped lead phosphate glass (Smith et al., 1986) in the temperature range between 1.2 and 2.1 K yielded spin recoveries profiles that followed a stretched exponential in time, varying as (1 − exp $(-t/\tau)^{0.5}$). The fitting parameter, $1/\sqrt{\tau}$, was found to vary with the square root of the Yb concentration and to increase linearly with increasing temperature. Proteins, by contrast, exhibit fairly exponential recoveries.

Alexander et al. (1985) have computed the temperature variation of a spin relaxation mechanism that involves two fractons (quantized vibrations on a fractal lattice). A major difference between this model and earlier ones is that these fractons are localized and thus relaxation can be stimulated only if the spin is within some minimum distance of two such fractons having the proper energy difference. This localization adds an additional parameter, dϕ, which characterizes the localization of the fracton wavefunction. It is expected to lie between unity and the chemical length dimension. For a Kramers' ion, the theory of Alexander et al. (1985) predicts a T^n temperature dependence with

$$n = 2m\{1+2\,d\phi/D\}-1 \tag{11}$$

where m is the spectral dimension and D is the mass fractal dimension. Setting n equal to 6.33 as a typical experimental value, and D equal to 2, based on protein X-ray data, we can vary dϕ between the reasonable limits of unity and the value of D = 2, and solve for the corresponding value of m, the spectral dimension. With $d\phi$ = 1, m must equal 1.83; for $d\phi$ = 2, m must equal 1.22. Thus even with the

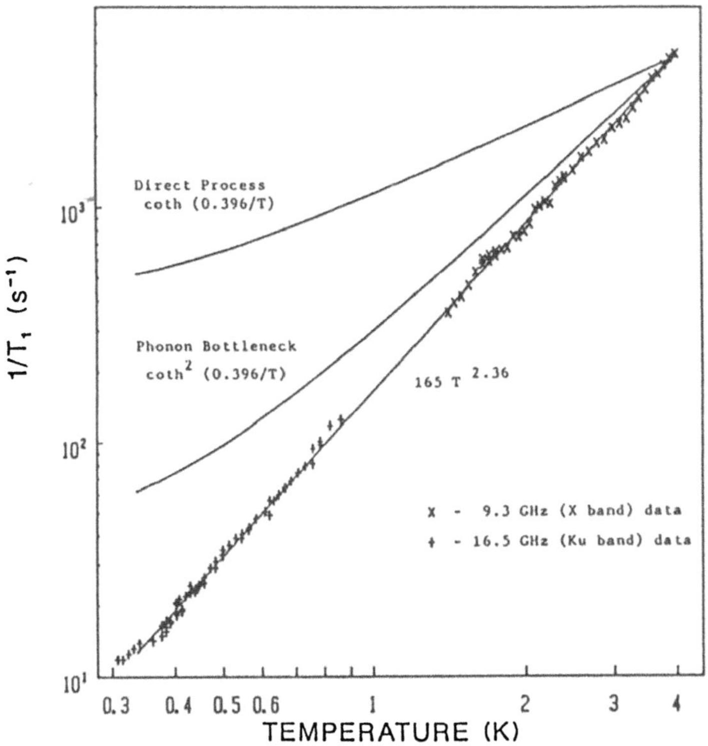

Fig. 11 — Electron spin lattice relaxation rates for amorphous silicon made by ^{28}Si ion implantation (Askew et al., 1986). The two solid curves, without data, correspond to the temperature dependences expected for an ordinary direct relaxation rate, and a phonon-limited direct relaxation rate involving a single phonon. A relaxation mechanism involving a distribution of localized two-level states, with a density of states that varies with energy as $E^{0.36}$, can explain the unusual temperature dependence.

two-fracton relaxation model, the spectral dimension of the biopolymer chain must be greater than unity. This is possible only with off-backbone connectivity.

More recently, Elber and Karplus (1986) have computed low-frequency modes of BPTI, myoglobin, cytochrome c-551, and a ferredoxin using an effective medium approximation (EMA). For each amino acid they included two interactions along the backbone and two main-chain hydrogen bonds at least six residues away. They evaluated force constant distributions and mass matrices for their EMA from the known protein structures. The force constant distribution along the backbone was never close to zero, while the distribution of cross-linking peaked at zero. They found that the resulting vibrational modes could be fit well to a spectral dimension of 1.6 for myoglobin and 1.4 for cytochrome c-551 and the ferredoxin. They conclude that the experimentally observed spectral dimensions reflect the distribution of interaction strengths between amino acids rather than any geometrical (fractal) effects.

In conclusion, there is a body of experimental evidence indicating that Raman spin relaxation rates in paramagnetic proteins follow a T^n temperature dependence, with $5 \leqslant n \leqslant 7$, that cannot always be explained on the basis of a low Debye temperature. Several models have been proposed to account for this. These include ones in which: (1) the spectral dimension of the protein equals its chain fractal dimension because of unspecified crosslinking bonds; (2) a relaxation mechanism involves localized fractons; and (3) a vibrational density of states can be described by a spectral dimension, but this results not from any fractal geometry, but rather from a distribution of the force constant between amino acid residues. Model 3 appears very promising, but it considers the observed correlations between m and $\bar{\mathbf{d}}$ to be completely fortuitous. Model 2 predicts some specific detail in the scatter of the measurements about a simple T^n law, which at present is not strongly supported by the experimental data. The requirements for the crosslinking of model 1 are not known at present.

This work was supported in part by the U.S. Public Health Service under NIH Grant GM-24488.

REFERENCES

J.P. Allen, Ph.D. Thesis, Physics Department, University of Illinois, Urbana, IL, 1982 (unpublished).

J.P. Allen, J.Y. Colvin, D.G. Stinson, C.P. Flynn, and H.J. Stapleton, Biophys. J. **38**, 299 (1982).

S. Alexander and R. Orbach, J. Phys. Lett. (France) **43**, L625 (1982).

S. Alexander, O. Entin-Wohlman, and R. Orbach, J. Phys. Lett. (France) **46**, L555 (1985).

T.R. Askew, H.J. stapleton, and K.L. Brower, Phys. Rev. B: Solid State **33**, 445 (1986).

F.C. Bernstein, T.F. Koetzle, G.J.B. Williams, E.F. Meyer, Jr., M.D. Brice, J.D. Rodgers, O. Vennard, T. Shimanouchi, and M. Tasumu, J. Mol. Biol. **112**, 535 (1977).

T. Bray, G.C. Brown, Jr., and A. Kiel, Phys. Rev. **127**, 730 (1962).

J.T. Colvin and H.J. Stapleton, J. Chem. Phys. **82**, 4699 (1985).

R. Elber and M. Karplus, Phys. Rev. Lett. **56**, 394 (1986).

J.P. Gayda, P. Bertrand, A. Deville, C. More, G. Roger, J.F. Gibson, and R. Cammack, Biochim. Biophys. Acta. **581**, 15 (1979).

N. Gō, T. Noguti, and T. Nishikawa, Proc. Natl. Acad. Sci. U.S.A. **80**, 3696 (1983).

J.S. Helman, A. Coniglio, and C. Tsallis, Phys. Rev. Lett. **53**, 1195 (1984).

R.C. Herrick and H.J. Stapleton, J. Chem. Phys. **65**, 4778 (1976).

C. Mailer and C.P.S. Taylor, Biochim. Biophys. Acta **322**, 195 (1973).

P.J. Muench, T.R. Askew, J.T. Colvin, and H.J. Stapleton, J. Chem. Phys. **81**, 63 (1984).

P.J. Muench, G.C. Wagner, and H.J. Stapleton, Bull. Am. Phys. Soc. **31**, 388 (1986).

R. Orbach and H.J. Stapleton, "Electron Spin-Lattice Relaxation" in *Electron Paramagnetic Resonance*, S. Geschwind, Ed., Plenum, New York, 1972

R. Rammal and G.J. Toulouse, J. Phys. Lett. (France) **44**, L13 (1983).

C.P. Scholes, H.T. Janakiraman, and T.E. King, Biophys. J. **45**, 1027 (1984).

D.L. Smith, H.J. Stapleton, and M.B. Weissman, Phys. Rev. B: Solid State (to be published) (1986).

H.J. Stapleton, Magn. Reson. Rev. **1**, 65 (1972).

H.J. Stapleton, Phys. Rev. Lett. **54**, 1734 (1985).

H.J. Stapleton, J.P. Allen, C.P. Flynn, D.G. Stinson, and S. Kurtz, Phys. Rev. Lett. **45**, 1456 (1980).

G.C. Wagner, J.T. Colvin, J.P. Allen, and H.J. Stapleton, J. Am. Chem. Soc. **107**, 5589 (1985).

21. The Role of Chaos in Biological Systems

F. Kaiser

With 5 Figures

ABSTRACT

For many years the possible influence of weak electromagnetic fields on biological systems has been investigated by both experimental and theoretical methods. In the present contribution the possible influence of deterministic chaos on the experimental results and on intrinsic biological order and function will be discussed.

We start with a brief discussion of the properties that a system in a deterministic chaotic state can display. The differences between deterministic irregularities and stochastic behavior are studied. It is shown that coherent oscillations are a possible basis for theoretical models and for biological rhythm as well. The interaction of these oscillations with external fields leads to a variety of periodic, quasiperiodic and irregular (chaotic) motion, exhibiting far-reaching analogies with respect to experimental results (e.g. sharp resonances, extreme sensitivity, and frequency and intensity windows).

The existence of deterministic chaos reveals a possible explanation of some further facts, e.g., failure of reproducibility, variation of effects in similar experiments, and partial absence of effects. Finally, two models are discussed, a nonlinear torsion pendulum and a specific model for coherent oscillations. The bifurcations are analyzed with respect to the externally supplied energy and its internal dissipation; the results presumably exhibit an universal behavior.

INTRODUCTION

The traditional way of thinking about the behavior of deterministic systems is based on the statement that given the initial conditions one knows how the system will develop in the far future. This statement has turned out to be false in many cases. In these cases small changes in the initial conditions do not lead to small changes in the outcome in the following sense: for infinitesimally differing adjacent starting points of trajectories one ends up with two completely different final states. Meanwhile it has become clear that a *sensitive dependence on the initial conditions* leading to a completely chaotic (irregular) behavior in time is a typical behavior of many systems. Such a property has been found in electronic circuits, in lasers and optical devices, in turbulent systems, in oscillating chemical reactions, in periodically driven biological rhythms, and in a number of other topics and disciplines.

The equations that can describe such a behavior, are sets of first-order, nonlinear differential or difference equations and nonlinear maps, to mention the most relevant modeling approaches. These nonlinear dynamical systems may be divided into dissipative systems and conservative systems; the latter include both classical and quantum systems. The present discussion is devoted to dissipative systems, since systems of this type are the most important ones for describing experimental findings, at least in biological systems. From a mathematical point of view, all dissipative nonlinear dynamical systems with more than two degrees of freedom (i.e. described by three or more autonomous first-order differential equations) can display chaotic motion and can lose their *final state predictability*. Whether they do depends on details of the internal nonlinearities. For dissipative nonlinear maps, a lower dimensionality can lead to chaos.

As early as 1892 Poincaré had discovered that strongly deterministic equations (of Hamiltonian type) for the motion of mechanical systems could display a chaotic motion (Poincaré, 1892). Not until 1963 was it found that a system of three coupled, first-order nonlinear differential equations can exhibit a chaotic behavior (Lorenz, 1963). Contrary to Poincaré's example (chaos in conservative systems), Lorenz discovered *deterministic chaos in dissipative systems.**

In the present paper the possible role of chaos in biological systems is discussed. On the one hand chaotic motion is viewed as a specific property of biological order and function under certain external or internal constraints and with respect to some modeling approaches. On the other hand both experimental methods and experimental results on biological systems must be reconsidered under the aspects of predictability, determinacy, and reproducibility.

COHERENT OSCILLATIONS AND CHAOS IN BIOLOGICAL SYSTEMS

Coherent Oscillations

Coherence means a static or dynamic order principle. Some degrees of freedom of the system under investigation exhibit phase correlations and/or space-time correlations. Examples are fixed phases of a set of coupled oscillators or the determination of a state at a space-time point (x_2, t_2) by the space-time point (x_1, t_1).

A well-known and important realization of temporal coherence is present in the laser. In this system oscillating atomic dipoles exposed to an external energy pump beyond a certain threshold are driven from a state of random oscillations into a state of a complete in-phase oscillation. Thus the cooperation of the atomic dipoles creates the coherence inherent in the laser light, i.e., the atoms operate in a state of self-organization. The establishment of certain coherent oscillations on a macroscopic scale, which might be important for biological systems, can be performed by nonlinear interactions between oscillating electric dipoles and their surroundings (e.g., rest of the system, heat baths,- - -). The random supply of energy may lead to a strong coherent excitation of a single mode supported by the whole set of dipoles. Again interacting subunits and their additional supply with external energy are required. A very simple model for this situation consists of a set of coupled oscillators, the stabilization and excitation of which are given by nonlinear

*For reviews of deterministic chaos see Collet and Eckmann, 1981; Eckmann, 1981; Ott, 1981; Haken, 1983;, Schuster, 1984; and Cvitanovic, 1984.

dynamics and external drives, respectively. The governing equations can exhibit a chaotic motion (irregular oscillation), i.e., a deterministic chaotic state in the driven, nonlinear and dissipative system occurs.

Oscillating phenomena are well known in biological systems, ranging from microseconds and shorter time scales to days (circadian rhythm) and even longer periods. From a rather fundamental point of view, Fröhlich has promoted the ideas that coherent states can be established in active biological systems (Fröhlich, 1968, 1980, 1984). Possible types of coherent vibrations are (1) highly excited single modes, (2) oscillation of highly polar metastable states, and (3) self-excited periodic processes (limit cycles).

Whereas the establishment of highly polar states requires a certain *amount of energy* to be put into the system ($E \geqslant E_{\mathrm{min}}$), the excitation of coherent vibrations is possible only if the *rate of energy flux* through the system is beyond a certain threshold ($S \geqq S_0$).

The possible excitation of coherent vibrations has been demonstrated with some simple model calculations (Fröhlich, 1968; Kaiser, 1977a). The coherent vibrations include a very large frequency range from extremely slow chemical or electrical oscillations to some beyond the gigahertz range in membranes and proteins.

Chaotic Behavior

The mathematical description of coherent oscillations each of which is assumed to model a certain biological function is intimately connected with the theoretical and numerical study of nonlinear dynamical systems. The occurrence of an irregular behavior (i.e. turbulent, chaotic, aperiodic, stochastic motion) in a large class of deterministic systems is accompanied by a quantitative universality. Coherent oscillations (e.g., of Fröhlich type) are thus linked with the fundamental problems occurring in nonlinear dynamics. Specific techniques and refined methods have to be applied to study the different types of behavior inherent in the dynamical systems (Schuster, 1984).

Coherence is inseparably connected with a profound deterministic concept, including a strong predictability, whereas chaos incorporates a strong stochasticity and indeterminacy. The transition from strongly periodic motions (coherent oscillations) to a chaotic motion and the possible coexistence of these is a surprising fact, since order and chaos seem to exclude each other. The solution of this problem is fascinating and well established.

The chaotic behavior found in nonlinear dynamic systems and in many experiments is due neither to an infinite number of degrees of freedom (three degrees can be sufficient) nor to external sources of noise (there are strong periodic stimulations) nor to the uncertainty stemming from quantum mechanics (a purely classical description is chosen). The source of chaos is a property of the nonlinear system itself. The nonlinear dynamics cause nearby trajectories to separate exponentially within a certain bounded region (basis of attraction), leading to a strange attractor in phase space.

The new type of stochasticity, sometimes referred to as turbulent statistics, is driven by fluctuations on a macroscopic scale instead of quantum or thermal fluctuations. It should be emphasized again that some specific properties are essential; very few degrees of freedom, classical systems, and strongly deterministic and nonlinear dynamics. Furthermore, besides nonlinear dynamical systems represented by ordinary differential equations (i.e., flow systems), low-dimensional discrete systems, i.e., nonlinear maps can exhibit deterministic chaos. In many cases it is convenient to study a d–dimensional flow system by the corresponding (d-1)-dimensional map) (Poincaré map).

Here we restrict discussion to chaos in time, that is, uncorrelated behavior in time. Spatial correlations are possible. The occurrence of chaos in space and time (fully developed turbulence) is still an open question.

The existence of a specific type of stochasticity (deterministic chaos) in nonlinear systems (either models or real objects) offers a lot of exciting questions, going finally to the fundamentals of physics. These questions include: the necessary and sufficient requirements for a system to become chaotic, the association of seemingly irregular measured signals with deterministic chaos, the preparation of a system into a specific initial state and its systematic perturbation (initial state sensitivity, and predictability of long-time behavior).

The detailed analysis of signals being measured permits one to separate stochastic signals having their origin in conventional noise from those originating from nonlinear dynamics. For direct calculation of correlations, trajectory divergences (Liapunov exponents), invariant measures, Hausdorff dimension, etc. are important methods. For example, a very low dimensionality (fractal in most cases, e.g. 4.2) shows that the underlying nonlinear dynamics is determined by only a few degrees of freedom (e.g. 5) and thus the irregular behavior is of deterministic origin; for details see Schuster (1984), where also many experiments are presented.

Chaos in Biological Systems

Biological systems exhibit many periodic states (biological rhythmicity) (Kaiser, 1980). Furthermore, there is increasing evidence that active biological systems can be driven from periodic states (e.g. coherent oscillations) to chaotic ones by external stimuli. Examples are periodic and chaotic activity in a molluscan neurone (Holden et al., 1982), chaos in sinusoidally stimulated neurons (Hayashi et al., 1982), and chaotic activity in enzyme proteins (Mandell et al., 1982). These examples belong to the class of externally driven periodic systems where the bifurcations from periodic to deterministic chaotic states occur on different routes via quasi-periodicity, period-doubling sequences, intermittency, etc. In these systems, the occurrence of deterministic chaos seems conclusive. However, there are experiments which in almost all cases exhibit only irregular motions. In order to classify such chaotic states a detailed analysis must be performed. Rapp et al. (1985) have done a correlation analysis on the dynamics of neural activity in the simian motor cortex. It shows that the spontaneous activity may be chaotic in the deterministic sense, exhibited by a strange attractor of very low dimensionality. Evidence of chaotic dynamics of brain activity during certain sleep cycles is reported by Babloyanz et al. (1985). Again a correlation analysis yields a low-dimensional chaotic attractor. Kaiser, (1985a) reviews some modeling approaches describing rhythmic activity of the brain and some related experimental facts.

Biological Relevance of Chaos

Deterministic chaos is a completely irregular motion. However, we have seen that there exist fundamental differences between this type of stochasticity and the conventional ones (thermal noise, quantum fluctuations). Some characteristics of deterministic chaos are: oscillation-diagrams and phase space diagrams show specific structures within the irregularities (U-sequences, two- and multi-band chaos,. . .), the power spectra exhibit a strong scaling behavior of the strength of different-frequency components, and also a specific sequence of the occurrence and disappearance of single-frequency lines.

An external noise source leads to further bifurcations and to the disappearance of periodic windows, resulting in an increase of irregularity. However, there are examples where the addition of noise creates either a decrease of irregular motion or even transition to regular behavior, i.e., besides *noise-induced disorder* there exists the possibility of *noise-induced order.* This again is a manifestation that deterministic chaos is completely different from noise. Nonlinear dynamic systems incorporate both order and chaos and order in chaos (coherence in chaos).

From this point of view biological order and function possibly require some new interpretations in those cases where an irregular motion has a strong deterministic origin:

1. only a few relevant variables determine the behavior, i.e., the corresponding active state is decoupled from thermal motion,
2. spontaneous patterns are based on a specific coherence (e.g., phase coherence of prestimulus and poststimulus activity of EEG),
3. "noisy"-looking signals are the results of synchronized components in certain rhythmic activities,
4. the loss of periodicity (i.e., aperiodic motion) may be an active part during biological development,
5. irregular motions are essential for certain types of order and function (e.g., decoupling from noise influences),
6. irregularities are essential functional components (as in gene reproduction, information storage ..) and not pathological signs.

In addition, a series of experiments has been performed where either active biological systems have been exposed to very weak nonionizing radiation or Raman measurements have been done. A number of unusual results have been found, ranging from extreme sensitivities, sharp resonances and strong frequency dependences via frequency and intensity windows, and threshold and saturation to some highly excited lines in Raman spectra. Though these results seem to present strong evidence for the existence of both nonthermal effects in irradiated biosystems and coherent oscillations in active biosystems (Fröhlich, 1980; Fröhlich et al., 1983; Adey, 1981; Grundler et al., 1983; Pohl, 1983; Kaiser, 1983a, 1985b), the experiments are the subject of highly controversial discussions.

The controversy is mostly based on the evident lack of reproducibility of a number of experiments. We do not here go into the details of experimental techniques applied to biological systems. However, some general features deserve further discussion in connection with nonlinear dynamics:

1. Periodic activity within biological systems can be driven by a weak external stimulus into a series of bifurcations; the resulting state either can be decoupled from the stimulus or can be driven into a more sensitive state,

2. The coupling of coherently oscillating subunits can create new states that are sensitive to the external fields. A blocking of an energy input might occur as well, i.e., the system decouples from the stimulus,

3. The extreme sensitivity of chaotic deterministic systems to the initial state presumably requires a precise preparation of the biological object, but this can hardly be performed; thus the actual state of the irradiated system may be completely different from experiment to experiment, which may lead to very different results,

4. The lack of final state predictability can lead to arbitrary results for irradiated systems.

It should be emphasized that a possible deterministic chaotic behavior in biological systems cannot completely solve the question of controversial experimental results, but it can give some explanations and remove some of the discrepancies. Furthermore, Fröhlich's concept of coherent oscillations in biological systems gets extended and it provides additional insight on biological activity, e.g., by the equivalence of rhythmic processes and of irregular motions.

NONLINEAR DYNAMICS

The transition from regular (i.e., periodic) motion to an irregular (i.e., aperiodic, chaotic) one has been investigated with the help of a large number of theoretical models. In a series of papers we have analyzed the bifurcations of externally driven nonlinear oscillators, including entrainment and phase locking, stability analysis, and onset of travelling waves. These calculations include an externally driven nonlinear torsion pendulum (Kaiser, 1985b), externally driven Van der Pol and generalized Van der Pol oscillators (Kaiser, 1983, 1983b, 1984, 1985b), and a biophysical system to model coherent oscillations (Kaiser, 1977b, 1978, 1983b, 1984, 1985b). The first model belongs to the class of systems which for vanishing external stimulus possess only a static attractor. The other examples are self-sustained oscillators (periodic attractors), which means that these systems exhibit internally created and internally stabilized oscillations (limit cycles).

In the present paper we restrict discussion to the external energy supply and to internal energy consumption (dissipation). From an experimental and biological point of view the distribution of the externally supplied energy within the system offers additional information for the bifurcation dynamics of driven systems.

Nonlinear Torsion Pendulum

The equation of motion reads (Kaiser, 1985b):

$$\theta \ddot{\psi} + \Gamma \dot{\phi} + A \phi - B \sin \phi = F_1 \cos \omega t$$

$(\cdot = \frac{d}{dt}$; $\phi =$ amplitude of oscillation, $\theta =$ moment of inertia, $\Gamma =$ damping con-

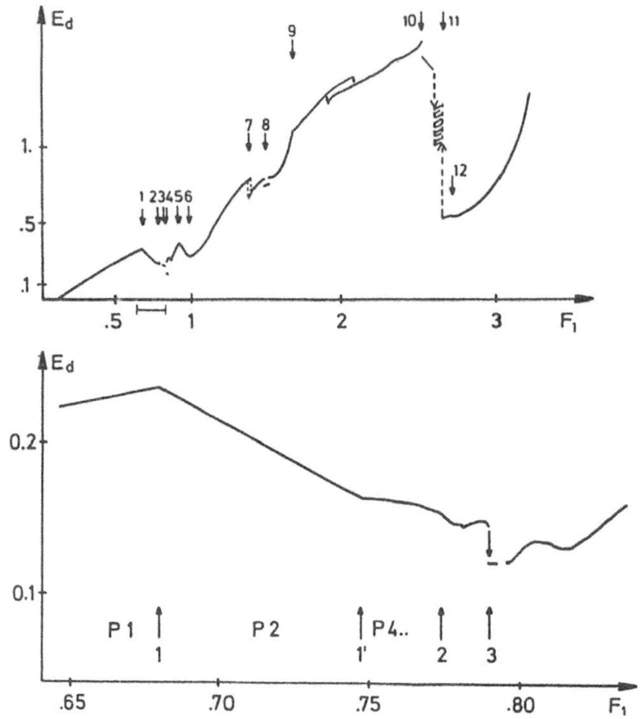

Fig. 1 — Dissipated energy E_d (= externally supplied energy E_D) as a function of the external periodic driver field F_1 for the nonlinear torsion pendulum. The lower picture is an enlargement of parts of the upper curve, marked on the F_1 axis.

The arrows point to different transitions:

1 $P1 \rightarrow P2$	5 $NP_3 \rightarrow P3a$	9 $NP_4 \rightarrow P2,..$
2 $P2'' \rightarrow NP_1$	6 $P3a \rightarrow P3s$	10 $NP_5 \rightarrow P2,..$
3 $NP_1 \rightarrow NP_2$	7 $P3s \rightarrow NP_4$	11 $Pn \rightarrow NP_6,\ P2,..$
4 periodic window	8 periodic window	12 $P2 \rightarrow P1$

The isolated dots belong to additional periodic windows. (Pn: periodic state with period $t = n\,\frac{2\pi}{\lambda}$; NP_i: different chaotic states; s = symmetric, a = asymmetric).

stant, A = coefficient of linear restoring force, B = coefficient of torque; $\theta, \Gamma, A, B > 0$). The right-hand side represents the periodic external drive. If all internal parameters and the driver frequency ω are fixed, the system runs through a series of bifurcations if the external field strength F_1 increases. In Fig. 1 the externally supplied energy per period t ($= 2\pi/\omega$)

$$E_D = \langle F_1 \cos \omega t \cdot \dot{\phi} \rangle_T$$

is given as a function of the driver field strength F_1 for the stationary state, i.e., where the externally supplied energy E_D equals the internally dissipated energy E_d. The diagram exhibits a rich variety of changes, each of which can be associated with a specific bifurcation. It turns out that one gets bifurcations from periodic to chaotic oscillations where the latter ones require less energy supply than the regular motions (e.g. transition 11 in Fig. 1). Furthermore, the transition from small (i.e. within 1 minimum) to large-amplitude oscillations is accompanied by a reduction of the energy supply and its dissipation (e.g. transition three in Fig. 1). Figure 2 shows the corresponding phase plane diagrams. It should be emphasized that for increasing F_1 the amplitudes increase, but for higher excitations they are driven back to small values. Already this simple model displays a multiplicity of effects,

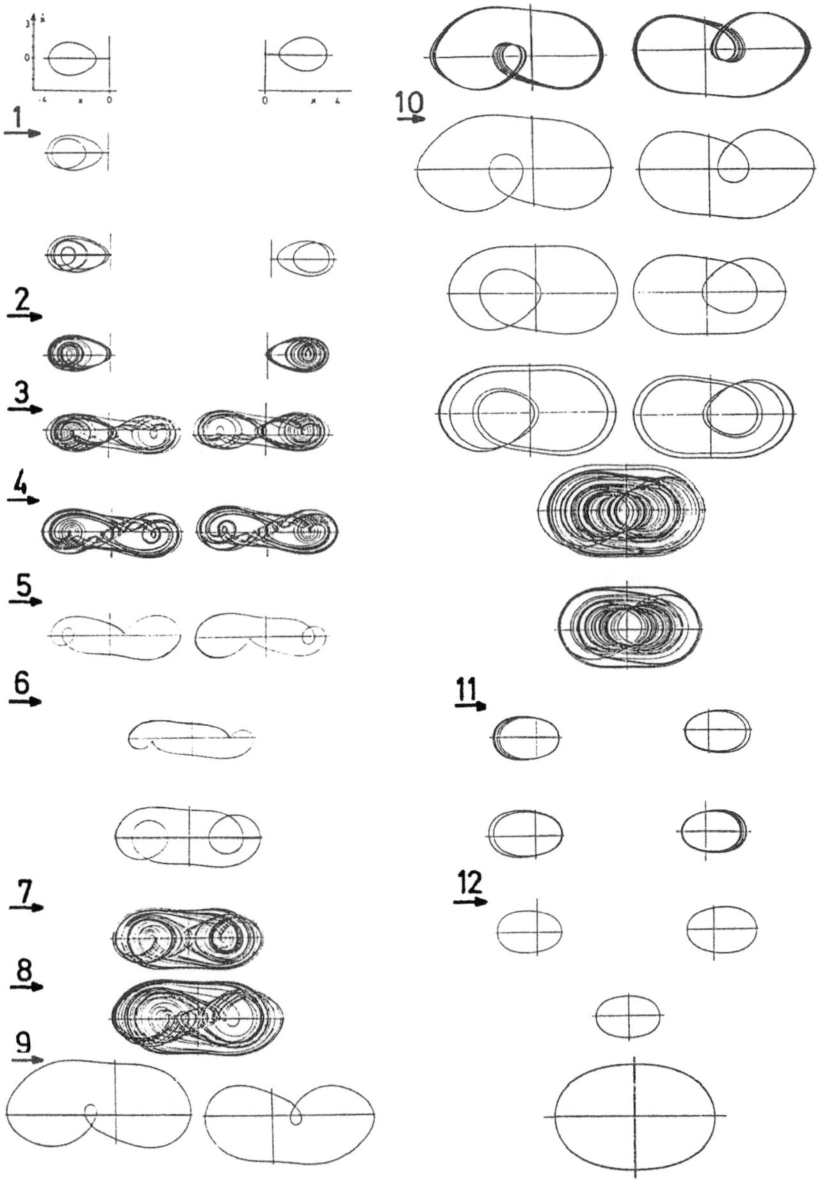

Fig. 2 — Phase plane diagram ($\dot{x}$ versus x) for a series of different states of Fig. 1. The arrows mark the transitions (bifurcations), which are already given in Fig. 1. F_1 increases from top to bottom in the left-hand side followed by the column on the right-hand side. The presence of two pictures in one column and in the same line indicates the existence of two different oscillations for the same parameter values, depending on the initial conditions (all pictures are drawn with the same scale).

which might be important for a better understanding of specific effects in irradiated systems. It is important to mention the existence of at least six different states of chaos.

Coherent Oscillation Model

The model has been suggested by Fröhlich, (1977). Its reduced version reads (Kaiser, 1977b, 1978):

$$\dot{x} = \gamma\, y + \alpha\, A\, x\, y + (c^2\, e^{-\Gamma^2 x^2} - d^2)\, x + F(t)$$

$$\dot{y} = -\beta\, x - \alpha\, A\, x\, y$$

$\gamma\, y$, $\beta\, x$ and $\alpha\, A\, x\, y$ result from a nonlinear chemical reaction with Lotka-Volterra dynamics; the third term on the right-hand side of the first equation describes collective gain processes and dielectric losses (modified Van der Pol dynamics). The external stimulus consists of a superposition of static and periodic fields, $F(t) = F_o + F_1 \cos \omega t$. The application of this combined field produces a great variety of oscillation pattern and bifurcation routes that are not found if only a periodic driver is operating (i.e., $F_o = 0$). In Fig. 3 the dissipated energy $E_d = E_D$ as a function of the driver field F_1 is given. Again there exist some bifurcations, where for increasing amplitudes the required energy supply is decreasing. A detailed analysis is required to investigate the jumps in the energy curve and to recognize their possible role for an explanation of experimental effects. It should

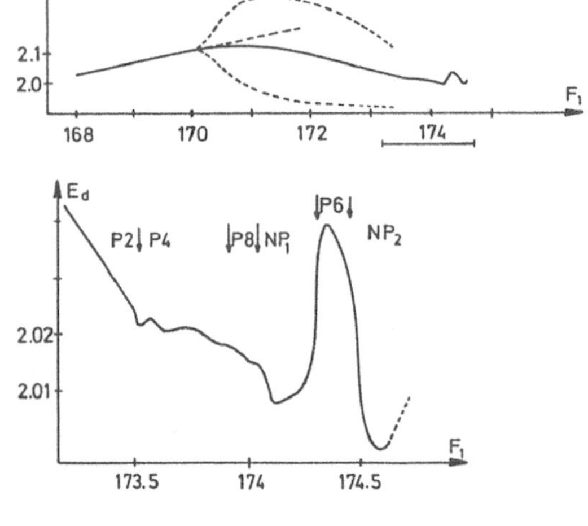

Fig. 3 — Dissipated energy E_d as a function of the external periodic driver field F_1 for the coherent oscillation model. The picture in the center is an enlargement of parts of the upper curve; the dotted straight line belongs to large $P1$ transients; the dotted curves are the separated $P2$-contributions (upper part; small amplitudes, lower part: large amplitudes).

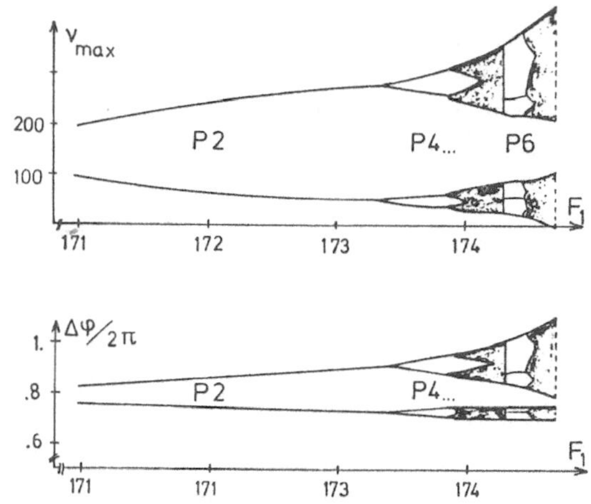

Fig. 4 — Bifurcation diagram for parts of Fig. 3. Maximum amplitude ν_{max} and phase shift $\Delta\phi$ as a function of the external driver field F_1. The shaded regions represent chaotic states.

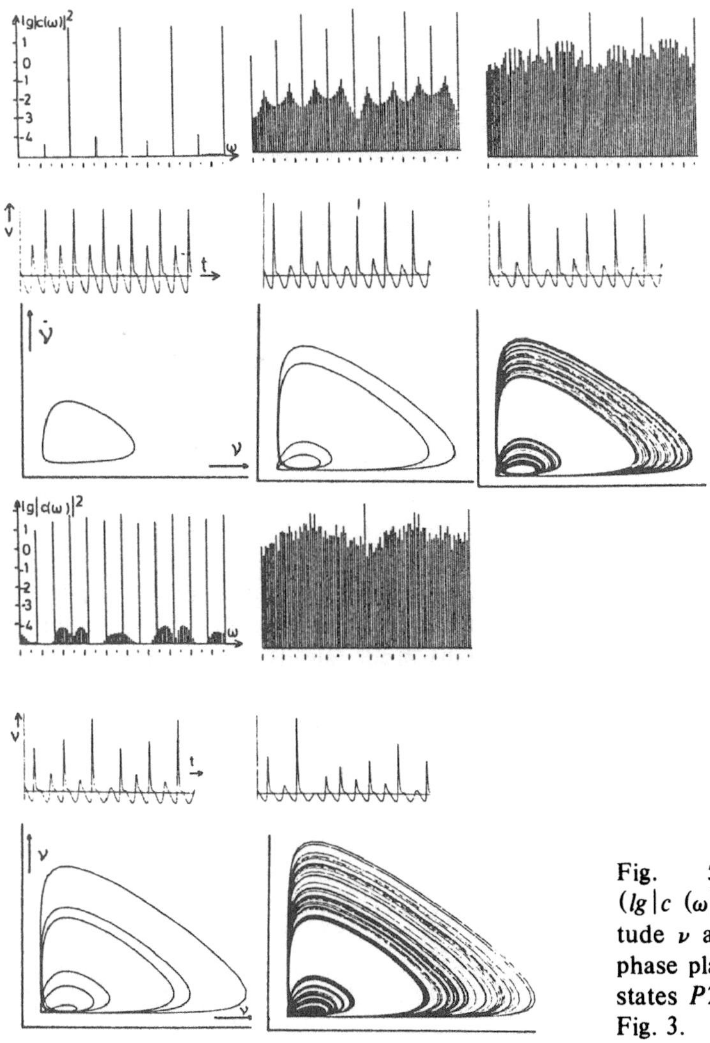

Fig. 5 — Power spectra $(lg|c(\omega)|^2 - \omega)$, oscillations (amplitude ν as a function of time t) and phase plane diagrams $(\dot{\nu} - \nu)$ for the states $P2$, $P8$, NP_1, $P6$ and NP_2 of Fig. 3.

be stressed that the transition from the first periodic state ($P1$) to the first period-doubled state ($P2$) is accompanied by a sharp decrease of E_d. At the transition point $P1 \rightarrow P2$ the system bifurcates into a state of a lower energy dissipation ($P2$), whereas $P1$ becomes unstable. The same behavior is given in Fig. 1. It has also been found for some other nonlinear oscillators. Presumably, a new type of universal behavior has been discovered. The amount of the externally supplied energy and its internal dissipation (per period) acts as a bifurcation parameter; perhaps a still unknown extremal principle for the dissipated energy in nonlinear dissipative systems exists. A future detailed analysis is required to clarify the situation.

In Fig. 4 the maximum amplitudes and the phase shifts for the bifurcating states are given. It is interesting to note that the larger the amplitude the smaller the phase shift. To give an impression of the different types of oscillations, power spectra, oscillation and phase plane diagrams for five different states are shown in Fig. 5.

CONCLUSIONS

The possible existence of irregular behavior on the basis of a deterministic chaotic motion necessitates a reinterpretation of those biological irradiation experiments in which either no effects have been detected or such effects could not be reproduced. Nonlinear dynamics displays an enormous variety of behavior for both experimental investigations and theoretical studies. Several different bifurcation routes that are found when only slight internal or external parameter changes are applied create qualitatively different final states. The loss of final state predictability in those cases where the system is extremely sensitive to the initial conditions makes it of doubtful value to run irradiation experiments with biological systems since their state is not known and can not be well defined. An extreme sensitivity during certain states of development may be followed by very low sensitivity. In addition, small changes of internal or external parameters can lead to drastic changes in the behavior of the system provided the system is near a bifurcation point.

The existence of nonthermal effects in irradiated systems and the evidence for coherent oscillations in active biosystems require further experiments and theoretical investigations, in particular with respect to the new results of nonlinear dynamics, e.g. deterministic chaos, loss of final state predictability, coherence in chaos, and macroscopic disorder.

REFERENCES

W.R. Adey, Physiol. Rev. **61**, 435 (1981).

A. Babloyanz, J.M. Salazar, and C. Nicolis, Phys. Lett. **111A**, 152 (1985).

P. Collett, and J.-P. Eckmann: *Iterated Maps on the Interval as Dynamical Systems*, Birkhäuser, Basel, 1980.

P. Cvitanovic: *Universality in Chaos* (a reprint selection), Adam Hilger Ltd., Bristol, 1984.

J.-P. Eckmann, Rev. Mod. Phys **53**, 643 (1981).

H. Fröhlich, Int. J. Quantum Chem. **2**, 641 (1968).

H. Fröhlich, in *Brain Interactions with Weak Electric and Magnetic Fields*, S.M. Bawin, and W.R. Adeys, Eds., Neurosci, Res. Progr. Bull. **15**, 67 (1977).

H. Fröhlich, Adv. Electron. and Electron Phys. **53**, 85-152 (1980).

H. Fröhlich, and F. Kremer, Eds., *Coherent Excitations in Biological Systems*, Springer, Berlin, 1983.

H. Fröhlich, in *Nonlinear Electrodynamics in Biological Systems*, W.R. Adey, and A.F. Lawrence, Eds., Plenum, New York, 1984 pp. 491-496.

G. Grundler, and F. Keilmann, in Fröhlich and Kremer (1983) pp. 22-37.

H. Haken, *Advanced Synergetics*, Springer, Berlin, 1983.

H. Hayashi, S. Ishizuka, M. Ohta, and K. Hirakawa, Phys. Lett. **88A**, 435 (1982).

A.V. Holden, W. Winlow, and P.G. Haydon, Biol. Cybern. **43**, 169 (1982).

F. Kaiser, Z. Naturforsch. **34a**, 134 (1977a).

F. Kaiser, Biolog. Cybernet. **27**, 155 (1977b).

F. Kaiser, Z. Naturforsch, **33a**, 294,418 (1978).

F. Kaiser, in *Nonlinear Electromagnetics*, P.L.E. Uslenghi Eds., Academic, New York, 1980 pp. 343-390.

F. Kaiser, in Fröhlich and Kremer (1983) pp. 128-133.

F. Kaiser, in *Biological Effects and Dosimetry of Nonionizing Radiation*, Plenum, Publ. Corp., New York, 1983b pp. 251-282.

F. Kaiser, in *Nonlinear Electrodynamics in Biological Systems*, W.R. Adey, and A.F. Lawrence, Eds., Plenum, New York, 1984 pp. 393-412.

F. Kaiser, in *The Living State - II*, R.K. Mishra, Ed., World Sci. Publ., Singapore, 1985a pp. 467-491.

F. Kaiser, in *Interactions Between Electromagnetic Fields and Cells*, A. Chiabrera, C. Nicolini, and H.P. Schwan, Eds., Plenum, New York, 1985b pp. 131-155.

E.N. Lorenz, J. Atmos. Sci. **20**, 130 (1963).

A.J. Mandell, P.V. Russo, and S. Knapp, in *Evolution of Order and Chaos in Physics, Chemistry and Biology*, H. Haken, Ed., Springer, Berlin, 1982 pp. 270-282.

E. Ott, Rev. Mod. Phys. **53**, 655 (1981).

H. Pohl, in: Fröhlich and Kremer (1983), pp. 199-210.

H. Poincaré, *Les Méthodes Nouvelles de la Mécanique Celeste*, Gauthier-Villars, Paris, 1892.

P.E. Rapp, A.M. Albano, I.D. Zimmerman, G.C. Deguzman, and N.N. Greenbaum, Phys. Lett. **110A**, 355 (1985).

H.G. Schuster, *Deterministic Chaos*, Physik-Verlag, Weinheim, 1984.

22. Non-Thermally Excited Modes and Free Energy Transduction in Proteins and Biological Membranes

D. B. Kell

ABSTRACT

The thermodynamic properties of aqueous globular proteins are considered, both when at equilibrium in a heat bath and during enzymatic catalysis. It is recognized that they exhibit thermally driven conformational fluctuations of great complexity and on time scales that may be much faster than the enzyme turnover time. Yet conformational states of proteins can store free energy under macroscopically isothermal conditions. This requires that they possess special, collective, non-thermally excited modes both during enzymatic catalysis and, in particular, during free energy conservation. Electron transport phosphorylation and related membrane processes provide well-defined examples of this latter behavior.

INTRODUCTION

It is indeed a singular honor and pleasure to be able to contribute to this Festschrift celebrating, on the occasion of his 80th birthday, the many achievements of Professor H. Fröhlich in physics and biophysics. For a young biochemist such as myself to have had the opportunity to enjoy intellectual contact with a physicist of genuine stature is inevitably an educative, revealing, and rewarding experience; that it has also been an outstandingly pleasurable highlight of my scientific career is due to the warm welcome and courteous, generous, and patient help I have always received from Professor Fröhlich during my studies of the salient features of his general theory of collective behavior in biological systems.

A fresh approach to a field of scientific enquiry by a brilliant outsider can often produce results of a special character, qualitatively different from those of Kuhn's (1970) "normal science". With such thoughts in mind, the Académie Francaise convened in 1967 a meeting, in the elegant surroundings of the palace of Versailles, at which they invited several of the outstanding theoretical physicists of the time to consider, from a physical perspective, the nature and problems of living systems and cellular organization. Whilst Schrödinger's (1944) classic essay in this direction had indicated that the apparent violation of the Second Law by living systems was accounted for by the realization that they are thermodynamically open systems, and the Brussels school (e.g. Nicolis and Prigogine, 1977) developed the idea of dissipative structures within the framework of a general treatment of the thermodynamics of irreversible processes, these treatments were essentially macroscopic in character. By contrast, Fröhlich's (1968, 1969) assessment at this meeting was of a more molecular or quantum mechanical character, and sprang from his knowledge of the unusual dielectric behavior of biological membranes and his earlier experience with nonlinear systems in general and (I assume) with superconductivity in particular.

The general idea, then as now, is that certain kinds of relatively well-defined physical systems can, by the *collective, nonlinear* behavior of ensembles of the particles of which they are constituted, display or develop a macroscopic order that would not result from, nor be predicted by, the linear superposition of their individual coordinates, velocities, and interatomic potentials. What Fröhlich (1968, 1969) suggested in particular, by extending certain general equations of microscopic physics, was that nonlinear interactions between the electric and acoustic modes of certain biological macromolecules could lead to the generation and stabilization of long-lived, polar modes, via a mechanism analogous to the Einstein condensation of a gas of bosons. Because my background is in bioenergetics, I shall concentrate upon what this means for *proteins*, especially those capable of free energy conservation or transduction. However, since proteins are highly sophisticated and complex arrays of atoms, we must inquire first whether they can or do exhibit such nonlinear behavior and stable, out-of-equilibrium states, and second, if they do, as to the types of mechanism by which this occurs. More explicitly, we must consider that the modes stabilized within biological systems by these nonlinear phase transitions must be highly resistant to the impact of the disordering tendencies of their internal thermal vibrations and those elicited by collisions with solvent and other molecules, allowing the system to store energy for times which are long compared to the normal cycle times of free-energy-transducing enzymes.

What I wish to do in the following, rather personal overview, therefore, is to summarize, within the framework of the above ideas, some of the key problems in bioenergetics as seen from a more biological or biochemical standpoint, and to indicate why I think that the generalized theory of Fröhlich concerning collective or coherent excitations in biological systems does provide a new, exciting, and persuasive formalism for the evolution of this subject. It is not my intention to review in detail either the theory and its mathematical foundation (Fröhlich, 1980) or some of the pertinent experimental evidence (Webb, 1980; Pohl, 1981; Fröhlich and Kremer, 1983; Del Giudice et al., 1984). Rather this paper is to illustrate, in very summary and elementary terms, some modern ideas concerning the structure, dynamics, and organization of isolated proteins, to extend this to situations in which proteins are catalyzing *favorable* chemical reactions, and finally to distinguish and discuss in a mechanistic fashion the very special biophysical features of some proteinaceous devices that are considered to take part in the storage and conservation of free energy in biology.

THE STRUCTURE OF ISOLATED GLOBULAR PROTEINS

The average three-dimensional ("tertiary") structure of a native, isolated, globular protein, both in solution and in the crystalline state, is determined purely by the primary sequence of amino acids which it contains. But a protein of molecular weight 20 kD can in principle possess or explore some 10^{80} conformational states (e.g., Jaenicke, 1984). However, since the Universe is "only" some 10^{17} s old (Barrow and Silk, 1983), even if we allowed our protein to assume these states at a rate of 10^{15} per second, it could not conceivably explore all of those available in passing from the unfolded to the folded state. Thus, given experimental realities, in the strictest terms, even an isolated protein in thermal equilibrium in a heat bath of solvent molecules cannot truly be considered as an ergodic system.

To allow progress, therefore, we consider macroscopic, "average" conformational states and substates, whose free energy difference $\leqslant kT$, a procedure that allows us to speak of "a" structure or conformation. Indeed, one may read in any biochemistry textbook that the energetically favorable transition from a "random coil" (unfolded state) to globular protein (folded state) may be accounted for by the formation of covalent, hydrophobic, dipolar, electrostatic, and van der Waals bonds between the amino acid units of the protein, and the changes in the angles and lengths of such bonds caused by amounts of energy equal to kT are both relatively small and are observable in the electron density map obtained by X-ray diffraction measurements (e.g., Ringe and Petsko, 1985). This in itself allows us to distinguish (with a "noise" amounting to kT) the conformation and free energy of a protein under a given set of conditions (of pH, temperature, ionic strength, etc) from those of the other molecules (solvents, ligands, etc) present, and of course, in this sense, the ("static") structures of many proteins are known to atomic resolution. What happens when we ask our protein to do something, for instance to catalyze a thermodynamically favorable reaction such as $A \rightarrow B$? To answer this, we must enquire not only into the "structure" but the *dynamics* of such proteins.

DYNAMIC BEHAVIOR OF AQUEOUS, GLOBULAR ENZYMES

One may discern two general types of approach that have allowed us to make progress in devising a mechanistic basis for the astonishing rate enhancements and specificity of which enzymes are capable. The more (physico-)*chemical* treatment (Jencks, 1980; Fersht, 1985) has shown how enzymes can, by virtue of their conformational flexibility, exploit the favorable free energy of substrate binding so as to stabilize the transition state, and has concentrated, as a result, on the interaction between the substrate and the protein side-chains to which it is bound at the "active site". A perhaps more genuinely dynamical and more purely (bio-)*physical* approach (reviews: Lumry and Biltonen, 1969; Careri et al., 1979; Lumry, 1980; Welch et al., 1982; Somogyi et al., 1984; and Welch, 1986) has stressed the view that the time-dependent, thermally induced conformational mobility *of the entire protein molecule* contributes significantly to the rate enhancements of chemical reactions catalyzed by enzymes, a view at least consistent with the fact that attempts quantitatively to imitate enzymatic activity with synthetic ligands of low molecular weight that mimic the active site alone have so far failed (Sinnott, 1979; Kell, 1982; Mutter, 1985). Obviously both treatments are important, but for the present purposes of developing a set of arguments, I shall lay stress mainly on the latter.

If we take this approach it might appear, from an equilibrium thermodynamic standpoint, that the following paradox arises: our enzyme is acting *isothermally* in a heat bath, yet we seem to be saying that it can not only exchange energy with the solvent but can in some way "use" this energy during its enzymatic cycle, violating the Second Law. The solution to this paradox (see e.g. Kemeny, 1974; Lumry, 1980; Cooper, 1984; Somogyi et al., 1984) is that such thermodynamic reasoning requires only that the *overall* ΔG for the enzyme-solvent interaction = 0 under these conditions, whilst the enthalpy and entropy may fluctuate considerably (owing to the molecular size of the protein molecule), *provided that such fluctuations ae compensatory*. As phrased by Kemeny (1974), therefore: "this does not mean that heat energy is turned into free energy but rather that the transduction of internal energy and heat exchange with the reservoir are part of the same mechanism." In other words, the enzyme may "borrow" heat energy from its surroundings but must "return" it after use, a phenomenon which abundant experimental evidence has shown is manifest in a huge variety of fluctuational modes about its "equilibrium" conformation (e.g., Gurd and Rothgeb, 1979; Englander and Kallenbach, 1984; McCammon, 1984). Indeed, it is worth mentioning that these fluctuations are probably crucial to the antigenicity of proteins and small polypeptide, protein-mimetic vaccines (Berzofsky, 1985; Williams and Moore, 1985), a topic of much current interest.

In devising treatments and mechanisms for this type of behavior, then, we must consider the *time scale* of protein fluctuations, for it is the time scale that tells us *how long* the protein will take to equalize any "excess" thermal energy between a number of degrees of freedom sufficient for it to be indistinguishable from kT. What we find especially (Careri et al., 1979; Welch et al., 1982) is that the exchange of energy at protein surfaces occurs (by collisions with solvent and buffer molecules, and in particular by proton-exchange equilibration) with characteristic frequencies of some 10^8 Hz. In contrast, the turnover time for a typical enzyme is say 10^{-3} s. Thus the view has evolved that whilst an enzyme is indeed a highly stochastically fluctuating molecule and, for a given free energy, possesses a *wealth* of conformational substates (at ambient temperatures), it is an "equilibrium chemodynamical machine" which, at every metastable, intermediate stage in its enzymatic cycle is, for practical purposes, in thermodynamic equilibrium with its surroundings. (This is not to say that it does not exhibit internal collective modes during transitions *between* macrostates in its reaction cycle (Ansari et al., 1985)). This is, at least to a first approximation, a reasonably satisfying picture for enzymes catalyzing the approach to equilibrium of a thermodynamically favorable chemical reaction. Our satisfaction is less than complete, however, when we come to consider enzymes that catalyze *free energy transduction*.

HOW CAN A PROTEIN MOLECULE STORE FREE ENERGY OVER LONG PERIODS UNDER ISOTHERMAL CONDITIONS?

We have seen that a typical enzyme, buffeted by its solvent environment, can only remain slightly out of equilibrium with its surroundings for a time scale that is short relative to its turnover time. In contrast, some enzymes can store free energy of a magnitude that is large relative to kT for a period that may be very long not only on these time scales but even relative to its "normal" turnover time. If we consider the synthesis of ATP from ADP and inorganic phosphate ($\Delta G^\circ = +31$ kJ/mol) "catalyzed" by an ATP synthase (normal turnover time ca. 10

ms), we find that biological systems containing an ATP synthase can store free energy (for instance, in photosynthesis, following absorption of a photon) as a "primary macroerg" (Blumenfeld, 1983) for periods of *seconds* before using it to make ATP. If such free energies are stored in protein conformational states (or in other states in equilibrium therewith), one must perforce seek a description of how this is achieved in the face of the same thermalizing environment that we have already admitted above would cause relaxation of the protein conformation to other conformations with a free energy equal to that in its (appropriately liganded) "ground" state $\pm\ kT$. This, in fundamental terms, is the key problem of principle in molecular bioenergetics (Somogyi et al., 1984; Welch and Kell, 1986). Obviously some mechanisms must be operative to restrict, in a kinetic sense, the decay of these conformational states to equilibrium, or, in Fröhlich's (1969) phrase, very strongly to excite a few modes of motion.

One way of restricting the number of kinetically available states is by putting the protein in an electrically (protonically) insulating membrane and maintaining an electrical potential or pH gradient across that membrane, a principle that forms the basis of Mitchell's chemiosmotic theory (Mitchell, 1966; Nicholls, 1982). Whilst there are nowadays many reasons to doubt that ATP synthesis in vivo occurs strictly according to this principle (Kell and Hitchens, 1983; Westerhoff et al., 1984; Ferguson, 1985; Kell and Westerhoff, 1985), Mitchell's ideas, and the experiments that they have stimulated, leave little room for doubt that electrogenic proton translocation reactions lie at the heart of ATP synthesis "catalyzed" by a membrane-located ATP synthase whose free energy is generally supplied by the exergonic reactions of electron transport. Bioenergeticists therefore now routinely refer to such devices as *proton pumps.* We are thus led to consider the possibility of long-lived, metastable, *out-of-equilibrium* states that may be attained when protons interact with proteins in general, and with the membranous systems of electron transport phosphorylation in particular.

FREE ENERGY STORAGE AND TRANSDUCTION IN PROTEINS AND BIOMEMBRANES BY SOLITONS AND OTHER OUT-OF-EQUILIBRIUM STATES

What we have learned from the above, and has in particular been brought home to me by my contact with Fröhlich and his writings, is that certain collective motions of ensembles of atoms can lead to the formation of a degree of macroscopic order under conditions in which normal frictional forces would have damped out any such ordering. Thus, in contrast to the impression one might gain simply by a general study of the more mainstream bioenergetic literature, it is at least plausible, and in my view the available evidence dictates, that proton pumps embedded in biological membranes also indulge in such collective motions as an (if not *the*) essential part of their biological activity. If this is so, one must be able to devise experiments which, from a bioenergetic standpoint, are both entirely novel and may be expected greatly to improve our understanding of electron transport phosphorylation.

For a variety of reasons, most authors who have considered the general problem of "high energy" states of proteins have suggested that solitary excitations (solitons) might possess the type of property required to account for the behavior of free-energy-conserving proteins. A soliton is a special kind of wave packet (phonon) which, because of the type of wave equation that it obeys (there are several different types), can carry energy over "long" distances in an essentially *dispersionless* fashion, without losing its energy by thermal exchange. It thus maintains its energy at a level greater than that expected on the basis of a Boltzmann distribution of vibrational energies at the ambient temperature. Solitons are thus characteristic of the non-thermally excited modes referred to in the title of this article, and are one of the internal modes of proteins generally for which Somogyi et al. have coined the general term *zymons*. It is not for me here to seek to distinguish the various proposals in detail (see e.g. Bilz et al., 1981; Blumenfeld, 1983; Davydov, 1983; Jardetzky and King, 1983; Ribeiro et al., 1983; Scott, 1983; Yomosa, 1983; Chou, 1984; Lomdahl, 1984; Careri and Wyman, 1984; Carter, 1984; Del Guidice et al., 1984; Lomdahl, et al., 1984; Somogyi et al., 1984 for an entrée to the literature), but what does come out of this work is that the existence of collective excitations in proteins is firmly established, both in well-defined theories and by experimental observation (mainly spectroscopic). What is not known with any degree of certainty at all is *exactly* which of the available models (if any) is actually used in vivo. Nor have I seen a treatment to date that is aimed specifically at considering the properties of solitons in membranes and proteins within the framework of Fröhlich's more general theory of coherent excitations. And neither has the exact role of the proton been established in this context. Most so-called localized coupling theories propose that protons move in concert with specialized modes of proteins (and perhaps lipids), and it is obvious that the *obligatory* vectorial movement of a proton as part of a solitary excitation will greatly affect the "normal" behavior of both the proton *and* the soliton (Kell and Westerhoff, 1985). In this context, it is extremely interesting that a recent dielectric study by Careri and colleagues (1985) has shown a highly cooperative channeling of protons (with a 7th-order dependence on the number of bound protons) from all over the surface of lysozyme towards the enzyme's active site. Such measurements show directly that there are indeed highly nonlinear interactions between the electric and acoustic modes of biological materials, as predicted by Fröhlich.

We ourselves have initiated a study of the dielectric properties of microbial membranes in the frequency range from ca 10 Hz-10 MHz. In the absence of protonmotive activity, the chief novel finding to date is that one may learn more about the movement of lipids and proteins than of protons (Kell, 1983; Harris and Kell, 1985; Kell and Harris, 1985). When is considered, the effect of membrane energization (i.e., proton pumping activity) on the dielectric behavior is considered, the search for possible resonant modes is bedevilled by the small number of *net* charges that are actually pumped (Kell and Hitchens, 1982; Hitchens and Kell, 1984) relative to the background conductivity in aqueous suspensions (Harris et al., 1984). Experiments at lower levels of hydration might ameliorate matters.

Since in the space available it is impossible adequately even to summarize the present status of the controversial subject of membranous proton-coupled free-energy-transducing systems, which has anyway been done elsewhere at some length (Kell and Hitchens, 1983; Westerhoff et al., 1984; Ferguson, 1985; Kell and Westerhoff, 1985), I wish finally to state the following point of view. This is that the enormous wealth of detailed (if often paradoxical) knowledge concerning protonmotive systems does indicate very strongly, when taken as a whole, that if collective excitations *sensu* Fröhlich are the norm in living systems, it is to electron transport phosphorylation that one might fruitfully first turn to describe them in a well-defined and clearly free-energy-conserving system. Whether such behavior is actually to be expected for *fundamental* reasons, as suggested in particular by McClare (1971), is at least plausible (Welch and Kell, 1986).

CONCLUDING REMARKS

There is now a growing realization that functional order in noncrystalline systems does not require a strict spatial order; more subtle types of behavior are possible, not only in inorganic systems such as lasers, superconductors, and superfluids but also in proteins and biological membranes. Whilst progress in such an interdisciplinary area is not without its difficulties, the acceptance that many types of nonlinear system obey fundamentally similar equations (Haken, 1977) can only hasten the day when biologists and physicists themselves indulge more fully and more regularly in collective behavior. That many are doing so already is due in no small degree to the ideas and efforts of Herbert Fröhlich.

ACKNOWLEDGEMENTS

I am delighted to be able to acknowledge my debt to H. Fröhlich for many rewarding discussions concerning the above ideas. Rick Welch has also contributed much to my understanding of these problems, and I thank him warmly. My work is supported by the Science and Engineering Research Council, U.K.

REFERENCES

A. Ansari, J. Berendzen, S.F. Bowne, H. Frauenfelder, I.E.T. Iben, T.B. Sauke, E. Shyamsunder, and R.D. Young, Proc. Natl. Acad. Sci. U.S.A. **82**, 5000-5004 (1985).

J.D. Barrow and J. Silk, *The Left Hand of Creation, The Origin and Evolution of the Expanding Universe*, Unwin, London, 1983.

J.A. Berzofsky, Science **229**, 932-940 (1985).

H. Bilz, H. Büttner, and H. Fröhlich, Z. Naturforsch. **36b**, 208-212 (1981).

L.A. Blumenfeld, *Physics of Bioenergetic Processes*, Springer, Heidelberg, 1983.

G. Careri, P. Fasella, and E. Gratton, Annu. Rev. Biophys. Bioeng. **8**, 69-97 (1979).

G. Careri, M. Geraci, A. Giansanti, and J.A. Rupley, Proc. Natl. Acad. Sci. U.S.A. **82**, 5342-5346 (1985).

G. Careri and J. Wyman, Proc. Natl. Acad. Sci. U.S.A. **81**, 4386-4388 (1984).

F.L. Carter, Physica **10D**, 175-194 (1984).

K.-C. Chou, Biophys. Chem. **20**, 61-71 (1984).

A. Cooper, Prog. Biophys. Mol. Biol. **44**, 181-214 (1984).

A.S. Davydov, "Energy transfer along alpha-helical proteins," in *Structure and Dynamics: Nucleic Acids and Proteins*, E. Clementi and R.H. Sarma, Eds., Adenine Press, New York, 1983, pp 377-387.

E. Del Giudice, S. Doglia, M. Milani, and M.P. Fontana, Cell. Biophys. **6**, 117-129 (1984).

S.W. Englander and N.R. Kallenback, Q. Rev. Biophys. **16**, 521-655.

S.J. Ferguson, Biochim. Biophys. Acta **811**, 47-95 (1985).

A.R. Fersht, *Enzyme Structure and Mechanism*, 2nd Ed., Freeman, San Francisco, 1985.

H. Fröhlich, Int. J. Quantum Chem. **2**, 641-649 (1968).

H. Fröhlich, "Quantum mechanical concepts in biology," in *Proc. 1st Int. Conf. on Theoretical Physics and Biology*, M. Marois, Ed., North Holland, Amsterdam, 1969, pp 13-22.

H. Fröhlich, Adv. Electron. Electron Phys. **53**, 85-152 (1980).

H. Fröhlich and F. Kremer, Eds., *Coherent Excitations in Biological Systems*, Springer, Heidelberg, 1983.

F.R.N. Gurd and T.M. Rothgeb, Adv. Protein Chem. 33, 73-165 (1979).

H. Haken, *Synergetics*, Springer, Heidelberg, 1977.

C.M. Harris, G.D. Hitchens, and D.B. Kell, "Dielectric spectroscopy of microbial membrane systems," in *Charge and Field Effects in Biosystems*, M.J. Allen and P.N.R. Usherwood, Eds., Abacus, Tunbridge Wells, 1984, pp 179-185.

C.M. Harris and D.B. Kell, Eur. Biophys. J. **13**, 11-24 (1985).

G.D. Hitchens and D.B. Kell, Biochim. Biophys. Acta **766**, 222-232 (1984).

R. Jaenicke, Angew. Chem. Int. Ed. Engl. **23**, 395-413 (1984).

O. Jardetzky and R. King, Ciba Found. Symp. **93**, 295-309 (1983).

W.P. Jencks, "What everyone wanted to know about tight binding and enzyme catalysis, but never thought of asking," in *Chemical Recognition in Biology*, F. Chapeville and A.L. Haenni, Eds., Springer, Heidelberg, 1980, pp 1-25.

D.B. Kell, Trends Biochem. Sci. **7**, 349 (1982).

D.B. Kell, Bioelectrochem. Bioenerg. **11**, 405-415 (1983).

D.B. Kell and C.M. Harris, Eur. Biophys. J. **12**, 181-197 (1985).

D.B. Kell and G.D. Hitchens, Faraday Discuss. Chem. Soc. **74**, 377-388 (1982).

D.B. Kell and G.D. Hitchens, "Coherent properties of the membranous systems of electron transport phosphorylation," in *Coherent Excitation in Biological Systems*, H. Fröhlich and F. Kremer, Eds., Springer, Heidelberg, 1983, pp 178-198.

D.B. Kell and H.V. Westerhoff, (1985), "Catalytic facilitation and membrane bioenergetics," in *Organized Multienzyme Systems: Catalytic Properties*, G.R. Welch, Ed., Academic Press, New York, 1985, pp. 63-139.

G. Kemeny, Proc. Natl. Acad. Sci. U.S.A. **71**, 2655-2657 (1974).

T.S. Kuhn, *The Structure of Scientific Revolutions*, Chicago University Press, Chicago, 1970.

P.S. Lomdahl, "Nonlinear dynamics of globular protein," in *Nonlinear Electrodynamics in Biological Systems*, W.R. Adey and A.F. Larence, Eds., Plenum, New York, 1984, pp 143-154.

P.S. Lomdahl, S.P. Layne, and I.J. Bigio, Los Alamos Sci. **10**, 2-31 (1984).

R. Lumry, (1980), "Dynamical aspects of small- molecule protein interactions," in *Bioenergetics and Thermodynamics: Model Systems*, A. Braibanti, Ed., Reidel, Dordrecht, 1980, pp 435-452.

R. Lumry and R. Biltonen, (1969), "Thermodynamic and kinetic aspects of protein conformation in relation to physiological function," in *Structure and Stability of Biological Macromolecules*, S.N. Timasheff and G.D. Fasman, Eds., Dekker, New York, 1969, pp 65-212.

J.A. McCammon, Rep. Progr. Phys. **47**, 1-46 (1984).

C.W.F. McClare, J. Theor. Biol. **30**, 1-34, (1971).

P. Mitchell, Biol. Rev. **41**, 445-502, (1966).

M. Mutter, Angew. Chem. Int. Ed. Engl. **24**, 639-653 (1985).

G. Nicolis and I. Prigogine, *Self-organization in Non-equilibrium Systems,* Wiley, Chichester, 1977.

D.G. Nicholls, *Bioenergetics. An Introduction to the Chemiosmotic Theory*, Academic Press, London, 1982.

H.A. Pohl, Coll. Phenom. **3**, 221-244 (1981).

A.A. Ribeiro, R. King, and O. Jardetzky, (1983), "Rotational solitons in folded polypeptide chains," in *Conformation in Biology*, R. Srinivasan and R.H. Sarma, Eds., Adenine Press, New York, 1983, pp 39-47.

D. Ringe and G.A. Petsko, Progr. Biophys. Mol. Biol. **45**, 197-235 (1985).

E. Schrödinger, *What is Life?*, Cambridge University Press, Cambridge, 1944.

A.C. Scott, "Solitons on the alpha-helix protein," in *Structure and Dynamics: Nucleic Acids and Proteins*, E. Clementi and R.H. Sarma, Eds., Adenine Press, New York, 1983, pp 389-404.

M.L. Sinnott, Chem. Brit. **17**, 293-297 (1979).

B. Somogyi, G.R. Welch, and S. Damjanovich, Biochim. Biophys. Acta **768**, 81-112 (1984).

S.J. Webb, Phys. Rep. **60**, 201-224 (1980).

G.R. Welch, Ed., *The Fluctuating Enzyme*, Wiley, New York, 1986.

G.R. Welch and D.B. Kell, "Not just catalysts. The bioenergetics of molecular machines," in *The Fluctuating Enzyme*, G.R. Welch, Ed., Wiley, New York, 1986.

G.R. Welch, B. Somogyi, and S. Damjanovich, Progr. Biophys. Mol. Biol. **39**, 109-146 (1982).

H.V. Westeroff, B.A. Melandri, G. Venturoli, G.F. Azzone, and D.B. Kell, Biochim. Biophys. Acta **768**, 257-292 (1984).

R.J.P. Williams and G.R. Moore, Trends Biochem. Sci. **10**, 96-97 (1985).

S. Yomosa, J. Phys. Soc. Jpn. **52**, 1866-1873 (1983).

23. Herbert Fröhlich, and the New Biophysics of Cooperativity

Herbert A. Pohl*

If one word is to be used for a description of the research created by Professor H. Fröhlich, that word is *cooperativity*. The phenomena of cooperative and collective phenomena have been much advanced by his insights, and recently, especially in biology. In commemoration of his 80th birthday on December 9, 1985 it is appropriate for us to honor this most productive scientist and friend.

In 1968, Herbert Fröhlich, already famous for his theoretical insights into solid state physics in such fields as superconductivity, semiconductor transport properties and instabilities, dielectric breakdown, and a host of others; turned his attention to cooperative phenomena in biology. In series of papers he opened up a new approach to biophysics. It has created a wealth of insights into how cells behave, especially in response to electromagnetic fields.

Said briefly, Professor Fröhlich pointed out that many of the various oscillators which comprise the living state will obey Bose-Einstein rules of cooperativity. This results in there being possible the existence of "boson-like condensations" leading to cooperativities appearing as the energy throughput reaches critical levels. The situation is rather like that of the operation of a laser. At a certain minimum power input, the laser begins to lase. Analogously, at a certain minimum energy throughput, as Professor Fröhlich showed, the cellular vibrators can "condense" to a cooperative state of interactions. The result is the appearance of long range ordering and sensitive interactions, observed, but unexpected by conventional classical approaches.

Some of these theoretical predictions have already borne fruit, and more is likely. For example, the astoundingly long range (up to 4 microns) interactions between live erythrocytes observed by Rowlands et al. can presently only be understood in terms of Froehlich interactions. Moreover the cooperative electromagneitc phenomena predicted by Fröhlich to occur in the experimentally difficult realm of about 100 GHz have been seen by Russian, German and American scientists, although this evidence is still developing. Terms such as "limit cycles," oscillating chemical reactions, cooperative degeneracies, Davydov solitons, "polar mode softening" and the like are beginning to invade the world of the biologists and biophysicists.

In broad brush strokes, he has pointed the way. A whole new science of vital oscillatory phenomena is evolving, due in large part to the stimulus and insights of Herbert Fröhlich.

*Deceased June 23, 1986.

24. Opposite Long-range Interactions Between Normal and Malignant Cells

W. Nagl and F. A. Popp

With 3 Figures

INTRODUCTION

Discussion on the origin and development of cancer has rapidly increased since the detection of the existence of oncogenes (for review see Kurth, 1983; Ochoa, 1985). Although carcinogenesis is assumed to be a multistep process, molecular biology has made it clear that oncogenes can contribute greatly to the development of a tumor. Normal cells of very different animals, such as worms, flies, other insects, birds, mammals, and humans contain similar oncogenes ("c-genes"). This indicates a high conservation of this set of genes, and an important function for them in growth and development, particularly in embryogenesis. Some retroviruses have appropriated oncogenes by genetic recombination from normal cells, so that retrovirus infection, introducing additional viral (and modified) oncogenes ("v-genes") into normal cells, as well as carcinogen-mediated activation of cellular oncogenes, may both lead to increased synthesis of oncogene-encoded informational proteins that convert normal cells to cancer cells. Some of these proteins are tyrosine kinases that catalyze the phosphorylation of the tyrosine residues of proteins. They find a location within the cell membrane and may alter the surface properties of the cell; others, evidently non-enzymatic in nature, are found in the cell nucleus and may influence gene activities in general. Further possible activities of oncogene are their mutation, their transposition, or their amplification (= extra replication, i.e. increase in number). The latter phenomenon is now commonly accepted as an important factor in drug resistance (e.g., Schimke, 1984).

In spite of all these new findings, the question of why, when, and where oncogenes become malignant remains unanswered. We are still missing some essential factors for solving both the problem of cancer and the problem of understanding normal differentiation and the living state in general (Rowlands, 1983). Biophysical approaches may lead to a deeper understanding.

In order to find a bridge between biochemistry and biophysical aspects, we introduced some years ago an electromagnetic model of cell differentiation (Nagl and Popp, 1983; Popp and Nagl, 1983). It is based on long-range interactions in biological systems as they have been postulated, first by Fröhlich (1968). Recently, our model found experimental support by Chwirot et al. (1985a, b, 1986), who examined sucessfully some basic predictions of the theory by using a new technique in preparing completely synchronized cell cultures (Chwirot, 1980). Since our model suggests opposite density dependence for cellular interactions within normal and malignant cell populations, it is applicable also to some questions of cancer research. In this paper we summarize the main aspects and present new experimental results confirming the validity of the basic approach.

EXPERIMENTAL BACKGROUND AND THE HYPOTHESIS

Our considerations originate from the phenomenon of a very weak photon emission of all active tissues that is called now "low-level luminescence". Recently, some excellent reviews on this field have been presented (e.g., Slawinska and Slawinski, 1983, 1985) so that a detailed description of the method and all the known results can be omitted here. Figure 1 displays an example of this quasi-stationary photocurrent from living plant tissue, whose intensity amounts to a few up to some thousand photons/ (s $\cdot$ cm^2), distributed over a spectral range from at least infrared (800 nm) to ultraviolet (about 200 nm).

Some elements of our approach are based on the following observations:

(1) There is evidence of a strong correlation between photon intensity and DNA conformations, as has been demonstrated by using ethidium bromide (Rattemeyer et al., 1981) or DNase (Chwirot et al., 1986) as tracer substances.

(2) The theory of unstable quantum systems predicts an exponential decay when under ergodic conditions a chaotic rescattering between source and decay products (e.g., photons) takes place (e.g., Fonda et al., 1978). All deviations from this law point to nonchaotic rescattering processes (Fain, 1981; Ngai et al., 1983) which

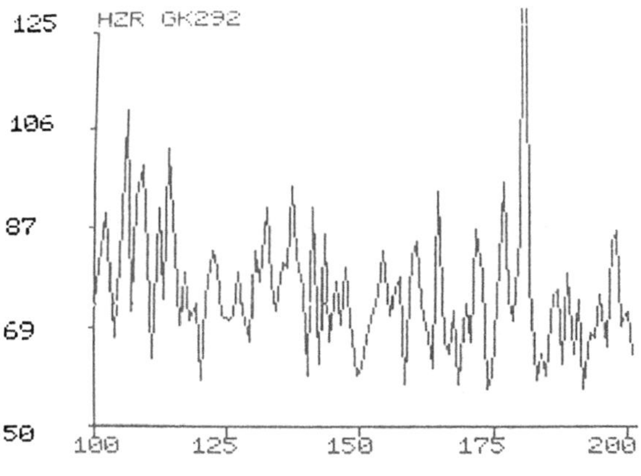

Fig. 1a — Low-level luminescence (in counts per second) from cucumber seedlings during time intervals arbitrarily chosen from 100 to 200 s.

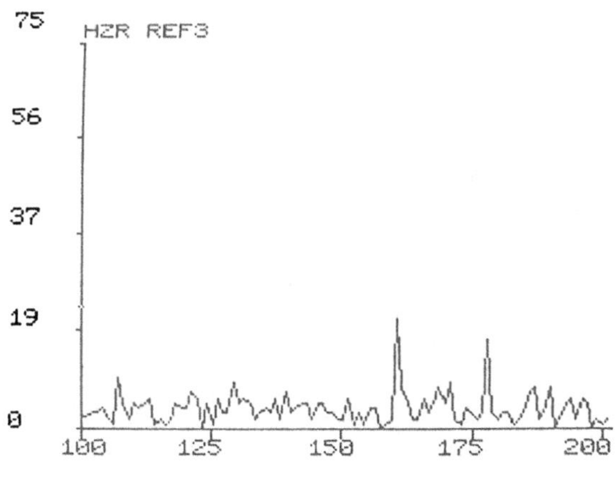

Fig. 1b — Dark-count rate (in counts per second) for the equipment used for measurements shown in Fig. 1a. Since it is about an order of magnitude lower than the photo count rate there, it can usually be neglected.

for a fully coherent field turn into a hyperbolic relaxation dynamics (Li and Popp, 1983; Li, 1986; Popp, 1985, 1986).

After exposure of living cells to weak light, one observes even with separation of narrow spectral parts with bandwidths of a few nanometers a decay function whose agreement with an exponential law (or more than one *independent* exponential) can be clearly rejected (Popp et al., 1981). Rather, it corresponds quite well to a hyperbolic relaxation (Popp et al., 1981; Chwirot et al., 1985a, b). In addition, the agreement with the hyperbolic law in synchronized cell cultures is correlated to the "biological" cooperativity within the system (Chwirot et al., 1985a, b; 1986).

On the basis of these and a variety of similar experimental results (Popp, 1979; Popp, 1984) we developed together with Li a model of "low-level luminescence" which suggests that exciplex formation in DNA linked to phonons in its lattice system (thus forming polaritons) is responsible for ultraweak photon emission from living tissues (e.g. Li et al., 1983; Popp, 1984). In fact, this model suffices to explain all the experimental data of low-level luminescence known, including the characteristic temperature dependence (Popp et al., 1984; Popp and Slawinski, 1986) and coherent rescattering between source (DNA exciplexes) and decay products (photons).

This model may also explain a variety of biological phenoma (Nagl and Popp, 1983; Nagl, 1985), and in particular it may give rise to new directions in cancer research. Roughly speaking, malignant growth according to this model is assigned to a definite loss of coherence in the intercellular mutual rescattering of photons and their molecular sources within the cells (Popp, 1979; Nagl and Popp, 1983). Let us repeat the main aspects of this argument in order to throw light onto the following basic experiments.

The classical approach to this problem can be based on the following relation:

$$x\ddot{x} = (1+n)\,\dot{x}^2 . \qquad (1)$$

For $n=0$ this formula is valid in the case of a classical oscillator with amplitude $x(t)$, providing thus an exponential function

$$x = x_0 \exp(\lambda t),$$

where λ is in general a complex constant, and x has to be taken as a complex value at any time t, too.

According to eq. 1, the kinetic energy (on the right-hand side) and the potential energy (on the left-hand side) are, aside from some phase-shifting, equally distributed over the ergodic system. The total kinetic energy always turns into potential energy and vice versa.

For coherent rescattering, on the other hand, a definite part of the kinetic energy of the decay products is restored in the form of potential energy of the source so that under ergodic conditions n becomes a constant $\neq 0$.

Consequently, the solution of eq. 1 takes the form

$$x = x_0\,(t+t_0)^{-1/n} \qquad (2)$$

where x and n are in general complex quantities, while t_0 is a real constant. Equations 1 and 2 hold also in a more refined quantum description of the problem (Li and Popp, 1983; Popp, 1985).

For a purely imaginary n, oscillations are obtained whose frequency drops down continuously, whereas for a completely real n the damped relaxation displays no oscillatory behavior.

According to our approach, n can be described in terms of a chemical potential μ due to the fact that the system is away from equilibrium. This corresponds in some way to Fröhlich's model (1968), where a chemical potential was introduced. At the same time, the frequency dependence of the chemical potential that has been investigated by our group corresponds to an "ideal open system" with always sufficient energy to stabilize it at a nonequilibrium phase transition (Prigogine, 1984). The relation between n and μ turns out to be (apart from a possible proportionality constant) very simple:

$$n = \frac{\mu}{kT} \tag{3}$$

where kT is the mean thermal energy of the surrounding heat bath. However, one should keep in mind imaginary parts of μ that describe long-range phase correlations which have been postulated by Fröhlich (1968). We found actual indications of such oscillatory behavior during "delayed luminescence" of *Bryophyllum daigremontanum*, while malignant tissue of the same plant failed to exhibit such oscillations under exactly the same circumstances (Popp et al., 1981). Recently, Chwirot et al. (1986) were also successful in finding evidence of those predicted oscillations by observing synchronized cell cultures after illumination with monochromatic light. Changing to white-light illumination caused the oscillatory behavior to disappear and the process turned into a hyperbolic decay with real n (or μ). Of course, according to our hypothesis n becomes real as soon as the imaginary parts of monochromatic contributions cancel out by random superposition. This by no means indicates a loss of coherence of the field.

Now, in applying these results on differentiation and related questions, we expect that for normal tissue

$$\frac{\partial \mu}{\partial S} > 0, \tag{4a}$$

while tumor cell populations satisfy the condition

$$\frac{\partial \mu}{\partial S} < 0, \tag{4b}$$

where S is the density of cells within the population and eqs. 4a, b are valid from at a least a definite $S = S_0$ on higher values of S.

METHOD, RESULTS, AND DISCUSSION

Recently, Schamhart et al. (1986) investigated hepatocytes and two corresponding malignant cell lines, hepatoma cell Reuber H35 (Pitot, 1964) and HTC (Thompson, 1966). While H35 is in some way premalignant, HTC displays a fairly high malignancy in all the significant characteristics, namely growth morphology and growth regulation, glucose and nitrogen metabolism, and mitochondrial development (Schamhart et al., 1979, 1984).

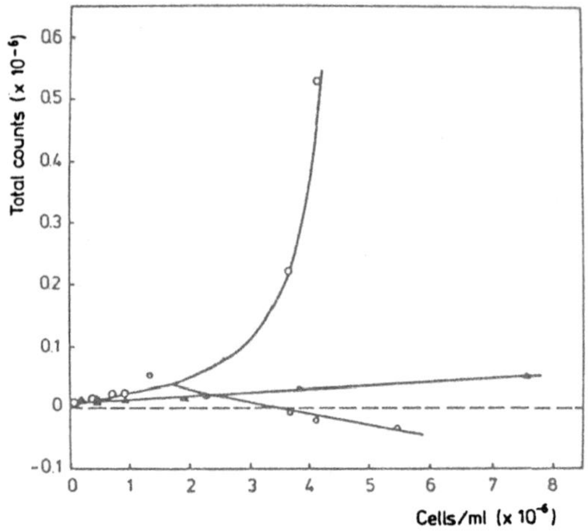

Fig. 2 — Total number of photons (counts) reemitted by hepatocytes (●●●) and hepatoma cell lines H35 (ΔΔΔ) and HTC (○ ○ ○) after a 5-min period of illumination in relation to the cell density compared to that of the medium.

Schamhart et al. illuminated 11-ml cell suspensions in a quartz cuvette with white light of a halogen lamp for 5 min. For that time the relaxation behavior no longer depends on the length of exposure. Immediately after excitation, the decay function of the cell suspension has been observed with the aid of our photocounting equipment (Ruth, 1977). The cuvette was always kept in darkness. The cells were continously spun with a magnetic stirrer in such a way that sedimentation was just excluded. Figure 2 displays a part of Schamhart's results.

These experiments were reproduced by the authors with precisely the same results. Thus, we can state that the S-dependence shown in Fig. 2 reflects well the conditions (eqs. 4a, b) of normal and malignant tissues, respectively.

In order to examine Schamhart's results for general validity, we applied the same method to normal human amnion cells (from the Institute of Anthropology and Human Genetics of the University of Heidelberg) and a corresponding malignant amnion cell culture, namely WISH cells (Hayflick, 1961). After trypsinization the cells were re-suspended in MEM without phenol red. The period of illumination was reduced to 3 min. where again no dependence of the decay behavior on this time interval could be observed. By use of a computer program for minimizing the deviations between the measured decay curves and a hyperbolic function due to eqs. 2 and 3, the value

$$k * ln\, x = (1 + (2kT)/\mu)$$

which corresponds to the emitted intensity $i \approx \dot{x}x$, was calculated and plotted versus the cell density S. The measurements were carried out three times with unsystematical change of S. From these repetitions the errors could be computed. Figure 3 displays the experimental results.

The observed results predicted by eqs. 4a and 4b are again in fairly good agreement with the experimental facts.

In most general terms, Schamhart's and our own results show evidence of essentially different mutual long-range interactions of normal and malignant cells, changing their optical properties sensitively, whatever may happen in detail. This can be seen by the following consideration:

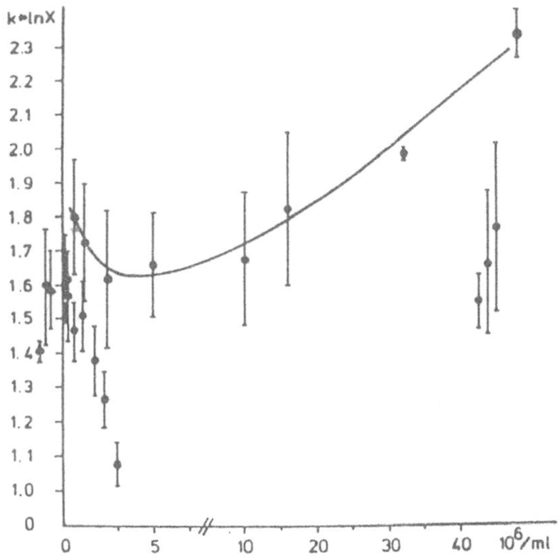

Fig. 3 — With increasing cell density in the medium, malignant cells display completely different decay behavior after exposure to light than the corresponding normal ones do. The decay parameter $k^* \ln x$, corresponding to the slope of a hyperbolic decay, increases with increasing cell density in case of malignant cells (upper curve), while for the corresponding normal ones this parameter decreases rapidly with increasing cell density in suspension (lower curve). Only the medium exhibits a linear response (see three measuring points at the right-hand side of Fig. 3).

If p is the photo count rate of a single cell, and S the cell density, we then obtain from a cellular layer parallel to the front plate of the quartz cuvette with area F at a distance x, an increase di of the measured intensity according to

$$di(x) = SFp(1 - W(x))\,dx \tag{5}$$

$W(x)$ is the probability of radiation absorption within the system on the way of length x between the layer and the counter. There is at first no reason to assume for $W(x)$ and deviation from the Beer-Lambert law:

$$W(x) = 1 - \exp(-\gamma x)\exp(-S\alpha x) \tag{6}$$

γ is an absorption coefficient of the device that is the same for all the experiments, and α represents absorption coefficient per unit of cell density for the cells within the medium.

After insertion of eq. 6 into eq. 5 and integration over the cuvette of length d we obtain

$$i_0 = \frac{SFp}{(\alpha S + \gamma)}\,(1 - \exp(-(\alpha S + \gamma)\,d) \tag{7}$$

where i_0 is the measured radiation intensity. The results shown in both Figs. 2 and 3 display correlations to i_0. Hence, differentiation of eq. 7

$$\frac{\partial i_0}{\partial S} = \frac{i_o}{S}\left(1 - \frac{S}{S + \frac{\gamma}{\alpha}}\right) + \frac{SFp}{\alpha S + \gamma}\,\alpha\, d \exp(-(\alpha S + \gamma) d)\,. \tag{8}$$

Since both terms of the right-hand side of eq. 8 are positive definite, this model could never explain first

$$\frac{\partial i_0}{\partial S} < 0$$

which is observed at higher cell densities of normal cells, and second

$$\frac{\partial i_0}{\partial S} >> 0$$

for malignant cells at higher densities. Of course, for $S \rightarrow \infty$ we have according to eq. 8

$$\frac{\partial i_0}{\partial S} \rightarrow 0 .$$

Consequently, the results shown in Figs. 2 and 3 cannot be explained in terms of individual properties of single cells. In spite of the rather large intercellular distances in these experiments, the optical behavior changes dramatically, being completely opposite for intercellular interactions in normal and malignant tissues. In this general model, either

$$\frac{\partial p}{\partial S} \neq 0$$

or

$$\frac{\partial \alpha}{\partial S} \neq 0$$

is valid, or both. At the same time it would be fantastic to explain these sensitive dependences in terms of photochemical events: the average intensity of emission amounts to a few photons per cell per day!

Let us, therefore, summarize as follows: Experimental results support the electromagnetic model of cell differentiation (Nagl and Popp, 1983). According to this approach we expect in fact a sensitive dependence of exciplex formation by DNA on long-range interactions between cellular units. These interactions may be responsible for excited states in different parts of the DNA depending on whether these states are occupied or not, thus determining, for instance, whether oncogenes are switched on or off. The bridge to biochemistry is then really accomplished. Further support for this link between molecular structure and photon couplings has recently been adduced from molecular data on the DNA conformations and published elsewhere (Popp and Nagl, 1986).

REFERENCES

W.B. Chwirot, Acta Physiol. Plant. **2**, 103 (1980).

W.B. Chwirot, R.S. Dygdala, and S. Chwirot, Cytobios **44**, 95 (1985a).

W.B. Chwirot, R.S. Dygdala, and S. Chwirot, Cytobios **44**, 239 (1985b).

W.B. Chwirot, R.S. Dygdala, and S. Chwirot, Cytobios (1986), accepted for publications.

B. Fain, Phys. Rev. A: Gen. Phys. **24**, 933, (1981).

L. Fonda, G.C. Ghirardi, and A. Rimini, Rep. Prog. Phys. **41**, 587 (1978).

H. Fröhlich, Int. J. Quantum Chem. **2**, 641 (1968).

L. Hayflick, Exp. Cell. Res. **23**, 14 (1961).

R. Kurth, Naturwissenschaften **70**, 439 (1983).

K.H. Li and F.A. Popp, Phys. Lett. **93A**, 262 (1983).

K.H. Li, Rep. Prog. Phys. (1986), in press.

K.H. Li, F.A. Popp, W. Nagl, and H. Klima, in *Coherent Excitations in Biological Systems*, H. Fröhlich and F. Kremer, Eds., Springer, Berlin, 1983.

W. Nagl, Int. J. Cytol. **94**, 21 (1985).

W. Nagl and F.A. Popp, Cytobios **37**, 45 (1983).

K.L. Ngai, A.K. Rajagopal, R.W. Rendell, and S. Teitler, Phys. Rev. B: Solid State **28**, 6073 (1983).

S. Ochoa, Naturwiss. Rdsch. **38**, 368 (1985).

H.C. Pitot, C. Peraino, P.A. Morse, and V.R. Potter, Natl. Cancer Inst. Monogr. **13**, 229 (1964).

F.A. Popp, in *Electromagnetic Bio-Information*, F.A. Popp, G. Becker, H.L. König, and W. Peschka, Eds., Urban & Schwarzenberg, München/Baltimore, 1979.

F.A. Popp, *Biologie des Lichts*, Paul Parey-Verlag, Berlin/Hamburg, 1984.

F.A. Popp, Proc. Int. Conference on Lasers '85, Las Vegas, Nevada, 1985, The Society for Optical & Quantum Electronics USA, in press.

F.A. Popp, in *Selected Papers on Selforganization and Dissipative Structures*, D. Reidel Publ. Co., Dordrecht/Boston, 1986, in press.

F.A. Popp, K.H. Li, and W. Nagl, Z. Pflanzenphysiol. **114**, 1 (1984).

F.A. Popp and W. Nagl, Cytobios **37**, 71 (1983).

F.A. Popp and W. Nagl, Polym. Bull. **15**, 89 (1986).

F.A. Popp, B. Ruth, W. Bahr, J. Böhm, P. Grass, G. Grolig, M. Rattemeyer, H.G. Schmidt, and P. Wulle, Coll. Phenom. **3**, 187 (1981).

F.A. Popp and J. Slawinski, (1986), in preparation.

I. Prigogine, private communication, 1984.

M. Rattemeyer, F.A. Popp, and W. Nagl, Naturwissenschaften **68**, 572 (1981).

S. Rowlands, J. Biol. Phys. **11**, 117 (1983).

B. Ruth, Dissertation (Exp. Phys.), Marburg 1977.

D.H.J. Schamhart, K.W. van de Poll, and R. van Wijk, Cancer Res. **39**, 1051 (1979).

D.H.J. Schamhart, A. Slawinski, and R. van Wijk, Cancer Res. (1986), in press.

R.T. Schimke, Cell **37**, 705 (1984).

D. Slawinska and J. Slawinski, Photochem. Photobiol. **37**, 709 (1983).

J. Slawinski and D. Slawinska, in *Chemiluminescence and Bioluminescence*, J. Burr, Ed., Marcel Dekker, New York, 1986.

E.B. Thompson, G.M. Tomkins, and J.F. Curren, Proc. Natl. Acad. Sci. U.S.A. **56**, 296 (1966).

25. Hopping Charge Carriers in Molecular Crystals and Biopolymers: The Fröhlich Connection

R. Pethig

With 3 Figures

INTRODUCTION

It is possibly the case for all scholars and researchers that the main thrust of their work is consciously, or subconsciously, influenced by a few key papers and books read in their formative years. For me these took the form of two small books (Fröhlich, 1958; Szent-Györgyi, 1960) and one research note (Fröhlich and Sewell, 1959). Szent-Györgyi's *Introduction to a Submolecular Biology* deals with the possibility that the phenomena of molecular charge-transfer interactions and electronic mobility might be of relevance for a full understanding of the subtleties of the living state. This fired off my interest in matters biological and directed me towards the interdisciplinary subject matter of biomolecular electronics. However, the basics of the methodologies of this work were derived from the works of Fröhlich.

From Fröhlich's book *Theory of Dielectrics* I was introduced to the concept of hopping charge-carriers and how these might be detected through dielectric measurements, and from his paper with Sewell the possibility that charge carrier mobility measurements might help in distinguishing between hopping charge transport and delocalized band-type conduction processes was firmly planted in my mind. It is therefore with great pleasure that, in this Festschrift in honor of Professor Fröhlich, I have the opportunity to describe some of our experiments on molecular crystals and biopolymers that were influenced by his writings and in doing so express my personal gratitude to him.

DIELECTRIC MEASUREMENTS OF HOPPING CHARGE CARRIERS

Figure 1 is a slightly modified version of the figure appearing on page 18 of Fröhlich's book on dielectrics. It represents the potential energy barrier confronting a charged particle "wishing" to jump from position A to position B, and the dotted line shows how this is modified by the application of an external electric field. This situation is shown by Fröhlich to be equivalent to a polar molecule of dipole moment $qs/2$ turning through 180°, where q is the particle's charge and s is the distance between the sites A and B. As shown by Lewis (1977) and outlined in Fig. 2, this model can more generally represent an electron hopping between two donor or acceptor sites i and j of a donor-acceptor pair DA. In this case the effective dipole moment is qr_{ij} and corresponds to the dipole $D^+ A^-$ reorienting through the angle $\theta_i - \theta_j$. However, the essential point is that the effective dipole moment is large compared to that exhibited by conventional molecular dipoles. If the

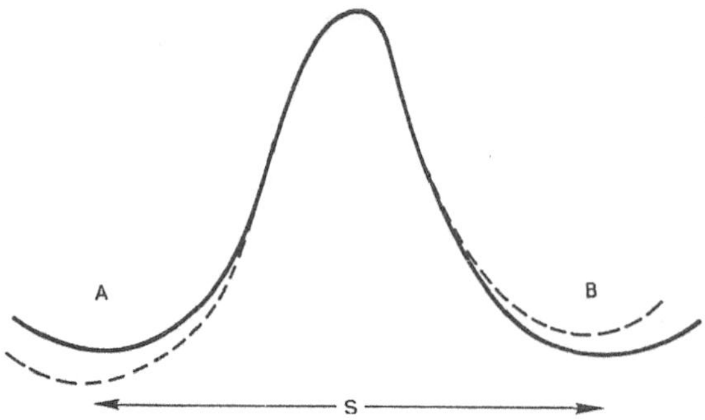

Fig. 1 — The potential energy barrier confronting a charged particle wishing to jump between the equilibrium sites A and B. The dotted curve shows the effect on an applied electric field. (After Fröhlich, 1958.)

distance r_{ij} is just 5 Å, for example, then the effective dipole moment is 8.0×10^{-29} cm or around 24 D units. Small polar molecules exhibit dipole moments of around 3 D or less.

The usefulness of this model has been demonstrated for organic crystals, charge-transfer complexes, and biopolymers. Figure 3 shows the dielectric loss peaks obtained for anthracene single crystals and polycrystalline β-carotene (Hayward and Pethig, 1975). Although the dielectric loss peaks are low in magnitude (for anthracene the maximum value for tan δ is 1.2×10^{-2} at 100 kHz) they extend over at least 12 decades of frequency. Dielectric theory (Fröhlich, 1958) shows that the area under the dielectric loss peak is proportional to the product Nm^2, where m is the dipole moment of the dipolar species of concentration N per unit volume. Since anthracene and β-carotene are non-polar the dielectric losses observed could be due to the presence of polar impurities or to charges hopping between localized sites. Organic polar molecules that might become incorporated into anthracene or β-carotene crystals would have dipole moments no larger than about 3 D, and this gives the lower limits of impurity concentrations of the order of 3% for anthracene and 20% for β-carotene. The anthracene had been crystallized after 80 zone-refining passes at a speed of 20 mm per hour, so that impurity contents of less than 50 ppm rather than 3×10^4 ppm would be more probable. An impurity content of 20% for β-carotene is also unreasonable. On the other hand, if we argue that for anthracene the dielectric loss is associated with the hopping of

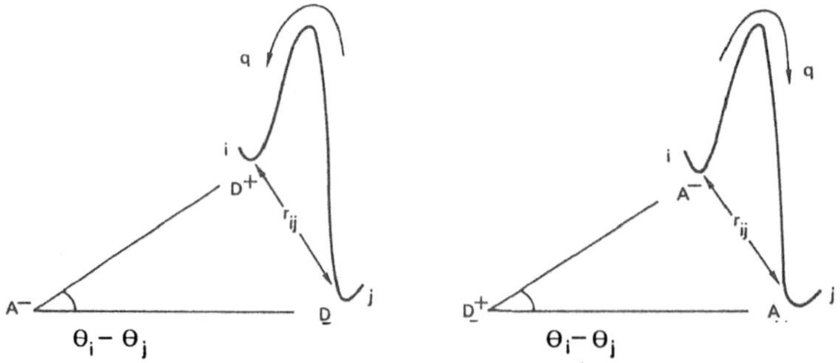

Fig. 2 — Electron transitions of either $D \pm D$ or $A = A$ associated with the dipole $D^+ A^-$ reorienting through an angle $O_i - O_j$. The effective dipole moment is qr_{ij}. (After Lewis, 1977.)

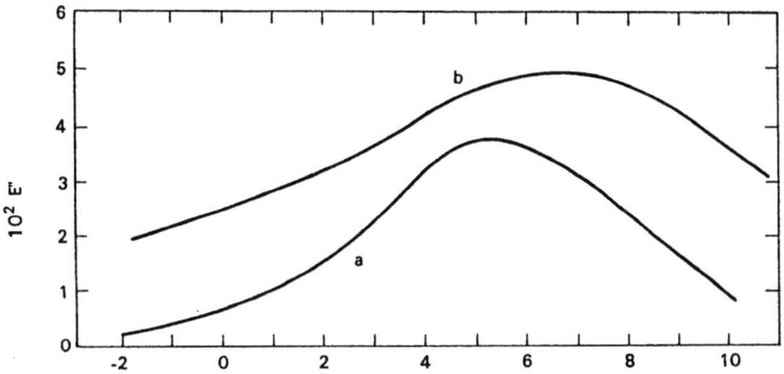

Fig. 3 — Dielectric loss peaks for (a) anthracene single crystal and (b) polycrystalline β-carotene. (After Hayward and Pethig (1975)).

electrons over distances of about three times the *c* dimension of the unit cell (i.e., a hop distance of the order of 30 Å) the equivalent dipole moment will be of the order of 150 D. The corresponding "dipolar" concentration would then be around 5×10^{22} m^{-3}, which is of the order of 10 ppm anthracene molecules. The localized states involved with this are most probably associated with crystal defects that give rise to local electron donor centers. An electron in the vicinity of an ionized donor will find itself in a Coulombic potential energy well, whose magnitude will exceed the electron's thermal kinetic energy over a volume of 10^{-23} m^3 about the donor site. This volume will contain of the order of 5×10^4 anthracene molecules. If the donor is associated with lattice defects then we would expect such defects to produce localized states. The dielectric loss is then envisaged as being the result of the hopping of electrons between such states in the vicinity of ionized donors. The broad loss peaks of Fig. 3 would conventionally be interpreted in terms of a wide distribution of dipolar relaxation times, and in the hopping model this is associated with the density and energy distribution of the localized states. The work on β-carotene was later extended (Lewis and Pethig, 1976) to show how the dielectric loss peaks could be used to determine the concentration and energy distribution of localized states in this crystalline material.

Attention was also directed towards charge-transfer complexes and biopolymers. A low-frequency dielectric dispersion was observed for the perylene + chloranil complex (Carnochan and Pethig, 1976) and this dispersion exhibited an apparent relaxation energy of 0.8 eV, which is very close to the observed semiconduction activation energy (Bone and Pethig, 1978). From other work the semiconductivity was concluded to involve the transport of electrons in thermodynamic equilibrium with extrinsic electron trapping sites. An analysis of the dielectric dispersion in terms of the conventional theories of dipolar relaxation, assuming that the dipoles were the perylene + chloranil charge transfer entities, provided a value of 2.7 D for the apparent dipole moment. This value was considered to be unrealistically high for the solid-state situation where relaxations of the dipoles would be hindered. A smaller dipole moment would also be expected for the complex, as indicated by the fact that the closely related naphthalene + chloranil complex in carbon tetrachloride solvent has a dipole moment of 1.1 D. If, however, the dielectric dispersion is analyzed in terms of electrons hopping between localized sites, then the required trapping site density is found to be 1.8×10^{23} m^{-3}, a value quite compatible with the semiconduction data. This result led to the conclusion that a direct relationship exists between the steady-state conductivity, resulting from the long-range percolation of hopping charges, and the dielectric properties associated with localized hopping. A formal development (Bone et al., 1981) based on experimental data obtained also for protein samples (Eden et al., 1980; Gascoyne et al., 1981) indicates that the dielectric relaxation time τ associated with the hopping charges is related to the steady-state conductivity σ by a relationship of the form

$$\tau = f(\epsilon)/\sigma$$

where $f(\epsilon)$ is a function of the limiting high- and low-frequency relative permittivity values that characterize the dispersion. Independent work by Rosseinsky et al. (1981) on mixed-valence salts has led to essentially the same conclusions.

CHARGE CARRIER MOBILITY MEASUREMENTS

My introduction into the solid-state physics of hopping charge transport was obtained by reading Fröhlich and Sewell's paper of 1959, and this also smoothed the way for a later appreciation of the basic features of polaron and soliton transport. In this paper a description is given of the difficulties that arise when the conventional energy band picture for electrical conduction is applied to the case where the charge carrriers have low mobilities. For normal band theory to be applicable it is necessary that perturbations arising from electron-phonon coupling and impurity scattering are sufficiently weak that the energy broadening (which must result from the Bloch states having a finite lifetime) is less than the conduction band-width W. Heisenberg's uncertainty principle leads to the condition that

$$W \geqslant \hbar/\tau$$

which Fröhlich and Sewell show to be equivalent to

$$\mu \geqslant \frac{qa^2}{\hbar}\frac{W}{kT}$$

where μ is the charge carrier mobility, a is an appropriate lattice constant, and the other symbols have their normal physical identities. For $W \simeq kT$, which is a band-width typical of many molecular crystals, and a lattice constant of a few Ångstroms, then at room temperature the borderline for band-type conduction is $\mu \gtrsim 1$ cm^2 V^{-1} s^{-1}. For mobilities lower than this there is a strong interaction with acoustic phonons and the charge carriers are transported by a thermally activated hopping motion from site to site instead of as a delocalized wave. In effect the charge carriers are randomly scattered at each site. Fröhlich and Sewell also discussed the displacement of lattice ions caused by electrons in neighboring sites, and this is essentially the concept of a polaron. Also, by describing how the absorption or emission of single acoustic phonons becomes impossible (in view of energy and wave vector conservation principles) for a Bloch state velocity lower than the velocity of sound, one of the essential characteristics of soliton transport was introduced to me by Fröhlich and Sewell long before such matters had to be taken aboard for biological systems (e.g., Balanovski and Beaconsfield, 1985).

Our first attempts to use this mobility criterion for investigating charge transport in molecular solids involved dc. Hall current measurements on anthracene crystals (Morgan and Pethig, 1971). The electron Hall mobility values, measured perpendicularly and parallel to the cleavage plane of the anthracene crystal structure, were 1.3 and 5.0 $cm^2 V^{-1} s^{-1}$, respectively. An analysis of the Hall effect for molecular solids having narrow energy bands is not straightforward, but these results have been taken to indicate that the band model is inapplicable for electrons in anthracene at room temperature (Pope and Swenberg, 1982).

As outlined elsewhere (Pethig, 1979) there is an extensive literature on the theoretical treatment of energy band structures and charge transport in biopolymers, especially for polypeptides and DNA. Some of the earliest work was in fact done by Szents-Györgyi's student, Solomon Laki and John Gergely, in the 1940's. Some of the theories developed (e.g., Suhai, 1974) predicted broad energy bands and electron mobilities as large as 10^6 $cm^2 V^{-1} s^{-1}$ for certain polypeptide structures. In this light, Hall effect measurements appeared to offer an ideal way for obtaining experimental validation, or otherwise, of such predictions. However, conventional dc and low-frequency ac Hall effect measurements on polycrystalline biological materials can be complicated by electrode effects and heterogeneities associated with intercrystalline boundaries and intracrystalline defects. These difficulties can be overcome by taking measurements at microwave frequencies, where the Hall effect manifests itself as a Faraday-type rotation of the plane of polarization of the microwaves passing through the test sample. Hall effect measurements at 10 GHz (Eley and Pethig, 1971) produced mobility values of less than 1 $cm^2 V^{-1} s^{-1}$ for dry cytochrome c, hemoglobin, and chloroplasts, but values some ten times this were derived for DNA, plasma albumin, and mitochondrial extracts. The accuracy of the mobility values was directly dependent on the measurement of the microwave resistivity of the test samples, and in fact these were later found to have been erroneously high for the samples deduced to have the highest mobilities. This brings the mobility values down to that corresponding to the hopping regime, or at best to the borderline between band and hopping conduction. Hall mobility measurements made later at 33 GHz (Pethig and South, 1974; Cross and Pethig, 1980) have tended to confirm this for a wider range of polypeptides, proteins, and DNA in the dry state. The basic reason for this large difference between experiment and some of the theoretical predictions appears to be the fact that the theories, of necessity, are bazed on idealized models which are not fully applicable to protein structures in general. However, steady progress in the theoretical modelling is being made and the existence of energy bands in biopolymers still appears to be a real possibility (J. Ladik, private communication, 1985). The influence of hydration also needs to be explored in more depth, for both the theoretical and experimental work.

CONCLUDING REMARKS

Professor Fröhlich's book on dielectrics, first published in 1949, influenced and educated several generations of dielectricians, and his paper with Sewell can be considered to be an important milestone in the development of the solid-state physics of charge transport in narrow band semiconductors. It is hoped that the full extent of the influence of just these two examples of Professor Fröhlich's contribution to science is indicated here, since our own modest studies have been made on systems that normally lie far outside those usually investigated by physicists and electrical materials scientists.

REFERENCES

Balanovski, E. and Beaconsfield, P., Solitonlike excitations in biological systems, Phys. Rev. A., 32 (1985) 3059.

Bone, S. and Pethig, R., Temperature dependence of the low frequency dielectric dispersion in the perylene + chloranil complex, J. Chem. Soc., Faraday Trans. I., 74 (1978) 720.

Bone, S., Eden, J., Gascoyne, P.R.C. and Pethig, R., Conduction and dielectric polarisation in proteins and molecular complexes, J. Chem. Soc., Faraday Trans. I., 77 (1981) 1729.

Carnochan, P. and Pethig, R., Low frequency dielectric dispersions in the Perylene + chloranil charge transfer complex, J. Chem. Soc., Faraday Trans. I, 72 (1976) 2355.

Cross, T.E. and Pethig, R., Microwave Hall effect measurements on biopolymers, Int. J. Quantum Chem: Quantum Biol. Symp., 7 (1980) 385.

Eden, J., Gascoyne, P.R.C. and Pethig, R., Dielectric and electrical properties of hydrated bovine serum albumin, J. Chem. Soc., Faraday Trans. I, 76 (1980) 426.

Eley, D.D. and Pethig, R., Microwave Hall effect measurements on biological materials, Disc. Faraday Soc., 51 (1971) 164.

Fröhlich, H., Theory of dielectrics, Clarendon Press, Oxford, 2nd edition, 1958. (1st. Ed. 1949).

Fröhlich, H. and Sewell, G.L., Electric conduction in semiconductors, Proc. Phys. Soc., 74 (1959) 643.

Gascoyne, P.R.C., Pethig, R. and Szent-Györgyi, A., Water structure dependent charge transport in proteins, Proc. Natl. Acad. Sci. USA, 78 (1981) 261.

Hayward, D. and Pethig, R., Frequency dependence of the conductivity of molecular solids, Phys. stat. sol. (a), 32 (1975) K177.

Lewis, T.J., The dielectric behaviour of non-crystalline solids., Dielectric and Related Molecular Processes, Chemical Society, London, Vol. 3 (1977) 186.

Lewis, T.J. and Pethig, R., The determination of localized energy states in -carotene by a.c. conduction studies, in Excited States of Biological Molecules, Ed. Birks, J.B., Wiley Interscience, London (1976) 342.

Morgan, K. and Pethig, R., D.c. Hall current measurements on organic crystals, J. Materials Sci., 6 (1971) 179.

Pethig, R., Dielectric and electronic properties of biological materials, Wiley, Chichester (1979).

Pethig, R. and South, R.B., Hall effect measurements at 33 GHz, IEEE Trans. Instrum. Meas., IM-23 (1974) 460.

Pope, M. and Swenberg, C.E., Electronic Processes in Organic Crystals, Clarendon Press, Oxford (1982).

Rosseinsky, D.R., Stephan, J.A. and Tonge, J.S., Site-transfer electronic conductivity, J. Chem. Soc., Faraday Trans. I., 77 (1981) 1719.

Suhai, S., Theoretical investigations of semiconductive properties in proteins. I. Electrical conductivity, charge mobilities, and free paths in polyglycine, Biopolmers, 13 (1974) 1731.

Szent-Györgyi, A., Introduction to a Submolecular Biology, Academic Press, New York (1960).

26. Cellular Molecular Processes Driven by Cell-Generated AC Electric Field

E. Del Giudice, S. Doglia, M. Milani, and G. Vitiello

INTRODUCTION

The postulation of long-range correlations at work in living matter has resulted so far in number of studies and discussions. Herbert Fröhlich is a pioneer in linking the emergence of collective dynamics of life to the electrical properties of biological components.

It is a matter of fact that an unordered mixture of cell components functions as a dielectric. Metabolic activity, however, as pointed out by Fröhlich in (1968, 1980), can transform the unordered mixtures of biocomponents into a medium where correlations are established that appear as "extraordinary" properties.

Coherent polarization modes appear when the energy flowing in the system exceeds a threshold and a nonlinear coupling exists between the set of electric dipoles (relevant macromolecules) and the heat bath (water and ions in the environment). Furthermore the polarization field acts upon the macromolecules, bringing them into highly polar metastable states; complex cooperative processes start, giving rise possibly to overall oscillations of the system.

Biomolecules, when hydrated beyond a certain threshold, exhibit low-frequency (fractions of Hertz) modes that point to the existence of zero-frequency modes responsible for the long-range ordering of the system (Celaschi and Mascarenhas, 1977; Hasted et al., 1981). Actually, it was suggested long ago that electrodynamics in a given medium assumes a peculiar form when the physical spectrum of the medium exhibits zero-frequency modes.

Schwinger (1962a, 1962b) has proven that electrodynamic theory for such a system, at a nonperturbative level, would involve a massive photon, because of the mixing between field modes and matter polarization modes. As a consequence the (usual) massless photon when traveling in the medium would acquire a mass whose square is equal to the residue of the pole of the polarization tensor at zero frequency; in the situation we are investigating, this residue is proportional to the vacuum expectation value of the polarization field.

"Photon mass" is elementary particle jargon; it means that an observer completely immersed in the system would see the photon as a massive object. From the outside, it is possible to see on phenomenological grounds the "photon mass" inside the system as a confinement process for the field within a region of finite size. The old result was originally derived in connection with the extraordinary electromagnetic properties of superconductors, but of course it is quite general (Anderson, 1984). Whenever a system exhibits in its physical spectrum a zero-frequency mode (this property is connected with the breakdown of some continuous symmetry and the Goldstone theorem) some degrees of freedom of the electromagnetic field combine with the degrees of freedom of the matter field. The field confinement produces quite strong field gradients on the boundaries and the corresponding forces could give rise to a matter compartmentation of the system.

In this contribution we work out the above arguments by showing that Fröhlich's model of coherent excitation of polarization modes for metabolically active systems agrees with the quantum field theory prediction about the electrodynamic response of a system where long-range correlations are autonomously sustained. We will analyze a specific example, namely the process of rouleau formation between metabolically active erythrocytes (Del Giudice et al., 1986) experimentally investigated by Rowlands and coworkers (1981, 1983a, 1983b).

FRÖHLICH MODES AND ELECTRODYNAMICS

We sketch in this section how quantum physics could account for some observed properties of biological matter.

Usually quantum physics is looked upon as synonymous with microphysics, but the experience of condensed matter physics in the last four decades shows that quantum physics has been successful in deriving the properties of macroscopic systems classically behaving. Detailed descriptions of the quantum theoretical approach to the dynamics of biological systems have appeared (Del Giudice et al., 1980, 1983, 1985). The approach can be schematized as follows.

Assumptions and Consequences

1. A living system is basically a dielectric that can be described as a set of electric dipoles.

2. We expect the Lagrangian of such a system to be exceedingly complicated. Therefore we will concentrate on its symmetries only.

3. Interactions among the system components are: *short range* (for instance, hydrogen bonding) and *long range*, namely electromagnetic.

4. The Lagrangian is invariant under the SU(2) group of dipole rotations.

5. What could be a reasonable ground state (vacuum) for the living system? A possibility would be one in which the dipoles would exhibit all possible orientations, leading to an isotropic macroscopic situation. This state would exhibit the same invariance as the Lagrangian and no symmetry breaking would occur in this case. We know from experiments (Celaschi and Mascarenhas, 1977; Hasted et al., 1981) that all the relevant biomolecules display a long-lasting electric polarization related to the hydration of the molecule. Oscillation modes with frequency less than 1 Hz and relaxation times longer than 10^5 s have been found. Highly polar metastable states have been theoretically predicted by Fröhlich (1968, 1980, 1985) as peculiar to living systems. We therefore

assume as ground state (vacuum) for a living system a state exhibiting nonvanishing electric polarization

$$<0| P | 0> = P \neq 0 .$$

6. The electric polarization P therefore appears as the order parameter in the living system. The SU(2) group of invariance of the Lagrangian is broken. The system displays a preferred direction of orientation.

7. The ground state lives long enough to be treated as a stable state in the mathematical formalism. On the other hand its existence depends on the previous ordering of the system (orienting fields possibly triggered by the metabolic processes). Actually dipole orientation in the ground state requires a certain amount of energy. In the framework of Davydov's theory of solitons (1979, 1982) on macromolecular chains the energy output of metabolic reactions is trapped on the chain in soliton form. Davydov has shown that a soliton is equivalent to an electric current flowing on the chain without any additional expenditure of energy. This current behaves therefore as a supercurrent and produces an orienting electromagnetic field without heat losses. The period of such a current is the lifetime of the soliton, which is expected to be very long, in agreement with the low-frequency modes observed experimentally for the biomolecule electret state. Moreover this dipole orientation seems to last very long owing to the observed very long relaxation time (Celaschi and Mascarenhas, 1977; Hasted et al., 1981). Actually the energy flowing through the system in form of the soliton current does not produce any thermal fluctuations in the system and metastability of the electret could arise as a consequence.

Coherent polarization modes arising from short-range interactions

When dipoles are totally oriented, short-range interactions dominate; we will start then by taking into account only such forces. In this case the Goldstone theorem holds. Of the three fundamental rotations of SU(2), two are broken and consequently two Goldstone modes appear, carrying electric polarization quantum numbers. Physical states are coherent with respect to the dipole wave quanta. The system components become correlated by these coherent modes at very long range. In the system the matter distribution is described by a coherent field (Del Giudice et al. 1983, 1985).

Electromagnetic field in the correlated medium

The appearance of a nonvanishing electric polarization in the ground state reduces the SU(2) symmetry to a U(1) symmetry. This residual symmetry also needs to be broken because otherwise well known theorems would bring disturbing long range electromagnetic interactions into the picture. The phase of the dipole field component along the polarization axis is then fixed. The U(1) breaking implies the appearance of a preferred helicity along the axis and of an additional Goldstone field Θ (x) (Del Giudice et al., 1980). However these quanta of Θ cannot appear in the physical spectrum; according to the Anderson-Higgs-Kibble mechanism (Anderson, 1958) they cancel out with the ghost component of the electromagnetic field $A_\mu(x)$.

As a final result the electromagnetic field acquires a longitudinal component giving rise to a "massive" photon.

After gauge fixing, Maxwell's equations transform into

$$\left[\frac{I\,\partial^2}{c^2\partial t^2} - \nabla^2 - M^2\right] A_\mu(x) = 0 \tag{1}$$

when

$$\hbar\,\partial_\mu\Theta(x) = qA_\mu(x) \tag{2}$$

M^2 is proportional to the order parameter P. The photon mass corresponds phenomenologically to the confinement of the electromagnetic field into a tube around the propagation axis whose radius is $\frac{\hbar}{c} \cdot \frac{1}{M}$. The bigger the order parameter the smaller the size of the confinement.

The field confinement in a correlated medium is governed by the linear eq. 1. However, eq. 1 is the final result of an infinite chain of nonlinearities appearing in the usual Maxwell equations.

The term M^2 in eq. 1 emerges as the sum of infinite multiphoton processes. In other words it is possible to describe the interplay between the conventional electromagnetic field and the space distribution of coupled oscillators as a "new electromagnetic field" characterized by noninteracting quanta, where all the nonlinearities add up to give a mass to the photon.

Self-focusing

A trend toward the result presented above can then be traced also by following the emergence of nonlinearities in the original Maxwell equations, through the introduction of field-induced source distributions. An interaction among photons mediated by the matter oscillators should then be taken into account.

It is possible to see that an incoming electromagnetic wave could undergo so-called self-focusing. This process can be understood as follows.

Since the nonlinear dependence of the field is taken into account, the refractive index of the medium would be

$$n = n_o + n_2\,E^2 + \ldots.. \tag{3}$$

If the constants in the above expansion are positive (this takes place when, for instance, the Kerr effect is at work), the cross section of the beam shrinks while it propagates in the medium.

It can easily be seen that when $n_2>0$ i.e. when the medium traversed by the wave is optically denser than the one outside, the incoming beam shrinks and after a critical path

$$R_{nl} = a/2\sqrt{n_o/n_2E^2}$$

(a is the beam radius) becomes confined within a thin waveguide ("filament") whose length is related to the field amplitude E in a complex way.

The geometric and diffraction losses of the propagating beam can be thus counteracted so that the propagation becomes very efficient. This mechanism is active only when the power of the beam exceeds a critical value

$$P_{cr} = \frac{(1.22)^2}{25\sigma n_2} \lambda^2 c$$

as can be derived on the basis of simple geometrical-optics considerations. Notice that the appearance of a critical power in self-focusing depends on the fact that the nonlinear response of the medium is triggered by a previous orientation, which in this case is performed by the field itself; since this orientational effect is an increasing function of the beam intensity, we obtain the existence of a critical power.

In the case of a biological system, the nonlinear response of the medium to the electromagnetic field is a consequence of a dipole orientation produced by a long-lasting dipole correlation triggered by metabolism. Fields are not required in this case to impose the ordering of the medium via its electrical anisotropy. They can however exploit the dynamical ordering produced by a different agent (metabolism). As a consequence we are not required to ask for high field powers. Quite low fields can then propagate in a highly nonlinear way.

Phenomenological Consequences

From eqs. 1 and 2 we deduce the following consequences (Del Giudice et al., 1980, 1983, 1985):

1. The above argument applies to electromagnetic fields whose strength is comparable to the correlation strength. For very intense such fields we recover Maxwellian propagation.

2. The electromagnetic field becomes the quantum wave field associated to a massive particle; this means that the field is completely carried by the propagating particle. In a sense the whole electromagnetic field becomes "radiative".

3. The field concentrates around the photon trajectory more and more with increasing M. Let us consider propagation through the water of a living system. In the extreme case where all the dipoles are correlated, $M=13.60$ eV (hydrogen ionization potential). The corresponding radius of the tube where the field is trapped is 146 Å. Outside this tube the field is vanishing. The effect is analogous to the Meissner effect in superconductors. The electromagnetic field filamentation has a stabilizing effect on the whole structure because

system components untouched by a filament are protected against the electromagnetic forces. Moreover field intensity remains constant along the filament, giving rise to a complete transparency of the medium.

4. Physical states are coherent states of the Goldstone quanta $\Theta(x)$. Equation 2 transforms this matter coherence into photon coherence. An externally supplied incoherent field whose intensity satisfies requirement 1 becomes coherent through propagation in a living medium.

5. Equation 2 implies the magnetic flux quantization into a living system. By integrating eq. 2 along a closed path we get

$$\phi(H) = n \cdot \frac{\hbar}{q}, \tag{3}$$

and Josephson-like effects can be expected.

6. Filamentation of the electromagnetic field produces strong field gradients on the filament boundaries. As a consequence frequency-selective forces appear. In a dilute solution where solute molecules are able to oscillate at frequencies υ_{ok} and the field oscillates at frequency υ, the force acting upon a molecule is

$$F = const \cdot \Sigma_k \frac{\upsilon_{ok}^2 - \upsilon^2}{(\upsilon_{ok}^2 - \upsilon^2)^2 + \Gamma^2 k} \cdot \nabla E^2 \tag{4}$$

where Γ_k is the damping of the k-th mode. When resonance occurs between the field frequency and a molecular mode in the solution (i.e. $\upsilon \simeq \upsilon_{ok}$) condensation of selected molecules can occur at the interface of the electric field filaments (Del Giudice et al., 1986).

This pattern compares with the picture emerging from the recent studies on cell cytoskeleton (Wolosewick and Porter, 1979). In the following we will discuss the problem of cell interactions in the specific case of red blood cells investigated by Rowlands.

ROULEAU FORMATION AMONG RED BLOOD CELLS

We discuss in this section some experiments by Rowlands and coworkers (1981, 1983a, 1983b) on the attraction of metabolically active erythrocytes. These experiments are interesting since the interaction involves not only the cells but also the interposed medium, where filamentary structures are formed according to a pattern that seems to fit into the scheme of the previous section. Let's summarize the main results.

1. A long-range attraction develops among metabolically active erythrocytes. The range of this attraction is about 12 μm, i.e. three times the cell radius. When the metabolic activity is suppressed by different methods the attraction disappears.

2. The attraction appears only when the medium where the cells are suspended contains special "transmitter" macromolecules. In the human plasma this role is played by fibrinogen. However the same role could be played in other suspending media by molecules such as polyethyleneoxide, and polyvinylpyrrolidone. The common feature to these molecules is their elongated shape.

3. The attraction is species specific.

4. When cells previously built into a rouleau are pulled apart, considerable resistance is met. By electron microscopy a thin material thread can be observed along the straight line connecting two adjacent cells. This fibril or "contractil" seems not to be attached directly to the cell membrane, but slides on the membrane surface when cells are rotated. Furthermore small end gaps between the contractil and the cell membrane are detected.

The above features point clearly to a metabolic origin of the process. In the framework discussed in the previous section, it can be seen that the region near the cell membrane is expected to present a number of peculiar properties. First of all, electric and magnetic fields originating from cell dynamics and connected with the external medium by membrane activity must be present. As a matter of fact there are many experimental reports about the presence of such fields (Pohl, 1983; Belinovska et al., 1985). Furthermore some ordering is assumed to be present in the water surrounding the cells (Clegg, 1983).

As in the intracellular case this ordering could be attributed to cell-correlated dynamics, specifically to the Goldstone bosons arising from a symmetry-breaking process and connected with low-frequency modes of the membrane fields (Fröhlich 1968, 1980, 1985). If water surrounding the cell is thus correlated, the electric and magnetic fields extending from the cell must follow the regime described above: they must confine themselves inside filaments. We (Del Giudice et al. 1980) have evaluated the maximal range of the correlation through Goldstone bosons at $T = 300$ K; it is expected to be of the order of some tens of micrometers, which compares with the range of the erythrocyte attraction. Two possibilities then arise:

1. The medium contains a molecule able to coresonate with the frequency of the field emitted by the cell. In this case according to eq. 4 this molecule is attracted and the electric field filaments become coated. This coating stabilizes the filaments and provides an improved transmission of the field for geometrical reasons.

2. No coresonating molecule is present. In this case the field filament cannot become coated at all and is not protected against the fluctuations of the medium.

The contractil formation process could be then a model system for understanding the more complex cytoskeleton. The extracellular character of the experiment would allow better control of the relevant parameters. It would be useful to study contractil formation in presence of an electromagnetic field of appropriate (low) intensity as an external control parameter.

A more detailed analysis of the Rowlands experiments has been provided in terms of an equivalent self-focusing model (Del Giudice et al., 1986); microscopic features of contractils have been explained in terms of the critical self-focusing parameters.

ACKNOWLEDGEMENTS

We are deeply grateful to Professor Fröhlich for all we have learned from him and for leading us to the enjoyable study of biological physics.

REFERENCES

P.W. Anderson, Phys. Rev. **110**, 827 (1958).

P.W. Anderson, *Basic Notions of Condensed Matter Physics: Frontiers in Physics*, Benjamin-Cummings, Menlo Park, CA, 1984.

E.J. Belinovska, W. Lech, and A. Wittling, Phys. Letter. **109A**, 124 (1985).

S. Celaschi and S. Mascarenhas, Biophys. J. **20**, 273 (1977).

J.S. Clegg, in *Coherent Excitations in Biological Systems*, H. Fröhlich and F. Kremer, Eds., Springer, Berlin, 1983, p 162.

A.S. Davydov, Phys. Scripta **20**, 387 (1979).

A.S. Davydov, *Biology and Quantum Mechanics*, Pergamon, New York, 1982.

E. Del Giudice, S. Doglia, M. Miliani, and G. Vitiello, "Electromagnetic Field and Spontaneous Symmetry Breaking in Biological Matter," submitted to Nuclear Physics (1986).

E. Del Giudice, S. Doglia, M. Miliani, and G. Vitiello, Phys. Lett. **95A**, 508 (1983).

E. Del Giudice, S. Doglia, M. Miliani, and G. Vitiello, Nucl. Phys. **251B**, FS13, 375 (1985).

E. Del Guidice, S. Doglia, and M. Miliani, J. Biol. Phys., in press (1986).

H. Fröhlich, Int. J. Quantum Chem. **2**, 641 (1968).

H. Fröhlich, Adv. Electron. Electron Phys. **53**, 85 (1980).

H. Fröhlich, Phys. Lett. **110A**, 480 (1985).

J.P. Hasted, H.M. Millany, and D. Rosen, J. Chem. Soc., Faraday Trans. 77, 2289 (1981).

H.A. Pohl, in *Coherent Excitations in Biological Systems*, H. Fröhlich and F. Kremer, Eds., Springer, Berlin, 1983, p 199.

S. Rowlands, L.S. Sewchand, R.E. Lovlin, J.S. Beck, and E.G. Enny, Phys. Lett. **82A**, 436 (1981).

S. Rowlands, in *Coherent Excitations in Biological Systems*, H. Fröhlich and F. Kremer, Eds., Springer, Berlin, 1983, p 145.

S. Rowlands, C.P. Eisenberg, and L.S. Sewchand, J. Biol. Phys. **11**, 1 (1983b).

J. Schwinger, Phys. Rev. **125**, 397 (1962).

J. Schwinger, Phys. Rev. **128**, 2425 (1962).

I.I. Wolosewick and K.R. Porter, J. Cell. Biol. **82**, 114 (1979).

27. Evidence for AC Fields from Living Biological Cells

H.A. Pohl*, W.T. Phillips†, and J.K. Pollock

With 4 Figures

ABSTRACT

Following the stimulating suggestion by Herbert Fröhlich in 1968, oscillating electric fields are now found to be produced by living cells. These natural oscillating electric fields of cells at frequencies in the order of 30 Hz or higher are evidenced by:

(1) direct instrumental detection and observation.

(2) the use of tiny particulate probes of either very high or very low dielectric permittivity, together with knowledge about nonuniform field effects (dielectrophoresis, or DEP for short).

(3) cellular spin resonance (CSR), the frequency-sensitive spinning of cells or small particles.

(4) the observation that certain species of fish emit radio frequency signals in the range of 1 to 60 kHz. This emission can hardly be imagined to be other than cellular in origin.

(5) the formation of rouleaux by red blood cells mediated by long-range interactions (up to 4 micrometers).

Two major conclusions emerge from these studies:

(I) That natural ac oscillations appear to be "universal" in that they have been detected in all types of cells examined, and especially in those capable of reproduction.

(II) The oscillations detected by DEP (mentioned above) appear to be maximal at or near mitosis, at least in yeast, where the phase of the life cycle is most easily followed.

The role of these natural ac fields of living cells is yet to be determined. Still under continuing study is the critical question as to whether they are essential or mere incidentals to cellular life.

---

*Deceased June 23, 1986.

†Present address, Dept. Of Radiology, Health Science Center, University of Texas, 7103 Floyd Curl Drive, San Antonio, TX 78284

INTRODUCTION

That living cells and even higher organisms produce detectable static or even occasional pulses of dc fields has been known and studied for several hundred years. Not so for ac fields, partly because of our relatively recent history of dealing with such fields. As a consequence of mankind's apparent need to use complicated circuitry to make ac fields, the existence of natural oscillating fields in and around cells has been but little considered. It has, in fact, only recently been reported. At least five techniques have been used to detect the natural ac fields of cells.

(1) The technique of micro-dielectrophoresis (μ-DEP) uses small dielectric particles to detect ac fields from cells (Pohl, 1980a).

(2) Cellular spin resonance (CSR) has been used to investigate the dielectric properties of cells, and to detect their ac fields. (Pohl and Crane, 1971; Pohl, 1978, 1980b,c, 1981a,b,c; 1983a,b,c; Pohl et al., 1981; Lamprecht and Mischel, 1981; Roy et al., 1981; Pohl and Braden, 1982; Mischel et al., 1982; Holzapfel et al., 1982; Arnold and Zimmermann, 1982; Mischel and Lamprecht, 1983; Mischel and Pohl, 1983; Glaser et al., 1983; Sauer, 1983; Sauer and Schloegl, 1985; Pohl, 1986; Rivera and Pohl, 1986). Particulate spinning in rotating electric fields had long been observed (Lampa, 1906; Lertes, 1921; Pohl, 1951; Heller, 1959; Cantril and Pohl, 1968) but its sharply resonant response to frequencies of the applied field was first described by Pohl and Crane in 1971. Perhaps the best theoretical analysis to date is that of Sauer and Schloegl (1985).

(3) Limited success in the direct instrumental detection of these fields from individual cells has been reported by two groups (Jafary-Asl et al., 1983a,b; Rivera et al., 1983, 1984).

(4) Direct instrumental detection of radio frequency signals from larger organisms such as electric fish of gymnoid species was reported by Westby and Kirschbaum (1982). It is difficult to imagine other than cellular sources for such radiation of these mormyrid fish.

(5) The newly discovered long-range interactions of erythrocytes has been found to mediate rouleaux stacking. These are best explained using the ideas of Fröhlich on metabolically generated short-wave radiations and their interactions (Rowlands, 1982, and this Festschrift; Rowlands et al., 1981, 1982, 1983a,b).

(6) The direct detection and observation of ac fields about cells is perhaps most prettily done in a "hanging drop" experiment (Rivera et al., 1985) by letting suspended dielectric particles form patterns according to their being either repelled or attracted to the region of most intense cellular field (cf. Figs. 1 and 2.)

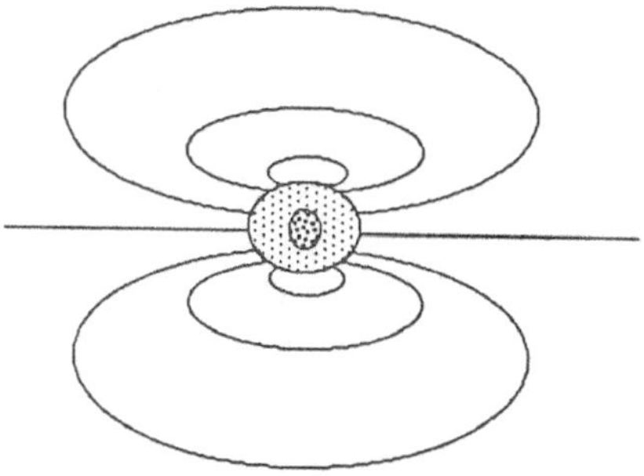

A typical single cell

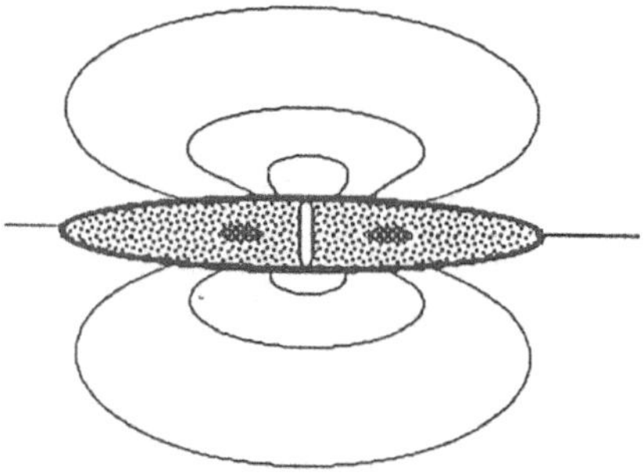

A Desmid algal cell

Fig. 1 — AC field lines about a cell.

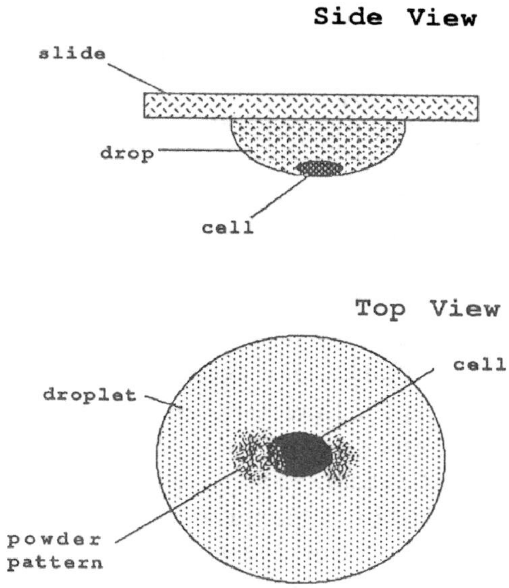

Fig. 2a — The hanging drop experiment to show AC fields about cells.

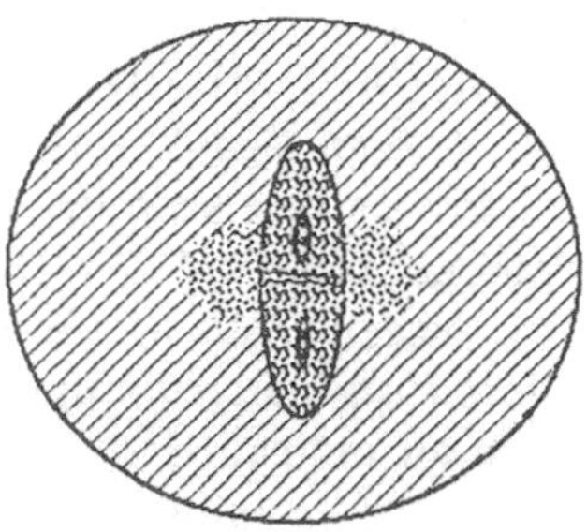

Top View of Hanging Drop Showing Cell and Associated Powder Pattern of a Highly Polar Powder

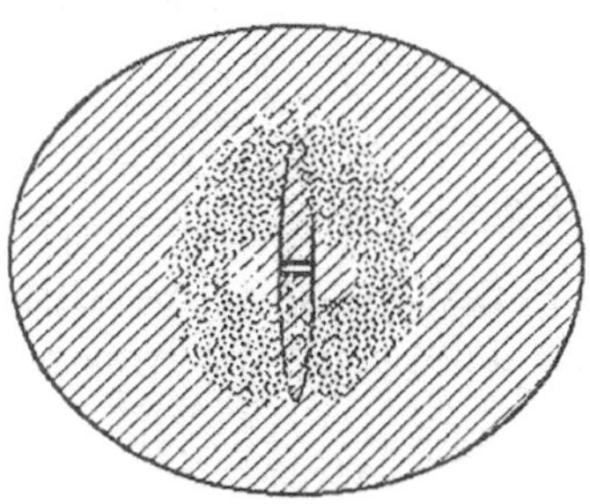

Pattern of a Weakly Polar Powder

Fig. 2b — Attraction and repulsion patterns about live algal cells (Simple dipole patterns).

Pohl and co-workers (e.g., 1981) used the μ-DEP experiments to show that ac fields exist in a wide variety of cell types: bacterial, fungal, algal, annelid, amphibian, avian, and mammalian. These studies indicated a phenomenon not only universal, but one that depends on the phase of the life cycle. In the yeast, *Saccharomyces cerevisiae*, the life cycle phase of the cell can be easily determined, according to the classical studies by Hartwell et al. (1973, 1974). Here, μ-DEP studies show that the intensity of the emitted ac fields appear to reach a maximum at or near mitosis (Pohl and Braden, 1982).

The CSR experiments confirmed the μ-DEP results when lone cells were observed to spin in response to an applied two-pole ac field (Pohl, 1982; Rivera and Pohl, 1986). One current explanation is that of an interaction between the externally applied field and an internal oscillating dipole. An alternative one is that the cells spin simply because of delayed polarization interactions, as would any material object (Sauer, 1983; Sauer and Schloegl, 1985). It must be said that the second explanation is the more likely, except in very few cases.

Rowlands and coworkers (1981, 1982, 1983a, 1983b) report that red blood cells form rouleaux (orderly stacks) at a much greater rate than expected from random diffusive collisions. The interaction has a surprising range of between 5 and 10 micrometers. It has been attributed to Fröhlich interaction of ac fields of very high frequency (ca. 100-300 GHz) produced by the cells (Del Guidice et al, 1982). Subsequent "after-contact" interactions in which thread-like processes are observed (contractils) upon separating once-united healthy erythrocyctes are described as involving both Fröhlich wave interactions and solitons (Del Guidice et al., 1982, and this volume.)

Direct instrumental detection of the minute ac fields produced by cells is quite difficult and has met with only limited success. Yeast cells (*S. cerevisiae*) have been reported to emit marginally detectable signals at 7 to 8 MHz (Jafary-Asl et al., 1983a, 1983b; Rivera et al., 1983) while an alga, *Netrium digitus*, was found to emit signals at about 35 KHz (Rivera et al., 1983). Power levels of the fields produced have been estimated at about 1 to 0.01 picowatts per cell.

Recently, a new technique has been developed that places a cell in a hanging drop along with polarizable powder particles. The powders form patterns under the influence of the fields generated by the cell and can give an indication of the size and shape of these fields. Rivera et al. (1985) first used the hanging drop technique to examine the electric field patterns about large algae (*Netrium digitus* and *Closterium acerosum*) and the fertilized eggs of the nematode *Monochoides changii* and of an amphibian, the African clawed frog *Xenopus laevis*. This work was confirmed in West Germany by Pohl and Lamprecht (1984) with the alga *Monorapidium griffithii*.

EXPERIMENTAL METHODS AND PROCEDURES

In this study we report the use of two techniques that use small dielectric particles to demonstrate the presence of natural ac fields about cells; the hanging drop technique, and μ-dielectrophoresis.

We have examined the following types of cells: BSC (African green monkey kidney line), VERO (African green monkey kidney line), AK-D (feline lung line), and fetal bovine kidney (FBK) cells (primary culture). These cells were all cultured in RPMI-1640 with 10 per cent horse serum. Gentamycin and L-glutamine were added to the growth medium. These cells were grown in culture until two days after they reached a complete monolayer. These postconfluent cells were used because younger cells did not seem to stand the procedure as well. Such cells were detached with 0.25% trypsin and resuspended in the growth medium described above to neutralize the trypsin. The cells were washed by repeated centrifugations followed by resuspension in deionized iso-osmotic solutions (i.e. ca. 300 mM dextrose, mannitol, or alanine of conductivities 15 μS/cm or less).

Suspensions of the several dielectric particles were prepared in the same iso-osmotic solutions used to suspend the cells. Particles of diameter ca. 1 to 2 μm were obtained by grinding the several pure powders in a small ball mill with alumina balls for a week. Particle sizing was done with the use of selective sedimentation through thin aqueous layers, followed by filtration through Millepore filters. The properties of the dielectric particles used are shown in Table 1. As can be seen, the first group are of high permittivity, and the second of low. Since the permittivity of the aqueous medium is about 80 in the low-frequency range (i.e., less than about 10 MHz), the members of the first group of dielectrics can be expected to be attracted to strong ac field regions in highly pure water, while members of the second group will be repelled. This is the normal action of DEP.

The hanging drop technique

The experimental procedure consisted of placing 2 μL of the particle suspension onto a microscope cover slip. A cell suspension was placed nearby on a micro-

Table 1. Dielectric properties of powders

Particle	Relative Dielectric Constant
$SrTiO_3$(reduced)	ca. 5000
$BaTiO_4$	2000
DP1A[a]	1000
$BaSO_4$	11.5
SiO_2	3.8
Al_2O_3	7.5

[a] DP1A is a copolymer of anthraquinone and pyromellitic dianhydride made by catalyzed Friedel-Crafts reaction. It has giant molecular orbitals enhancing electronic delocalization and internal dissociation to produce a very large polarizability (nomadic polarization; cf. F. Gutmann and H. Keyzer, *Organic Semiconductors,* Part B, Plenum Press, New York, 1984.

scope slide and several cells were micropipetted out under direct view through an inverted microscope. These cells were added to the 2-μL particle suspension. The "loaded" cover slip was then turned over and placed in the pocket of a hanging drop culture slide. The resulting patterns of particle distributions about the cells in this hanging drop were observed with a top-view microscope fitted for video camera filming. The patterns usually developed discernibly in 5 to 20 min.

The effects of altering the conductivity of the medium were determined by examining the powder patterns produced by the FBK cells developed using strontium titanate in media having conductivities of 30, 200, or 3,000 S/cm. The heightened conductivities were obtained by adding culture medium to the cell and the particle suspensions just before mixing the latter as they were placed upon the cover slip. As is known (cf. Pohl, 1978) the effective relative dielectric constant, i.e. the absolute value of the complex relative permittivity, of water at frequencies lower than 100 kHz exceeds that of the titanates when its conductivity exceeds about 1000 μS/cm. This can readily be calculated from the formula:

$$e_{\text{eff}} = \{e' + (\sigma / (\omega\epsilon_0)^2\}/2$$

where e_{eff} is the effective relative dielectric constant, e' is the relative (in phase) dielectric constant, σ is the conductivity, ω is the angular frequency $= 2\pi f$, where f is the circular frequency, and ϵ_0 is the absolute permittivity of free space.

There are several simple experimental precautions necessary for the proper development and interpretation of the powder patterns. First, mechanical vibration of the developing powder patterns in the hanging drops must be minimized. Second, it is necessary also to minimize the presence of detritus or of clumps and particle agglomerates in the powder particle preparations. Moreover, actively lysing cells can temporarily perturb the developing powder pattern. An exudate from a cell can form a film about the cell that is particularly evident at the air-water interface. This latter causes modification of the powder patterns, especially to form those in which the particles are held away from the cells; but it is not expected to be able to imitate both attractive and repulsive patterns. The cells were shown to remain viable in the nonionic medium containing particles for at least 3 h. This was shown both with Trypan blue "vital" staining and by reculturing the cells.

Micro-dielectrophoresis (μ-dep)

Here the cells are simply mixed with the small particles as above, and let settle on a slide. In a few minutes the cells subside and the average number of particles, n, seen to be associated with a cell (minimum of 40 cells) is determined by counting, as well as the number of particles, p, per unit volume in the liquid around the cells. The ratio n/p is a measure of the particle gathering (or repelling) by the cell. The statistics of the counting is evaluated, and the average n/p determined. Again as in the hanging drop technique, the medium must be of low conductivity, yet be iso-osmotic to the cell for best results. The typical behavior of living cells with such dielectric particles is shown diagramatically in Figs. 3 and 4. Summing up the results of numerous experiments, for which the standard error is in the order of about 10% of the observed mean value, gives the results shown in Table 2.

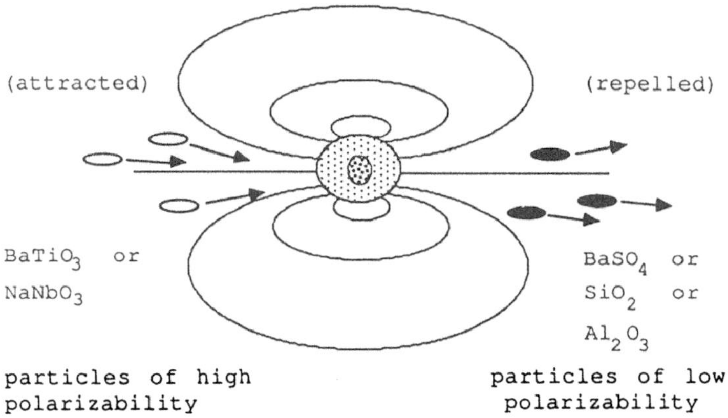

Fig. 3 — Micro-dielectrophoresis.

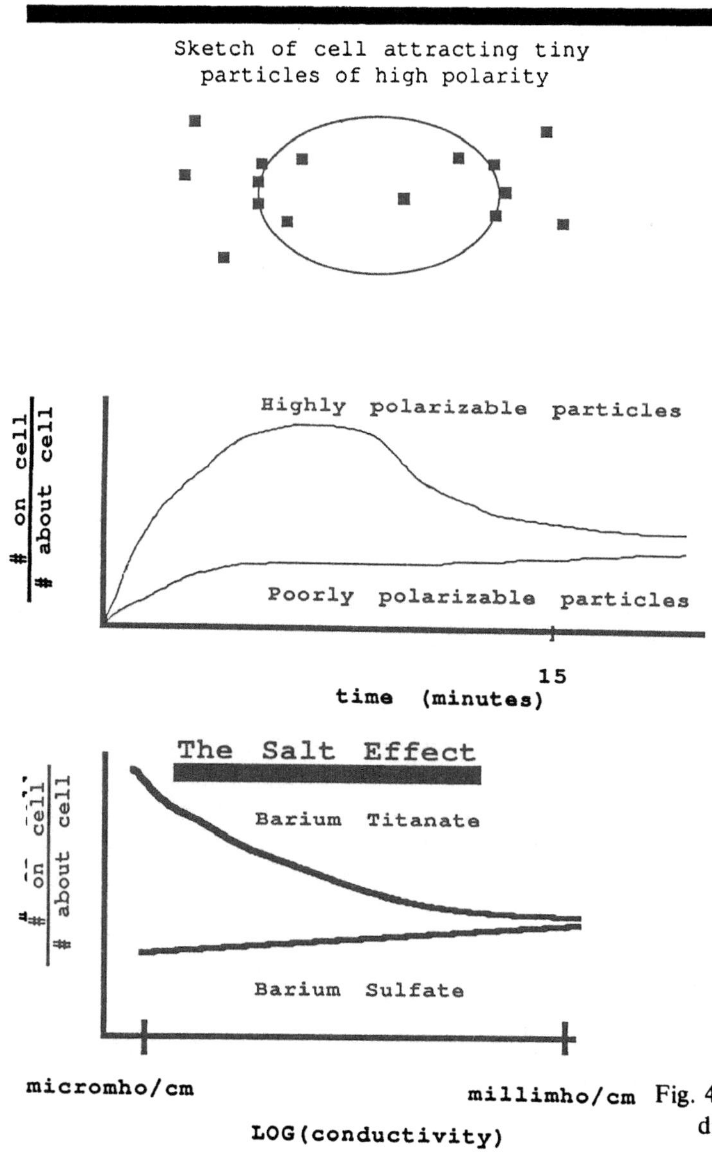

Fig. 4 — The formation of micro-dielectrophoresis patterns.

Table 2. Particle-gathering ability of cells

Cell	Ratio of Polarizable to Nonpolarizable Particles $(n/p)_{polar}$ to $(n/p)_{nonpolar}$
Bacteria: *Bacillus cereus* (synchronized)	4.0
Fungi: *Saccharomyces cerevisiae* (yeast)	2.0
Nematode: *Monochoides changii* (worm)	2.0
Amphibian: *Xenopus laevis* (clawed frog)	2.0
Aves: Chicken red blood cells	2.0
Mammalian:	
Mouse "L" fibroblasts	2.0
Mouse ascites (tumor) fibroblasts	2.2
Mouse fetal fibroblasts	2.0
Cell Types	
Mouse "L" cells (rapidly dividing/ slowly dividing)	
Mouse ascites tumor	2:1

RESULTS AND DISCUSSION

If a cell is emitting a natural ac electric field, the small size of the cell means that the field is diverging, as shown in Fig. 1. In the hanging drop experiment, sketched in Fig. 2, the living cell is seen to manifest such a divergent electric field by acting upon the suspended dielectric particles. The experiment somewhat resembles that in which one sees the familiar pattern of iron filings about a magnet. It is the dielectrophoretic analog of this lines-of-force (magnetophoresis) experiment.

The cells acting on the particles in the hanging drops form powder patterns that fall into three categories: attraction, repulsion, and random. The patterns are best shown in aqueous media of very low conductivity. Highly polarizable particles such as the titanates or the polymer DP1A are seen to be attracted to the live cells ... mammalian, avian, amphibian, annelidal, algal, fungal, or bacterial. The much less polarizable particles such as silica form different, repulsive-type patterns about such cells. Only random patterns develop about dead cells. It is moreover observed that of the highest conductivity, the attraction pattern of the strontium titanate particles reverts to repulsive, as expected if the effective dielectric constant of the medium exceeds that of the powder (cf. Fig. 2b).

By means of a pulsed dc electrophoresis experiment using Pt electrodes, all the cells and particles were tested for their surface electric charge. All were found to be charged negatively. Therefore the powder pattern formation about the cells could not be explained by simple *static* electrical forces. The existence of both repulsions and attractions for particles that are very similar chemically would also seem to exclude the possibility that the patterns are due solely to exudates from the cells. The attraction pattern seen at low conductivity is normal, but modified, we would say, by the presence of a film of exudate holding back the bulk of the surrounding powder particles, and letting the smaller particles skate over the film to gather normally at the cell under the influence of the natural ac field. In the higher-conductivity cases there is also evidence that an exudate from the cell holds off the highly concentrated particles, but some of them slip over the thin exudate film and still gather at the cell. The interpretation we would give for behavior observed in the last case, that of the highest-conductivity medium, also includes the effect of both ac fields and an exuded film. Here there appears to be an exuded film about the cell and restricted to the water drop surface. But even those particles which might skate over the film are seen not to lie near the cell because of the repelling action of the field at this high conductivity, at which the effective dielectric constant of the water exceeds that of the strontium titanate.

The experiments taken together show that the patterns formed by the cells and particles in a hanging drop are as expected for cells producing ac fields. The relevant observations follow.

(1) Live cells are seen to attract particles of high dielectric constant to themselves. If the orientation of the cells happens to be such that the dipolar oscillations lie in a plane more or less perpendicular to the line of sight of the observer, then a dipolar pattern of attraction (or repulsion) can occasionally be observed.

(2) Live cells are seen to repel particles of low dielectric constant from themselves.

(3) Dead cells have little or no effect on the distribution of particles of either low or high dielectric constant other than that of showing a volume of exclusion.

(4) The patterns become less well defined as the suspension conductivity is increased.

(5) The presence of added ions in the aqueous medium can increase the effective dielectric constant so as to evoke repulsive patterns even with highly polarizable particles such as reduced strontium titanate.

These facts are in accord with the theory for dielectrophoretic behavior of dielectric particles in the presence of nonuniform ac fields and can be understood in those terms.

We conclude that the powder patterns formed about living cells give evidence of the presence of ac fields.

The μ-DEP experiments lead to similar conclusions. Considering the above observations together, we can ask the critical question: are the observed particle patterns due to ac electric fields?

It is concluded that the observed effects are indeed due to ac fields about the cells and particles because:

1. Particles of high dielectric constant *collect* near the cell whereas those of low dielectric constant are *repelled* from near cells by long-range forces.

2. The observed effects cannot reside in mere chemical differences, because barium sulfate, aluminum oxide, and silica (all of low dielectric constant) are seen to be *repelled*, whereas particles of high dielectric constant such as barium titanate, strontium titanate, sodium niobate, and the organic polymer DPIA particles are all *attracted*.

3. The salt effect (seen for both NaCl and KCl) destroys both the attractions and repulsions, even at low ionic strength (ca 100 μM), in accord with the theory for dielectrophoresis. The salt added to the water increases its effective permittivity above that of even the titanate or niobate particles, so that the particles are electrically "invisible" (cf. Fig. 4).

4. Tests of the zeta potential show that both the cells and the particles have ionic double layers with the negative charge innermost. Ergo there cannot be a simple dc or static charge attraction accounting for *both* the attractions and the repulsions observed.

SUMMARY

The bioelectric fields of mammalian cells evoke pattern formation among tiny particles near them. The particle patterns form in the manner predicted for the action of natural cellular ac fields acting on dielectric particles through the phenomenon of dielectrophoresis. Of the particles tested, those with low dielectric constants were repelled from the cell, while those with high dielectric constants were attracted. This occurred although electrophoretic mobility (zeta potential) studies showed that all the particles and cells in this study were negatively charged. Distinctive powder patterns are present only about living cells. The patterns are sensitive to the conductivity of the suspension, becoming less pronounced as the conductivity increases in accordance with dielectrophoretic theoretical expectations. The observed patterns of the particles are due neither to static charge nor to chemical effects, but by *natural and nonuniform ac fields of the cells* acting on the dielectric properties of the powders and of the aqueous medium (dielectrophoresis).

Separate evidence indicates that the frequency of the observed fields lies in the range of about 30 to 10,000 kHz. The results reinforce those obtained by other methods of detecting natural ac fields in several laboratories here and abroad, so that the existence of natural ac fields, as originally postulated by Fröhlich (1968, 1977, 1980, 1982) can be considered well confirmed. Their function, if any, still remains unknown. There is much to learn.

ACKNOWLEDGEMENT

This research was sponsored and supported by the Pohl Cancer Research Laboratory, Inc.

REFERENCES

W.M. Arnold and U. Zimmermann, Z. Naturforsch. **37**, 908 (1982).

M.J. Berridge and P.E. Rapp, J. Exp. Biol. **81**, 217 (1979).

H. Block, E. Kluk, J. McConnell, and B.K.P. Scaife, J. Colloid Interface Sci. **101**, 320 (1984).

J.M. Cantril and H.A. Pohl, J. Electrochem. Soc. **115**, 700 (1968).

C.S. Chen, (1973), *On the Nature and Origins of Biological Dielectrophoresis*, Ph.D. Thesis, Oklahoma State University, Stillwater, OK 74078.

A.S. Davydov, Sov. Phys. JETP **51**, 397 (1980).

E. Del Guidice, S. Doglia, and M. Milani, Phys. Scripta **26**, 232 (1982).

H. Fröhlich, Int. J. Quantum Chem. **2**, 641 (1968).

H. Fröhlich, in *Brain Interactions with Weak Electric and Magnetic Fields*, S.M. Bawin and W.R. Adey, Eds., Neurosci. Res. Progr. Bull. **15**, 67 (1977).

H. Fröhlich, Adv. Electron. Electron Phys. **53**, 85 (1980).

H. Fröhlich, Bioelectromagnetics **3**, 45 (1982).

A.A. Furedi and R.C. Valentine, Biochem. Biophys. Acta **56**, 33 (1962).

R. Glaser, G. Fuhr, and J. Gimsa, Studia Biophys. **96**, 11 (1983).

L.H. Hartwell, R.K. Mortimer, J. Culotti, and M. Culotti, Genetics **74**, 267 (1973).

L.H. Hartwell, J. Culotto, J.R. Pringer, and B.J. Reid, Science **183**, 46 (1974).

J.H. Heller, (1959), Digest of 12th Ann. Conf. on Electronic Techniques in Medicine and Biology, Institute of Radio Engineers/American Institute of Electrical Engineers/Instrument Society of America, p 56.

C. Holzapfel, J. Vienken, and U. Zimmermann, J. Membrane Biol. **67**, 13 (1982).

A.H. Jafary-Asl and C.W. Smith, Ann. Rep. Conf. on Elec. Insul. and Dielec. Phenomena, IEEE Publ. 83 **CH 1902-6**, 350 (1983).

A.H. Jafary-Asl, S.N. Solanki, E. Aarholt, and C.W. Smith, J. Biol. Phys. **11**, 15 (1983b).

A. Lampa, Akad. Wiss. Wien, Sitz. Ber. **115**, (2a); 1659 (1906).

I. Lamprecht and M. Mischel, Naturwissenschaften **68**, 501 (1981).

P. Lertes, Z. Phys. **4**, 315 (1921).

M. Mischel and I. Lamprecht, Z. Naturforsch. **35c**, 111 (1980).

M. Mischel and I. Lamprecht, J. Biol. Phys. **11**, 43 (1983).

M. Mischel and H.A. Pohl, J. Biol. Phys. **11**, 98 (1983).

M. Mischel, A. Voss, and H.A. Pohl, J. Biol. Phys. **10**, 227 (1983).

H.A. Pohl, J. Appl. Phys. **22**, 869 (1951).

H.A. Pohl, J. Appl. Phys. **29**, 1182 (1958).

H.A. Pohl, *Dielectrophoresis, The Behavior of Matter in Nonuniform Electric Fields*, Cambridge University Press, London, 1978.

H.A. Pohl, "Micro-dielectrophoresis of dividing cells," in *Bioelectrochemistry*, H. Keyzer and F. Gutmann, Eds., Plenum, 1980a, pp 273-295.

H.A. Pohl, Int. J. Quantum Chem. **7**, 411 (1980b).

H.A. Pohl, J. Biol. Phys. **7**, 1 (1980c).

H.A. Pohl, J. Theor. Biol. **93**, 207 (1981a).

H.A. Pohl, J. Bioenerg. Biomembranes **13**, 149 (1981b).

H.A. Pohl, Coll. Phenom. **3**, 221 (1981c).

H.A. Pohl, Int. J. Quantum Chem., Quantum Biol. Symp. **9**, 399 (1982).

H.A. Pohl, J. Biol. Phys. **11**, 59 (1983a).

H.A. Pohl, J. Biol. Phys. **11**, 66 (1983b).

H.A. Pohl, Int. J. Quantum Chem. **10**, 161 (1983c).

H.A. Pohl, "Cellular Spin Resonance Spectrometer," U.S. Patent No. 4,569, 741. Issued Feb. 11, 1986.

H.A. Pohl and T. Braden, J. Biol. Phys. **10**, 17 (1982).

H.A. Pohl, T. Braden, S. Robinson, J. Piclardi, D.G. Pohl, J. Biol. Phys. **1**, 133 (1981).

H.A. Pohl and J.S. Crane, Biophys. J. **11**, 711 (1971).

H.A. Pohl and I. Lamprecht, (1984), Umschau **6**, 366 (1984).

H.A. Pohl and M. Pollak, J. Chem. Phys. **66**, 4031 (1977).

H.A. Pohl and J.J. Wyhof, Non-Cryst. Solids **11**, 137 (1972).

J.R. Pringle and L.H. Hartwell, in *Molecular Biology of the Yeast Saccharomyces,* J.N. Strathern, E.W. Jones, and J.R. Broach, Eds., Cold Spring Harbor Laboratory, 1981.

P.E. Rapp, J. Exp. Biol. **81**, 281 (1979).

S.I. Reed, (1980), Genetics **95**, 561 (1980).

S.I. Reed, (1984), in *The Microbial Cell Cycle*, P. Nurse and E. Streiblova, Eds., CRC Press, Boca Raton, FL, 1984.

H. Rivera, J.P. Biscar and H.A. Pohl, (1983), private communication.

H. Rivera and H.A. Pohl, (1986), "Cellular Spin Resonance," in *Modern Bioelectrochemistry*, F. Gutmann and H. Keyzer, Eds., Plenum, New York, 1986, Chap. 15.

H. Rivera, J.K. Pollock, and H.A. Pohl, Cell. Biophys. **7**, 43-55 (1985).

S. Rowlands, in *Coherent Excitations in Biological Systems*, H. Fröhlich and F. Kremer, Eds., Springer, Berlin, 1983a, p 145.

W. Rowlands, C.P. Eisenberg, and L.S. Sewchand, J. Biol. Phys. **11**, 1 (1983b).

S. Rowlands, L.S. Sewchand, and E.G. Enns, Can. J. Physiol. Pharm. **60**, 52 (1982).

S. Rowlands, L.S. Sewchand, R.E. Lovlin, J.S. Beck, and E.G. Enns, Phys. Lett. **82A**, 426 (1981).

S.C. Roy, T. Braden, and H.A. Pohl, Phys. Lett. **83A**, 142 (1981).

F.A. Sauer, in *Coherent Excitations in Biological Systems*, H. Fröhlich and F. Kremer, Springer, Berlin, 1983, p 134.

F.A. Sauer and R.W. Schloegl, in *Interactions between Electromagnetic Fields and Cells*, Chiabrera, C. Nicolini and H.P. Schwan, Eds., Plenum, New York, 1985, p. 203.

S. Schmidt and P. Ortoleva, J. Chem. Phys. **71**, 1010 (1979).

J.E. Treherne, W.A. Foster, and P.K. Schofield, (1979), *Cellular Oscillators,* J. Exp. Biol. **81** (review volume).

M. Westby and F. Kirschbaum, J. Compar. Physiol. **145**, 399 (1982).

28. On Morphogenesis in Living Systems

F. W. Cummings

With 3 Figures

ABSTRACT

A proposal is given for a description of single-cell-thickness surfaces that are undergoing morphogenetic movements such as in the early development of living systems. A two-dimensional "middle" surface must obey an equation which is independent of the coordinates, a fact which is useful in its determination as the Gauss curvature, a second-order partial nonlinear equation for the metric coefficient in conformal coordinates. This equation is thought of as coupled to a "pattern"-determining equation, which describes a combination of a set of biochemicals in terms of a single dominant one. A suggestion is made that this master pattern biochemical may be the ubiquitous calmodulin, a suggestion independent of the formalism given. The coupled system of two equations for pattern and form is then discussed as a model for gastrulation, the movement of the single-cell-thick sheets observed in numerous organisms. It is seen that a simple assumption for the Gauss curvature as a function of pattern, having one minimum, suffices to describe interesting aspects of early growth movements.

It is a deep pleasure and honor to contribute to a volume honoring such a good friend and constant inspiration as Herbert Fröhlich.

One of the most fascinating and challenging questions is that of the development of the pattern and form of living organisms. It is well understood that DNA plays a central and crucial role in development, but the fact that every cell contains the instructions (DNA) to form the complete organism tells one at once that there is more to be said on the subject than merely that there is a "genetic program" which gives rise to the various parts or organs of a multicellular organism. Furthermore, it is well known also that even single-celled organisms exhibit many of the remarkable regenerative and pattern-forming properties of multicellular ones, indicating a possible common origin (see Goodwin, 1980, for a review and references pointing out these strong similarities). In any case, the question is "How does the information coded into a linear sequence (i.e. DNA) get translated into a form and pattern existing in three-dimensional space?"

Since it is not known what molecular or cellular processes are involved in the spatial organization of organisms, it is perhaps best as a starting point to ask, "How can it happen, within the constraints of our present knowledge?" rather than the usual "How did it happen?". Of course, the latter question is largely unanswered at present, although long strides in this direction have been made in the last decade. It must also be stated that an extreme position is adopted in this present work, viz., that there are universal mechanisms at work; different species, perhaps different classes of organisms, elaborate on a common theme. To put it another way, we will assume that there is another universal which has yet to be uncovered besides DNA—perhaps not as strictly universal as DNA, but akin to it in its power to organize our understanding of the development of living organisms. In Wolpert's concept of positional information, cellular interactions are the basis of pattern formation and the pattern must be interpreted by a nuclear genome in every part of the field; this interpretation by the cell leads to cell differentiation (Wolpert, 1971; Wolpert and Lewis, 1975). (This interpretation has been criticized by Goodwin (1980) because single-celled organisms often exhibit pattern-forming and regenerative strategies strikingly similar to those of multicellular organisms.) Although Wolpert's positional information concept allows for any conceivable interpretation of the pattern, we will develop a model in which the response of the organism is greatly restricted. Cell differentiation in a multicellular organism can be viewed as a closing down of a part of the DNA, usually irreversibly, as if certain keys on a piano were nailed down; what fraction of the full potential a given cell displays depends on which genes are active and which are turned off. Each cell is characterized by a specific pattern of active and inactive genes that undergoes sequential changes as development goes ahead. The genome of a higher organism includes as many as 10^4 genes, and it seems that each gene is not controlled individually, but in groups, with a 'master' gene controlling the action of a sizable group. In recent years, genes have been identified that play a very special role in development, controlling both the timing and the spatial organization of tissues in the embryo precisely. Genes controlling the timing of events were discovered in a small worm (nematode, actually, *Caenorhabditis elegans*). This worm has 959 cells and its development has been studied to the point that the fate of each cell can be traced to a single gene. The fate of every cell in this nematode is determined by its ancestry, and the chronological pattern of events clearly has a particular significance. Control of the sequence of developmental events is of particular importance in an organism such as this where the developmental plan relies on invariant cell lineages. Most organisms do not develop this way, and *C. elegans* lies at one extreme of a continuum. At the other end are vertebrates whose embryonic cells may end up in almost any adult structure. The fate of a cell of this kind is

These master genes (dubbed the "homeobox"; see McGinnis et al., 1984; Gehring, 1985) that control development show a very surprising duplication of 59 out of 60 of their amino acids in vertebrates and invertebrates, which diverged in evolution about 500 million years ago; this gives impetus to the idea of universality. It is speculated that the master genes produce proteins which bind to the DNA, where and how being a current topic of much research. What is apparent from this is that these master genes must themselves be regulated. This is the subject of concern in the present paper: how the genes that control large groups of genes are themselves regulated.

Because of the striking similarities of development shown in different species, as well as the emerging universal features of the control genes mentioned above, it seems reasonable to adopt a "universalist" outlook in our attempt to find what controls the controllers.

The organization of the present work is as follows. In the first Section, a description of the surface of a developing organism will be proposed, to lay the groundwork for deriving an equation that describes the surface (actually the middle surface), which changes under the impetus of patterns of biochemicals in this surface, the changes involving the DNA directly or indirectly. After the simple description of a surface, the second Section develops an equation or equations which are to couple to that of the surface metric. This is approached initially by a brief consideration of a particular and well studied biological system (the insect imaginal disc) which suggests certain general characteristics of any description of a biochemical system crucial to development. The actual systems or biochemicals involved in the "pattern field" are unknown, tempting one to speculate. At the close of the second Section, a biochemical, a protein, is proposed as the key element of the pattern function, the controller of the control genes, although the core of the article is independent of this specific speculation. The "toy" model suggested in the second Section (which may help point the way to a more complete or correct one), is examined in the third Section. This set of equations coupling morphological changes to biochemical pattern changes is examined as a model of both gastrulation (in echinoderms such as *Synapta*), and segmentation in insects; these two processes are seen as variations on a single theme. Time dependence is introduced as being impelled, at least initially, by the change with (developmental) time of the total area, an input to the model, when cell proliferation as a function of position is also introduced.

A SIMPLIFIED DESCRIPTION OF MORPHOLOGY

The early development of multicellular organisms typically occurs morphologically by the folding of sheets of cells (epithelia) of one-cell thickness. The cellular sheet can be likened to a surface made of a large number of cans with one surface (the apical surface) locally forming a plane and all cans joined to their neighbors at their sides. The cells are deformable, however, so that there is in general a gradient of radius-to-height ratios of the cells as we move in the surface of the sheet. Furthermore, the ratio of the apical surface area of a cell to its basal area also varies as we move in the plane of the surface. In the following we will be restricting ourselves to consideration of the middle surface; if a cell has height τ, then the middle surface is the plane that bisects each cell at a depth $\tau/2$ below the apical cell surface. The thickness of the cell sheet will be given by a function $\tau(u, v)$ where the coordinates u and v give the position on the middle surface. (This description, although couched in terms of cells for picturesque purposes, is also an appropriate one for single cell surfaces, since it does not in the end include any cellular parameters.) A method of treating the middle surface appropriate to the developing organism is now discussed.

In order to describe any surface relative to an external coordinate system (Cartesian x,y,z, say), specification of both the so-called first and second fundamental forms is required, or six numbers in general at each point (see, e.g., DoCarmo, 1976). The first fundamental forms are the three metric coefficients, tensor components g_{11}, g_{12}, g_{22} and b_{11}, b_{12}, b_{22}, each a function of the coordinates

u,v of the middle surface. Likewise, the three second forms are also tensor components. In the usual application in physics or engineering, forces or torques are applied externally and the resultant deformation of the surface is required. Interest here is rather in a situation in which all forces are in the surface, generated by the organism; that is to say, they are self generated forces that can be viewed as acting, at least to a first approximation, in the middle surface. We will suppose that the forces are always in equilibrium; a conceptual division of the forces into two will be made, so that the forces responsible for the cohesion of the sheet (those that hold a given little square of middle surface intact) are balanced by forces that act to stretch or compress the little square of middle surface at any time. Numerous cells will make up any little square, and these as a group will exert no net shear. The (adiabatic) deformation of the entire surface by forces that act essentially in the plane of the middle surface is what is required here, so we will not need to consider the second fundamental forms, since there are no "external forces" comparable in magnitude to the (balanced) "internal forces". The times over which morphological changes occur will be presumed to be long compared to the times over which the in-plane forces come to equilibrium. Then the orientation or position of the organism relative to an external coordinate system is irrelevant. Again, this is to be contrasted to the usual mechanical engineering application, when the external or applied forces are a crucial element, and the second fundamental forms are thus most relevant. It would seem most beneficial for the developing organism to have a situation where the forces shaping it are under its own control, so to speak, so that its development is not subject to the vagaries of external forces. Thus in our present description of the morphogenetic movements of our (in general closed) middle surface (hereafter called the "surface") only the three metric coefficients that give the distance in the surface between neighboring points will be retained. Also it is to be realized that whenever a part of the developing organism is isolated for consideration, and then the remaining part or parts are in essence considered as external forces, the present description is invalid.

In order to have the discussion as simple as possible, and yet as general as desired, choice of a particular class of coordinates will be useful, the so called conformal coordinates. In these special coordinates, the two diagonal elements of the metric tensor are taken equal, and the nondiagonal term is set to zero, so that there is only one function which describes the geometry of the surface. Any surface is thus locally divided into little squares, and the size of the squares varies with coordinates u, v. The area of any small region is given by ΔA where

$$\Delta A = g \Delta u \Delta v, \tag{1}$$

and

$$(ds)^2 = g\left[(du)^2 + (dv)^2\right]. \tag{2}$$

Here $(ds)^2$ is the squared distance between the points u, v and $u + \Delta u$ and $v + \Delta v$ and g is the single metric coefficient (hereafter called 'the metric') containing all relevant information on the surface. It is easily verified that any surface can be covered by an infinity of different coordinates, since any analytic function (two functions obeying the Cauchy-Riemann conditions) represents a change of coordinates that preserves the conformal character typified by specification of one metric coefficient. It must be added that some surfaces require covering in patches, when one metric is not sufficient to cover the entire surface.

The Gauss curvature of the surface at any point is the product of the two principal inverse radii of curvature, and are given in conformal coordinates by the simple expression (see, e.g. DoCarmo, 1976).

$$K = -\frac{1}{2g}\left(\partial_u^2 + \partial_v^2\right) \ln g. \qquad \left(\partial_u \equiv \frac{\partial}{\partial u}\right). \tag{3}$$

It will be assumed in what follows that all expressions are given in the conformal coordinates unless specifically indicated otherwise. The simplest form of the metric for a sphere, involving as it does only one variable, is

$$g = R^2 sech^2 u, \tag{4}$$

with u going to $\pm\ \infty$ at the north and south poles, respectively. The *sech* (squared) has a bell shape, so that the little squares are largest at the equator, and decrease rapidly in size as the poles are approached. Use of this g in the Gauss curvature, the product of the two principal curvatures, gives at once that K is $1/R^2$ as expected. It is easily verified that the argument of the *sech* function may be replaced by the real or imaginary part of any analytic function, and K remains equal to $1/R^2$. This is a reflection of the fact that K is an invariant, (as is $(ds)^2$), so that its value is unchanged when coordinates are transformed.

It should be pointed out that the present description is incapable of distinguishing between any two surfaces that can be transformed one into the other by operations which do not either stretch or tear the surface. Thus, for example, all right cylinders of the same total surface area are equivalent for our purposes, and all have zero Gauss curvature. The sphere is uniquely the surface with positive constant curvature. (The surface of negative constant curvature resembles a "saddle of revolution," and has a metric of $(sech^2 u)$ (see Cummings, 1985). It is interesting to note that this property of indistinguishability will be a desirable feature of a model developed later.

Now that we have decided that we are to take the single function g to describe our surface, the question is "What equation can we write for g that will couple surface to pattern?" Elementary considerations tell us that we do not have far to look, as our options are very limited. Coordinates are unknown to the organism, being solely of the observer's own free choice. This means that when we write down an equation for g which describes observable properties of the organism (e.g., its surface) this equation must describe the same surface no matter what coordinate system we choose; that is, as we say, the expression must be covariant. We will further wish an equation which is linear in the highest derivative, so that the solution may be unique. Since the first covariant derivative of the metric tensor vanishes, we turn at once to the second derivative. It can be shown that the Gauss curvature is uniquely the covariant expression which involves the second derivative of g linearly. Then we take for our equation determining the metric g

$$-\frac{1}{2g}\left(\partial_u^2 + \partial_v^2\right) \ln g = K(\eta, \tau). \tag{5}$$

Here it has been indicated that K is to be a (covariant) function of the biochemical(s) $\eta(u, v)$, and also of the sheet thickness $\tau(u, v)$.

It will be apparent to anyone who has looked at Einstein's General Theory (Adler et al., 1975) that what we have here is a two-dimensional analog of it. The Gauss curvature is determined by biochemicals, which form a pattern on the surface, so that the pattern function(s), (and perhaps also the sheet thickness t, presumably also determined by the pattern function(s)), play the role of the matter-energy tensor in Einstein's theory. Here the equation is not "two-tensor equals two-tensor", but only "scalar equals scalar" since there is no effective difference in two dimensions between the curvature scalar, the curvature tensor, or the Ricci tensor. The next task is to examine possibilities for the form of the equations governing the pattern.

Before going on to this next issue, several remarks are in order concerning the form of the "source" term in the equation for the curvature, that is, the form of the function $K(\eta,t)$. It must be apparent that phenomenology must enter any description of biological phenomena, and at this point we expect to see phenomenology in an unabashed state. The curvature K at any developmental time is expected to depend on the particular organism under discussion. Possibly it is at this point in the description that the particular properties of the genome of the organism enters most forcefully, and one organism will differ from another to a large extent by having a different form for the function K. Put another way, the pattern functions of different organisms are expected to elicit different responses from the different genomes, and thus result in a different cellular deformation in a given region. In a simple model, which we discuss later, it is seen that it is the interplay between the parameter(s) of the pattern field and the shape of the K function which determines morphology as a function of time. A minimum model in which a rather simple functional form for K (a function with one minimum) as a function of a single pattern function will lead to interesting morphological changes similar to those observed in several organisms.

Time will enter the present model parametrically through the dependence of the total area of the middle surface on time. To clarify how this surface can change in time, imagine the following situation common to the very early development of echinoderms, for example. Cells, at first few in number, form a sphere-like structure in which the thickness of the surface is a non-negligible fraction of the radius. If the volume of the total cellular mass stays constant as cell division proceeds, then inasmuch as a single-celled surface is being formed of smaller and smaller thickness, it is clear that, as time goes on, the area of the middle surface will increase. It will be seen later that this allows time to enter the analysis with the aid of the Gauss-Bonnet theorem of surfaces. (This theorem says that the integral of the Gauss curvature over the entire surface must equal a constant, the value of the constant dependent only on the topology of the surface: 4π for a surface topologically equivalent to a sphere, 8π for a surface equivalent to a sphere with one handle (or a doughnut), and so on. As has been implied, the topological equivalent of a given surface is any surface which can be obtained from the original by stretching and bending, but not tearing or puncturing.)

REMARKS ON THE PATTERN FUNCTION(S)

There has been extensive work on the theory of nonlinear diffusion-reaction equations, and their possible relevance to the patterning in developing organisms (see, e.g., Meinhardt, 1982; Murray and Oster, 1984, and references therein). We will not attempt to summarize these works, and will state only that such systems are capable of giving almost arbitrarily complex patterns. Although it seems very likely that such reaction-diffusion schemes play some (unknown at present) role in development, there has been as yet no experimental evidence that such schemes are actually those made use of in developing systems, although there is a good bit of circumstantial evidence for this. At any rate, the underlying interpretation of such systems in microscopic (biochemical) terms is still an open question. One possibility from the present point of view is that some such set of coupled diffusion-reaction equations are the appropriate ones and are coupled to the surface as we have indicated, that is, by giving K as a function of one or more of the biochemicals, e.g., calcium. From this point of view, the present work then amounts to a suggestion for the coupling of pattern formation to geometry or morphology. In the reaction-diffusion equations, the metric enters by way of the Laplace operator, where

$$\nabla^2 = \frac{1}{g}\left(\partial_u^2 + \partial_v^2\right). \tag{6}$$

We take a different path here, and suggest that the coupled set of chemicals be simplified, and replaced by one single linear equation, the motivation being that since the actual chemicals involved have not been identified at this time, taking a toy model that mimics the main features of the more complicated nonlinear coupled set will allow us to see general features without resorting to tedious machine calculations. It will be assumed that the curvature K depends mostly on one pattern function, and further that there is one dominant parameter (a length) which governs the pattern behavior (and thus the curvature). Much of the nonlinear behavior of a more complete future system of equations will be included by fiat, as will be detailed later.

A number of pertinent and interesting points may be illustrated by a brief summary of the results of surgical intervention experiments on the imaginal disc of the developing fruit fly, *Drosophila*. These experiments exemplify the "regeneration-duplication" motif observed more loosely across a wide range of organisms; we mention specifically the imaginal discs because they have been studied most extensively and carefully in this respect (Bryant et al., 1981). Imaginal discs of insects are the origin of the adult insect's wings, legs, and genital and other appendages. It is composed of a single layer of cells, similar to a deflated balloon. After surgical bisection, a disc fragment is first allowed to grow by implantation in an adult fly, after which it is transplanted to a larval host where it undergoes metamorphosis with the host, when it has no opportunity for growth. Upon metamorphosis, the fate of the various structures is then revealed, or the various parts are differentiated. The often-noted phenomenon of parallel regeneration and duplication is most conveniently described by reference to the wing or leg disc of the fly.

The principal results of such a procedure may be most easily illustrated by reference to a sphere instead of a folded balloon. Imagine a sphere with some pattern on its surface. Next imagine surgically removing a cap piece from the sphere, i.e., one including the north pole. This pattern (of various biochemicals, presumably) represents the presumptive fate of the cells after metamorphosis to the adult. When the two pieces of the sphere are cultured, the larger of the two will regenerate itself, i.e. will reproduce the original sphere with the original pattern, while the smaller of the two will duplicate itself in mirror symmetry. If a cap piece is removed from both north and south pole regions, then the two caps will duplicate while the annular midregion will regenerate the entire sphere. Cuts that bisect the sphere and that are made near the midsection will often result in both pieces regenerating or both pieces duplicating, as well as the regeneration-duplication result. Thus there are four different outcomes to any surgical bisection. It may be worth noting that immediately after bisection and before the parts are replaced by cell division, the wound edge will "purse-string" and create a closed structure as soon as possible.

Thus, it would appear that the boundary edges are determining, at least in part, the piece that is being grown. In the case of a regenerating and a duplicating piece, in a sense, each piece grows the same replacement piece. Not only must the correct pattern be replaced, but the concomitant geometry must also; it is not sufficient to replace the excised cap with a planar circular piece where the patterns match on the boundary; the geometry of the regenerated or duplicated piece must also be that of a spherical segment. So any model must minimally be required to lead to the regeneration-duplication results outlined above. In Cummings (1985) the same equations as in the present paper were discussed from a different point of view. There, adhesive molecules were proposed as the protein underlying the pattern function, whereas in the present paper the pattern of any adhesive molecules is seen as determined by a more primitive (and ubiquitous) protein.

If it be imagined that there is only one predominant pattern function, then it must also be supposed that, in a multicellular system, the genome is making binary (or possibly trinary) decisions at each point in developmental time, depending on the pattern function (and perhaps even also on the metric or geometry). Supposing that there is one predominant pattern function, we see that it must obey an elliptic partial differential equation, one whose solution is given uniquely once the conditions on the boundary are given. We presume here that the values on the boundary stay fixed, and the new parts are replaced by cell division and growth of new tissue, or as is said, the system is "epimorphic". The simplest elliptic equation which satisfies our conditions that the pattern and geometry be replaced properly is the Helmholtz equation, as this is the simplest one that contains a length. It is readily seen, for example, that the Laplace equation will not suffice since one finds a solution for any geometry with a pattern function matching the boundary conditions.

In keeping with our aim to not complicate matters beyond what is needed to see a possible scheme of development, a "derivation" of our toy model will now be given, as well as a speculative identification of the pattern function with a specific biochemical. Imagine a surface (the middle surface) that has been covered with a set of orthogonal u,v coordinates. Also imagine a biochemical (a protein) that can exist in two states, which we imagine can exist in two states of conformation or folding. Let us call these C^+ and C, where C^+ gives the number per unit area of the excited or active state, i.e., folded. The simplest assumption that yields a nontrivial result is to suppose that C^+ changes in a given little square bounded by $u \pm \Delta u$ and $v \pm \Delta v$ proportionally to the deviation of its value at time t in this square from the average of its value in the neighboring squares, and also inversely proportionally to the area of the square. This will have the result of having C^+ smooth itself out as well as producing possible consistency with the boundary conditions. Besides this 'smoothing' or diffusion term, we mimic the autocatalytic effect expected in similar circumstances by further supposing that C^+ will decrease in the square in question proportionally to the difference of the density of unexcited chemical C from some average (for example, one-half of the total constant density). If it can be supposed that the chemical is distributed uniformly, so that the sum of the excited and unexcited states is a constant, then we write these relationships as

$$C^+(t+\tau) = C^+(t) + D\left\{< C^+> - C^+(t)\right\}/\Delta A - \epsilon(C - C_0). \tag{7}$$

$$C^+ + C = const. = C^{+0} + C_0,\ \Delta A = g\Delta UA\Delta v, \tag{8}$$

and

$$< C^+> = \text{average of } C^+ \text{at neighboring sites.}$$

The proportionality constant D has dimensions of area, and ϵ is dimensionless. As we go to the limit of infinitesimals, the first term on the right goes into the Laplacian operator, and using the value of C given in the second equation in the first leads to

$$\nabla^2\eta + k^2\eta = 0, \tag{9}$$

$$\eta = C^+ - C_0^+ . \tag{10}$$

Here we suppose that we are always in the steady state, i.e., that the processes above equilibrate at a rate much greater than the rates usual for developmental processes, typically minutes or even hours. Thus we have taken C^+ at time t to equal C^+ at time $t + \tau$. The equation is manifestly covariant, that is, independent of choice of coordinates, since it involves the Laplacian operator, and the equation may be written in conformal coordinates as

$$\left(\partial_u^2 + \partial_v^2\right)\eta + k^2 g\eta = 0, \tag{11}$$

showing its coupling to the equation for the "geometry" explicitly through g.

Several points must be emphasized here. This equation is to be thought of as a linearized version of a more exact equation, or set of coupled equations. Since the equation is linear, the amplitude may be anything. We avoid this problem by postulating that the amplitude will attain a maximum (say unity) wherever possible. We thus hope to circumvent much (and we feel unnecessary, given all other uncertainties) mathematical complexity of equations that properly show the growth and stabilization of the amplitude. This has been worked over in much detail since Turing first suggested the system of reaction-diffusion equations as a source of pattern in development. The effect is obtained here by fiat, and the result is an equation that can be dealt with more intuitively. The excited state C^+ will form into "islands" with one dimension the size of the "wavelength", which we view as a parameter that dominates the development and is determined by the biology, i.e., the DNA. It is taken as fixed during one particular developmental time period, although it can change from one period to the next. This will hopefully become clearer when we discuss certain features of development using our present simple "Ansatz", consisting of the two coupled equations, eqs. 5 and 9. As the simplest assumption, it is further supposed that the curvature K of eq. 5 depends only on the pattern function η and the time, and not explicitly on the thickness of the cell sheet, except insofar as this may be represented in the functional form of K.

To repeat, we think that in view of the present state of uncertainty or ignorance, a more complex mathematical formulation is unwarranted at the present time. Already the present simple model presents great difficulty in solution, mostly because of the highly nonlinear nature of the equation for geometry g.

To end this section we present a speculation as to the possible biochemical nature of the pattern function(s), a discussion which can stand apart from the previous ones, but which attempts to give a specific biochemical identity to the C^+ and C above.

Calmodulin is a protein consisting of a single chain of 148 amino acids. The protein is folded into four roughly matching domains, each with a calcium (Ca^{2+}) binding site. The affinity of the sites for calcium is such that whether the ion is bound or not depends on its concentration in the intracellular environment. Calmodulin is a "tough" molecule, withstanding low pH and boiling water, conditions that disable most proteins. A highly relevant point is that it is the most ubiquitous protein known, having been found in all organisms above the level of bacteria, as well as in every kind of cell in all organisms. It is apparently a component of all eukaryotic (nucleated) cells. Another ubiquitous component of living systems is the ion Ca^{2+}, the ever-present partner of calmodulin. The ion is also abundant in all living systems, and its concentration is between 10^4 and 10^5 times as high in the extracellular fluid as in the cell interior. A cell cannot tolerate a high internal Ca^{2+} level, because the ion combines with essential organic phosphates such as ATP. On binding calcium, the calmodulin molecule takes a more compact shape and becomes activated. Calmodulin is found both in the plasma membrane and in the cytoplasm of the cell. In its active state in the plasma membrane, it is known to "turn on" an enzyme that produces cyclic AMP, a sort of universal activator of the cellular processes. In the cytoplasm, it causes the enzyme (phosphodiesterase) to degrade cyclic AMP. Thus it acts both to activate and to deactivate cyclic AMP. There is a cellular pump which pumps excess calcium out of the cell, and calmodulun has been implicated as an activator of the enzyme that does this, at least in some cells (probably in all cells), thus regulating the intracellular concentration of the ion.

In the terms of our very primitive model of an autocatalytic system of several biochemicals, calmodulin is suggested to play the central role. The excited state is the conformation in which four calcium ions are bound. If we imagine a sphere covered with excited proteins uniformly distributed, except for the presence of a few excess ones in the excited state at the north pole, then we suppose that these few will (auto)catalyze the excitation of others in its neighborhood until the top half of the sphere is covered in a graded way from north pole to equator with excited protein, in such a way that the density of excited to unexcited proteins is zero at the equator. Half of a given surface will then finally be covered predominately with excited and half with unexcited protein, these grading smoothly into each other, this *provided* that the relationship between the length k and the surface size is correct. The integral over the surface of the variable η will be required to vanish in any allowed pattern, expressing the conservation of protein. As an example of the simplest nontrivial pattern, a sphere will be covered with a pattern function η of the form

$$\eta = \tanh u \,(= \cos\theta \text{ in spherical polars})$$

where the second expression on the right is in the more familiar spherical polar coordinates, while the first is given in the conformal coordinates. The maximum will be chosen to be $+1$, so that the minimum will be -1 and the maximum will be assumed to be achieved whenever the condition on the integral just mentioned can be achieved, i.e., that the surface, in the simple case just mentioned a sphere, is of sufficient area that the condition can be satisfied. For example, in the case of the sphere, the minimum sphere size will be related to the length k so that $(kR)^2 = 2$.

It is to be noticed that the "Helmholtz" equation (9) determining the pattern for a given geometry always has the solution $\eta = 0$. It is an illustration of the mimicry of an autocatalytic process by fiat, as mentioned earlier, we require the function to take on its maximum and minimum, here specified by fiat also as ± 1, whenever this is allowed by the conjunction of pattern and geometry, i.e., whenever the Helmholtz equation can be solved on the given geometry. Thus much complicated nonlinear mathematical fancy-stepping and tedious numerical work is short-circuited, the motivation being to a large extent that the underlying processes themselves are largely unknown.

Whenever the correct match between pattern and surface cannot be achieved, for example in the sphere just mentioned when its radius is too small to accommodate the $\cos\theta$ pattern, it is supposed that the pattern function is zero, that is, half of the proteins are activated and half deactivated in any small area (or to be more general, each half is equal to its average value). It is also to be noted that the selection of the position of the "north pole" as a special point must be determined by processes not considered here, e.g. gravity. Thus determination of the animal and vegetal pole of the undivided egg is not considered.

It is to be stressed that this is not an eigenvalue problem where the "length" k is to be determined, but is such that the sphere radius giving rise to a possible pattern (to speak in the terms of the simplest geometry), is determined by the biologically fixed k, fixed at least over the time of development of a single field, (after which it may take on a different value).

Two points would seem to bear repeating. First of all, the simple model described above is not to be taken literally. It is proposed as a toy model to emphasize the salient points of the scheme of "pattern coupled to geometry". Of course there must be several chemicals at least involved in the reactions leading to the autocatalytic action of the key chemical (if calmodulin is the key protein, the ubiquitous calcium ion must of course be involved too), which in turn actuates the relevant cell processes (e.g., by leading to the activation of adenyl cyclase, the production of cyclic AMP, and thus the activation of the various cell-specific protein kinases that in turn phosphorylate the various cell-specific enzymes). Various final proteins will thus be involved in the cell differentiation.

The interaction between the key pattern function η and the curvature K is in any case to be thought of as indirect, in the sense that a chain of chemical processes set off by the activation of the key chemical (calmodulin, say), will put in operation the structural machinery of the cell (e.g., the synthesis of microtubules from tubulin). This may occur via the mediation of the DNA either directly, i.e. by the activation of the chemicals already produced earlier by the DNA, or by induction of the nucleus to produce the relevant geometry-changing factors. It would seem that this process of pattern-to-morphology change is a prime candidate for the phenomenological representation of curvature function as a function of the key activator, thus shortcutting numerous biochemical interactions, and a large amount of ignorance.

A MODEL OF GASTRULATION AND SEGMENTATION

Animal eggs (we will focus on echinoderm eggs, small ones in which morphogenetic movements easiest to follow) usually show evidence of a polar axis before fertilization, such as the eccentric position of the nucleus. Subsequent cell divisions following fertilization lead to a blastula, a hollow ball consisting of a single spherical layer of smaller cells, one cell thick (see for example Berrill, 1983; Trinkaus, 1983). The polar effect is seen in the direction of early cell cleavages (for example, the first cleavage in most small eggs is in a plane through the polar axis) but more interestingly for our purposes, its effect is also seen in the transformation of the blastula into the gastrula. Growth of the blastula is accomplished by successive cell divisions in a plane of its middle surface so that cells remain side by side; after each division the daughter cells assume the same cuboidal shape as before except on a smaller scale. Thus the wall thickness diminishes while the aggregate cell surface enlarges with each successive cleavage. When a certain middle-surface area is attained, consisting of some thousands of cells, gastrulation is brought about by an inpushing, or invagination, from the center of the vegetal pole into the internal cavity, or blastocoel. Prior to gastrulation, and determinative of it, some factor becomes progressively concentrated toward the vegetal pole. Gastrulation is a process whereby the organism is established as a sort of tube within a tube—or as endomesoderm within ectoderm. One result of excising a region about the vegetal pole (called the vegetal plate), is that this isolated piece continues the invagination. The inpushing deepens and the rim of the isolated plate rolls up and closes over the central evagination to form a comparatively small gastrula. This shows (see, e.g., Berrill, 1983) that the forces act in the plane of the vegetal plate. What also happens sometimes is that the isolated piece exo-gastrulates, that is, performs the same folding except that inner surface exchanged for outer; thus any model must not be overly deterministic with regard to this process, and must allow both gastrulation and exo-gastrulation. It is this early stage of animal development that we will model, from the (say) sixteen-cell stage through gastrulation.

The process of blastula formation will be picked up at the (say) sixteen-cell stage, when it has begun to be identifiable as a hollow ball of cells. As cell division proceeds and the total area of the blastula increases, according to the model above, the pattern function will remain zero until a certain critical area of middle surface is approached, when the pattern will more or less rapidly approach the cos θ distribution. This pattern will have a maximum of one at the vegetal pole, grade to zero at the equator, and graded further to minus one at the animal pole. Thus the observation that the blastula grows as a hollow ball of cells until a critical size is reached is seen to be consistent with the model. According to the Gauss-Bonnet (G-B) theorem of surfaces, as already stated, the surface integral of the Gauss curvature is a constant, 4π. Thus during *this* stage of growth, the curvature will have the form, a constant positive curvature (thus that of a sphere), being assumed,

$$K_0(\eta, t) = 1/R^2(t) = 4\pi/A(t)$$

Here $R(t)$ is the sphere radius and $A(t)$ is the total area of the sphere. This form for K clearly satisfies the G-B theorem, and it is in this way that growth enters the model, i.e., via the total area. There is also another term in the Gauss curvature, a term that depends on the pattern function η and is taken as zero for η equal to zero. This term $K_1(\eta)$ is shown in Fig. 1, where we treat it as a function with one minimum in the interval of $+1$ to -1. This function has been assumed to have a vanishing area integral to satisfy the G-B theorem, which necessitates that K_1 be negative over part of the interval. The total Gauss curvature K, the sum of the η-independent and the η-dependent terms, is shown in Figure 2. What is now apparent is that the total area will grow so that the function will begin to become large at one pole and small and negative at the other. At this point, the function $K(\eta, t)$ will have a negative part as required by the G-B theorem. This shape will be that of a gastrulating organism, having a sphere-like geometry where the region in the vicinity of one pole has been invaginated. To carry the analysis further would require the solution of the pattern equation (eq. 9) on a similar structure (i.e., on a sphere with an invagination) with the area increased to the stage when the next pattern with more than one "island" of positive pattern is allowed. Or as another possibility, the next stage could be one with a different value of the "wavelength" k^{-1}. At any rate, this next stage is too difficult to attempt to describe at our present stage of insight.

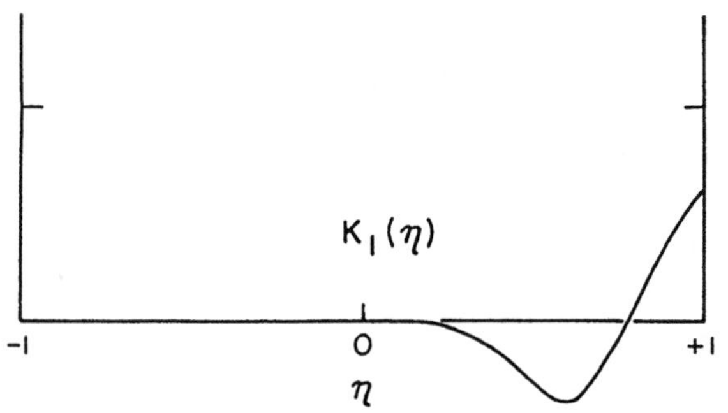

Fig. 1 — Schematic of $K_1(\eta)$ vs. η.

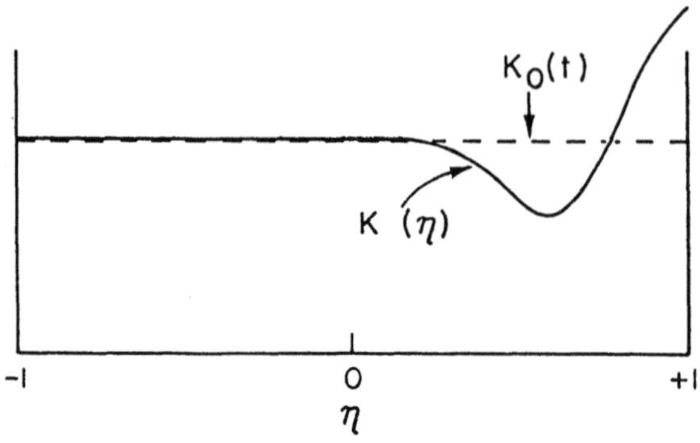

Fig. 2 — Schematic of $K(\eta, t)$ vs η.

Insect segmentation is a process that appears at first sight to be entirely different from the gastrulation of the echinoderms described above. After the sperm cell unites with the egg nucleus, synchronized cycles of nuclear division beginning soon after fertilization double the number of nuclei about every 10 min, so that after the eighth division there are 256 nuclei, at which time the nuclei begin to migrate to the cortex, or periphery, of the egg. When there are 512 nuclei, the first cell membranes are formed at the cells of the posterior end, which cells ultimately give rise to the germ cells of the adult fly. The other nuclei continue to divide in the cortex. When there are about 6,000 nuclei in the periphery, membranes separate the nuclei into a monolayer of cells called the blastoderm. Early in embryogenesis, even before cell membranes have been formed around the nuclei, a gene is being expressed in alternate sections, or segments. Thus the nuclei, without membranes, have a sensor which tells them their position in the cortical cytoplasm (see Gehring, 1985). The segmentation pattern of *Drosophila* has thus already been established at the blastoderm stage, and groups of cells in the blastoderm have been assigned to one segment in the larval and adult body, although the assignments at this stage are by no means complete. After the segments are laid down, each cell in the segment is assigned to either the anterior or posterior half, and the segment is thereby divided into two compartments. Then follows the distinguishing of the imaginal discs from one another. It is the process of segmentation that we will correlate with our model.

Note first of all that it is required that the autocatalytic process of pattern formation must be operative in the absence of cell membranes, and our speculation that calmodulin is the primary patterning protein must be so constrained; this says that it is not only as a membrane-bound component(s) that the pattern function(s) operate.

The blastula of the fruit fly can be viewed as an ellipse of revolution about the long axis. It is supposed that the first stage of development, during the migration of the nuclei to the cortex, consists of the formation of a pattern analogous to that of the echinoderm blastula above. The anterior end (as contrasted with what was ‘to become’ the posterior end in the previous case) develops a pattern of positive η that is graded to zero at the mid-circle and goes to negative values as the posterior end is approached. This will lead to an invagination of the head region (as observed), for the same reason as before. The second field will then be laid down, characterized by a larger value of k and a smaller (by a factor of about 11)

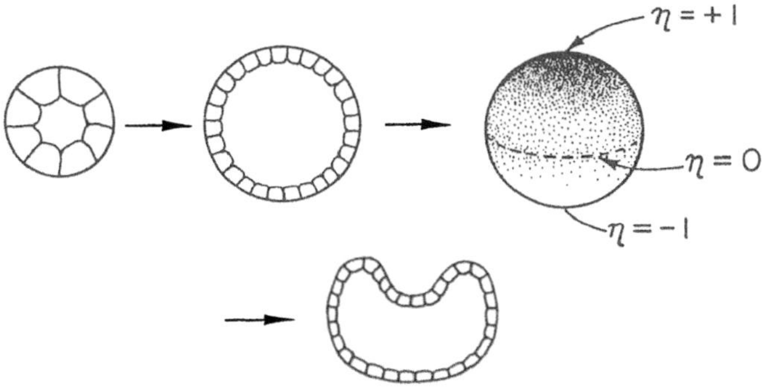

Fig. 3a — Schematic of gastrulation.

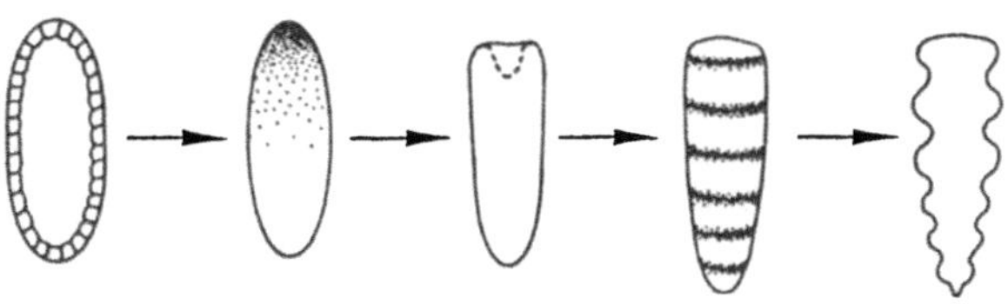

Fig. 3b — Schematic of insect segmentation.

wavelength. The ellipse with the retracted head segment parts is not girdled by bands that alternate from $+1$ to -1 as the length is traversed. The positive-value bands will invaginate as shown in Fig. 3, for the same reason that this occurs in the head region, i.e., that there exist now negative-curvature regions in alternating bands corresponding to the various segments. Notice also that the sausage without invaginations has a nearly zero curvature except near the ends. The value of k, assumed to be under DNA control, can continue to increase as we progress from field to field, and the coupling of pattern to geometrical change can also vary from field to field, a field being defined as having a single value of k, or wavelength.

REFERENCES

R. Adler, M. Barzun and M. Schiffer, *Introduction to General Relativity* 2nd Edition (McGraw-Hill, New York) 1975.

N.J. Berrill, *Developmental Biology, McGraw-Hill, New York, 1983.*

S.V. Bryant, V. French and P.J. Bryant, Science **212**, 993-1003 (1981) (and references therein).

F.W. Cummings, J. Theor. Biol. **116**, 243-273 (1985).

M. DoCarmo, *Differential Geometry of Curves and Surfaces,* Prentice Hall, Englewood Cliffs, NJ, 1976.

W. Gehring, 1976 Sci. Am. **253**, No. 4, 152B-163 (Oct., 1985).

B. Goodwin (1980) in *The Eukaryotic Cell*, Soc. for General Microbiol. Symp 30, Cambridge Univ. Press, pp.

W. McGinnis, R.L. Garber, J. Wirz, A. Kuroiwa and W. Gehring, Cell **37**, 403-408, 409-414 (1984).

H. Meinhardt, *Models of Biological Pattern Formation*, Academic, New York, 1982.

J.D. Murray and G.F. Oster (1984) IMA J. Appl. Math. Med. Biol. **1**, 51-60 (1984).

J.P. Trinkaus, *Cells into Organs; the Forces that Shape the Embryo,* 2nd ed., Prentice Hall, Englewood Cliffs, NJ, 1983.

L. Wolpert, Nature, **231**, 329-330 (1971).

L. Wolpert and H. Lewis, Fed. Proc., **34**, 14-20 (1975).

29. Generalities: Living Systems and Dielectrics

C. W. Smith

INTRODUCTION

In October 1973, Professor H. Fröhlich, FRS, retiring from the Chair of Theoretical Physics at Liverpool University, joined the Department of Electrical Engineering at Salford University, whose Charter did not require him to become "Emeritus" at such an early age. It is this which began the writer's active participation in the Fröhlich co-operative phenomenon, and it is to celebrate the 80th birthday of the still young Fröhlich that this contribution to his Festschrift is respectfully dedicated. When we started talking about dielectrics and the physics of biological systems, he warned me that it might only take 10 weeks to solve the problems, but equally it could take 10 years. It has now been running for more than 10 years, very interesting and enjoyable years too, and the more we continue to find out about the physics of nature the more wonderful nature appears.

Some two decades ago, Fröhlich had been asked to give a talk on a physicist's view of biology (Fröhlich, 1969). Although what he had to say probably horrified the biologists, it delighted the physicists, as can be seen from the reported discussions, which also raised some very fundamental questions. Since that time, biological questions have in some respects kept appearing as distractions from work on his other interests in theoretical physics.

In 1973, the writer was measuring the dielectric properties of biological materials in general and enzymes in particular. It was then clear to Fröhlich that from the point of view of atomic and molecular physics and chemistry, biological materials function in a most systematic manner even though they are extremely complicated systems. Although the molecular biologists had established the structure of many macromolecules and found that enzymes possessed active sites where catalyzed chemical reactions could take place according to the laws of chemistry, but at a greatly enhanced rate, there was nothing to explain their enormous catalytic power. Furthermore, the molecules could still only be represented in a rigidly frozen state, not in a dynamic interacting condition. The "tour-de-force" was to show the detailed spatial atomic distribution of the lysozyme molecule with its active site binding to the bacterial cell wall trisaccharide (Kelly et al., 1979). There is no simple connection between the molecular structure and the enzymatic activity, nor anything relating to the philosophical purpose of such systems. An enzyme-substrate system can be considered as an amplifier having a gain of 10^9 to 10^{10}; the characteristics of such an amplifier will be determined by its feed-back control-loop, that is, its regulatory system.

The search for a physical characteristic common to most biological materials led Fröhlich to their dielectric properties (Fröhlich, 1975) and the realization that the high electric field on biological membranes, of the order of 10^7 V/m, puts all the molecules in a membrane into the region of nonlinear dielectric polarization. From the point of view of physics, he expressed this in terms of nonlinear coherent excitations and showed that they would lead to long-range interactions on a very frequency-selective basis, very necessary for the control of an enzyme system—there are typically 3000 such systems in every living cell.

ENZYMES

The specificity of enzyme action, which is not confined to living systems, has often been modelled on a mechanical "lock and key" principle, but this still leaves many features unexplained. The analogy of a magnetic card "key" would overcome the energy transfer problem, but it is still necessary to consider how the enzyme "key" can find the substrate "lock" and fit it without jamming (competitive inhibition). This requires an information exchange at a distance, which in turn implies radiation and an interaction as between coupled Fröhlich oscillators. In other words, the "key" must "see" the light coming through the "lock keyhole". Fröhlich (1978) considered such a system in physical terms, as represented by two highly polarizable dielectrics at a distance larger than their dimensions and capable of giant dipole oscillations. He showed that there would be a strong interaction if they both oscillated at the same frequency, which for elastically bound particles of molecular dimensions came out at the order of 10^{13} Hz. The frequency corresponding to ambient (300 K) thermal energy (kT) is 6.25×10^{12} Hz. Today, this is still outside the region where precise dielectric measurements can be made on living materials.

It is important to remember the parallelism that exists between chemical structure and electromagnetic (or acoustic) frequency. Chemical bonds can be identified by measurement of their characteristic vibrational, rotational, librational wave numbers; without this parallelism, chemical analysis by spectroscopy just would not work. As a corollary to this, there must be as many refinements of frequency spectrum as there are chemical structures. Since the electric field in biological membranes is sufficient to take dielectrics into their nonlinear region, these nonlinearities should give rise to components at the various sum and difference frequencies and further complicate the observed pattern of frequencies.

A further point of importance concerning any measurements expressed as a frequency spectrum is that the time sequence of the signals from which the spectrum was generated has been lost. It is not possible to recover the original music or computer program from its time-averaged spectrum. If any observed features at frequencies in a dielectric spectrum arise because nature is producing a series of time-sequential control signals like a computer program, there is no way of deducing this from the spectrum.

At distances up to about 10 nm from any single electronic charge, typical of opposite charges on opposite sides of a biological membrane, the electrostatic field due to these charges will be of the same order as the membrane field. The membrane thus provides a bias field that will insure that fields arising from charge oscillations on the membrane will be of the same frequency and not twice the frequency. Two pairs of electronic charges interacting across a biological membrane will have an electrostatic energy of attraction greater than the thermal energy at 310 K, and thus will remain stable against thermal diffusion.

The enzyme lysozyme is a compact globular protein with dimensions about 3 × 3 × 4 nm, so that the whole of this enzyme and any of the substrate that may be in the active site would clearly be in the non-linear dielectric region due to all the electric charges present on adjacent molecules and on the membrane of the cell being lysed, and hence the simple linear summation of the electric field vectors cannot be presumed.

ENZYME MEASUREMENTS

The first thing of importance concerning the dielectric properties of materials such as enzymes is that while the enzyme is in the form of a dry lyophilized powder it may have a relative permittivity little greater than unity, but once the humidity exceeds a few percent the permittivity and loss start to increase significantly. At high values of humidity they may exceed a million in the lower part of the frequency spectrum (Rosen, 1963). Such measurements are susceptible to errors resulting from electrode polarization; however, the experimentation can be managed since there are techniques whereby such polarization effects can be eliminated (Shaw, 1942).

Our preliminary dielectric measurements (Ahmed et al., 1976) did little more than show that moist proteins behaved like moist ferroelectrics (Mansingh and Bawa, 1974). Dielectric measurements of both lysozyme and ribonuclease as a function of temperature showed discontinuities at temperatures that corresponded to documented changes in the activity of the molecules or their crystal structure. This encouraged our belief that we were not looking at artifacts.

The method of introducing moisture into the system did appear to be important. Specimens that had been humidified by exposing the enzyme (as a lyophilized powder) to water vapor at 6 °C for more than 12 h gave greater dielectric increments and hysteresis effects than those specimens which had been humidified by the addition of an equivalent amount of liquid water.

Unexpectedly, the 0.14-T magnetic field from a permanent magnet gave a reduction in both the permittivity and loss by about 40% for the humidified powders of lysozyme, ribonuclease, and ovalbumin. Similar results were obtained with an alternating (50-Hz) magnetic field for lysozyme and trypsin, and subsequently with various concentrations of lysozyme solution. Fröhlich immediately wanted to know whether diamagnetism or paramagnetism was involved in these magnetic effects. Experiments using an improvised torsion balance, constructed from a suspension galvanometer burned out in a teaching laboratory, showed that diamagnetism was definitely involved. As Fröhlich pointed out, this can come only from the equivalent of a short-circuited current loop.

Further measurements showed that magnetic fields of the order of 0.06 T could cause, in dilute solutions of lysozyme in water, an increment in the magnitude of the diamagnetic susceptibility 10^4 times as high as that expected for an ordinary diamagnetic material. The effect disappeared above critical fields of 0.08 T (Ahmed et al., 1975). These results were discussed on the basis of a model in which attached to each molecule was a small superconductive region with linear dimensions smaller than the London penetration depth. A comparable (closed) physical system would be superconducting colloidal mercury (Schoenberg, 1940).

More precise magnetic susceptibility measurements were made possible through the courtesy of our Department of Pure & Applied Physics, who gave access to their magnetic susceptibility balance (Booth, 1966). These measurements confirmed the preliminary results in respect of lysozyme, namely that there were anomalously high values for diamagnetic susceptibility up to a critical magnetic field strength, beyond which the values tended to those expected for water. There was also considerable hysteresis on temperature cycling.

These effects were eventually interpreted in terms of an analogy between stimulated charge-hopping conduction on lysozyme-coated contaminant cells and the "bucket brigade" memory circuit in which charge will continue to circulate so long as the pump frequency is present, that is, so long as the cell remains alive. The charge circulating in such a "bucket brigade" memory would simulate a persistent current loop of the necessary coherence to produce the observed diamagnetic effects (Smith, 1985).

A further and so far unexplained magnetic anomaly was observed in some measurements. Fluctuations appeared in the measured force on lysozyme solutions (40-50 °C) contained in quartz cells (flame sealed); these fluctuations had an amplitude of $\pm 5\ \mu g$ wt. and a periodicity of 20 s, and reversibly ceased above 0.11 T. These must be associated with the live contaminant.

Since lysozyme contains the α-helix structure, magnetic measurements were made on three polyamide materials containing, one, two, and three α-helices respectively to determine whether there were any effects due to an α-helix per se. None showed any anomalies, nor did any of the new quartz test cells.

It was found to be important to use newly made quartz test cells for each experiment. These were tested individually in a prior test run. In the early work, attempts to repeat measurements, using either fresh lysozyme or sodium chloride for a calibration, with a previously used glass or quartz cell all produced a null effect. This is because lysozyme forms a film that adheres tenaciously to glass and quartz. This film gives an overwhelmingly large effect that appears in both the calibration and test measurements and masks any smaller effects. The only technique that was successful in removing the lysozyme deposit and its associated magnetic effects was an extended bake in air above 200 °C. Ordinary detergents and sonication were ineffective.

Other laboratories did attempt experiments related to this work, but without published success (Sorensen et al., 1976; Chu et al., 1976; Careri et al., 1977). For our part we did not succeed in *failing* to obtain the effects until we installed sterile laboratory facilities, with flame sterilization of the quartz cells, autoclave sterilization of all liquids, TUV sterilization of the lyophilized lysozyme powder, and all handling and flame sealing carried out in a TUV-flooded enclosure. Then we obtained magnetization measurements for quartz, water, and lysozyme that were reproducible within experimental error and without any anomalies.

In this context it is essential to study all laboratory artifacts and understand them. Results of measurements on living systems that do not show evidence of biological variability are probably instrumental artifacts. In 1922, Fleming studied the effect of what was later known as lysozyme, on an airborne coccus of some kind drifting into his laboratory; this led to the discovery of bacteriolytic enzymes. At the beginnings of nuclear magnetic resonance, as Fröhlich has often recounted,

Bloembergen in 1949 managed to detect NMR signals from Gorter's original 1936 sample of LiF. The "Bloembergen Dirty Effect" as it became known, showed up the important point which Gorter had missed, namely that paramagnetic impurities could shorten NMR relaxation times so they became measurable. In view of the lack of progress in the understanding of the ways in which living systems make use of electromagnetic fields and frequency, it seems that too much power has been used on too pure materials, and as will be seen later, with too little biological stress.

While the above described dielectric and magnetic measurements were being carried out, we were also looking at the effect of electromagnetic fields on the lytic activity of lysozyme. There were many complicated effects and it was soon realized that the standard assay protocol for lysozyme represented a good optimization for reproducible results but not for investigating electromagnetic effects. Consulting biochemists usually drew the reply, "Why don't you find yourself an enzyme with a sensible substrate, or at least use a synthetic substrate?" The "substrate" *Micrococcus lysodeikticus* obtained as a freeze-dried powder is still alive, and the saline in the assay protocol used does not kill it. Further investigations showed that the enzymatic activity, that is, the rate at which lysozyme lyses these cells, is affected by the nutrients available to the substrate and the phase of the cell division cycle. The lytic activity is also photosensitive; even as little light as that in the beam of the spectrophotometer makes a great difference to the reaction rate compared with running the reaction in total darkness. The lysozyme solution can "remember" for long periods of time the electromagnetic and magnetic fields to which it has been exposed. We could expose lysozyme solution in the Electrical Engineering building and perform the assay in the Chemistry building as readily as performing the whole experiment in a single location. The widespread use of magnetic stirrers (approx. 0.01 T, steady field from the encapsulated magnet and an additional alternating magnetic field of a few hertz from the stirrer unit when operating) ensures that many magnetic anomalies are saturated and remain unobserved. This improves apparent consistency of the assays, of course.

A delay time, dependent on the substrate concentration, was found at the start of the lytic reaction of lysozyme. This situation has been discussed by Fröhlich (1980) as one of the consequences to his theoretical treatment of the excitation of coherent electric vibrations by random metabolic energy.

The lysozyme reaction was also found to be sensitive to proton magnetic resonance conditions, as described in a later section. The lysozyme assay was run in the usual way, but when the cuvette containing a partly lysed substrate was exposed to the combination of a steady magnetic field and a radiofrequency field that exactly satisfied the proton magnetic resonance condition, the lysis was halted completely, although it resumed when the cuvette was returned to the spectrophotometer. It was necessary to expose the cuvette to the steady magnetic field, that is, to split the energy levels, before switching on the radiofrequency field. Because the resonance is so sharp, it was also necessary to scan the frequency slowly and manually, to be certain of regularly encompassing the resonance condition (Jafary-Asl et al., 1983).

As commonly occurs in living systems under good homeostatic control, larger effects (often biphasic) are obtained when the system is stressed. A convenient way of doing this for the lysozyme *Micrococcus lysodeikticus*, enzyme-substrate system is by use of the competitive inhibitor *N*-acetyl-*D*-glucosamine. A complicated pattern of frequency-dependent effects on the activity was found between 50 kHz and 300 MHz; the effect of the inhibitor could be enhanced or cancelled depending on the frequency (Shaya and Smith, 1977). Inasmuch as the *Micrococcus* substrate was alive, it is interesting to note that this type of frequency dependence was again encountered several years later when we were investigating the radiofrequency emission from dividing yeast cells. It was then interpreted in terms of acoustic and surface tension wave resonances on the cell membrane (Jafary-Asl and Smith, 1983).

THE JOSEPHSON FREQUENCY VOLTAGE RELATION

In the report of the preliminary evidence for collective magnetic effects in lysozyme (Ahmed et al., 1975), we suggested that not only the enzyme but also water and ions may play a role in the establishment of superconductive regions. This could give rise to a layer structure and, with an appropriate distribution of ions, to electric fields and to an ac Josephson effect, that is, to electrical vibrations that might be in resonance with the electric vibrations in the molecules. The large-scale zero resistance phenomena of superconducting metals were not envisaged. The long-range forces resulting from coherence phenomena need not give rise to particle aggregation or micelle formation but may represent a coherence in operation among separated units over the whole volume cooperatively concerned. It was thus that we carried out some experiments to see whether we could detect any Giaever-type (1965) effects in thin film capacitors having lysozyme as the dielectric.

The most important result we obtained was a set of current-voltage characteristics for a lysozyme dielectric junction, on which 40-nV voltage steps appeared in the presence of 9-GHz microwaves (in darkness, inside a waveguide). The Josephson frequency corresponding to this would be 20 MHz; when this frequency was coupled loosely into the system, the junction resistance increased by a factor of 15 and the steps increased by a factor of 7.5 to 300 nV. Although no mechanism was suggested, it did seem worthwhile applying the Josephson frequency-voltage conversion of 500 MHz/μV wherever appropriate (Ahmed and Smith, 1978).

The next occasion that this conversion proved useful was during experiments on "pearl chains" of yeast cells collected between electrodes by dielectrophoresis (Jafary-Asl et al., 1983). We were looking for changes in electrical activity of a synchronized culture of *Saccharomyces cerevisiae* (normal diploid strain) through the cell division cycle. Current-voltage characteristics were obtained for conduction along the "pearl chains." The yeast cells were grown synchronously in well deionized sucrose (0.25 M) which was sterilized after making up, since a deionizer can give 1000 cells/ml in its effluent. These bud off from cultures established within the ion-exchange resin. "Pearl chains" from aliquots of synchronously dividing yeast suspension were tested at 5-min intervals from the start of division. Steps in voltage were detected only when the cells were near the time of cytokinesis. The smallest voltage steps detected were 15 nV, for which the corresponding Josephson frequency would be 7.5 MHz. There was also a distribution of larger voltage steps commencing at about 150 nV (corresponding to 75 MHz) and extending beyond 1 μV.

Since spectrum analyzer coverage was available for the 7.5-MHz frequency region, the synchronous yeast cell cultures collected by dielectrophoresis at electrodes separated by only 1 μm (the yeast cells were about 5 μm diameter) were examined very carefully in an electrically screened laboratory and with the culture in total darkness. At the time when the synchronized cells would be undergoing cytokinesis (4 h after starting the synchronous culture at the temperature of the laboratory), a signal appeared somewhere between about 7 MHz and 9 MHz. It came up through the instrumental noise and usually reached an amplitude of a few tenths of a microvolt. It commenced with a bandwidth of about 2 MHz. Then over a few minutes it narrowed down, the narrowest recorded value being 50 Hz width; it then broadened and disappeared into instrumental noise (Smith, 1984). The narrowness of this emission requires that it consists of coherent photons (Smith et al., 1986).

Emissions were also found in the region of 75 MHz and below 1 MHz. These are interpreted, as has been mentioned, in terms of acoustic and surface tension wave resonances on the cell membrane (Jafary-Asl and Smith, 1983). The 7.5-MHz emission is less easy to explain; it may arise from a parametric oscillation produced as the septum forms between mother and daughter cells (Smith, 1986). The overall pattern of these emission frequencies resembles the pattern of resonances obtained after exposing lysozyme to radiofrequency fields (Shaya and Smith, 1977). It may also be associated with the rate of ATP hydrolysis (Webb, 1984).

From micro-dielectrophoresis experiments, Pohl has shown that the preference of a wide range of cells for polar particles is maximal at or near mitosis and attributed this to electrical oscillatory radiofrequency phenomena (Pohl, 1983). Because the polar particles used were probably also piezoelectric, these experiments may not differentiate between electric and acoustic oscillations; however, since the particles were counted at the highly polar cell membranes where the electric and acoustic fields are equivalent, this is of no great consequence; but it would be interesting to know whether any of the high-permittivity particles used were not piezoelectric, as this would separate the ranges of the two possible types of oscillation.

Nonthermal resonant effects have been found in the growth of yeast cultures exposed to 42-GHz microwaves. The result was a biphasic dependence on frequency, with resonance bands of 8 MHz full width at half maximum (Grundler et al., 1983). It appears that a frequency of 8 MHz is rather fundamental for yeasts.

ONSET CONDITIONS FOR MAGNETIC FIELD EFFECTS

While trying to sort out the unreproducibility of the initial work on lysozyme we found that larger effects were obtained when the experimental conditions approached those needed for synchronous cell division. It was then found that the substrate was alive. We therefore also commenced work using that much investigated living system, the bacterium *Escherichia coli*.

To be quite certain that we were not going to miss anything between spot measurement intervals, we used batches of 18 cuvettes situated in a gradient of magnetic field so that no value of field went untested, although the resolution would only be that corresponding to the 1-cm dimension of the cuvette in the field gradient. Over a thousand cultures were grown and their mean generation time measured under carefully controlled conditions. The standard deviation of the controls was reduced to 0.5%. The results for magnetic fields of square waveform and frequency 50 Hz and 16.66 Hz showed a marked threshold effect and strong indications of a periodicity in field strength above this threshold. The application of the F-ratio test indicated a probability of less than one in two million that the effects were due to chance (Aarholt et al., 1981).

The magnitude of the effect was only about 4%, so the work put into getting good controls was essential. These results demonstrate an important feature of the response of biological systems to an external stress when they are under good homeostatic control. The effects do not get larger as the stress is raised; they become more complicated. Biological systems are nonlinear and often discontinuous in response. The oft-used argument that a null effect at a strong stress (e.g. an electric or magnetic field) implies an even smaller effect at a weaker stress, is not valid.

The above realization also gave us the clue as to how to obtain larger effects than the 4% then seen; the system needed to be biologically stressed. When *E. coli* is stressed by feeding lactose as its sugar, it transcribes an enzyme β-galactosidase which is gene controlled but is normally repressed by a protein that binds very strongly to a specific site on the DNA located just outside the structural gene. This repressor protein comes off in the presence of lactose (Davies and Walker, 1979). This is known as the *lac* operon system of *E. coli.* The relative rates of β-galactosidase synthesis were measured as a function of the strength of a 50-Hz square-wave magnetic field using the field gradient magnet as previously. The results showed a fivefold reduction in the rate at 0.3 mT ± 5% and a two-and-a-half fold increase at 0.6 mT ± 10%. There was no effect at cell concentrations above 10^8 cells/ml (Aarholt et al., 1982).

If spot readings had been taken at 0.1-mT (1-gauss) intervals instead of using the gradient of magnetic field, only one nonzero measurement would have been obtained. Clearly, we must take care to measure living systems with sufficient precision.

MAGNETIC FLUX QUANTIZATION

It was becoming apparent from experiments on a number of different living cells of various sizes, that the critical magnetic field strength for the onset of the various magnetic effects corresponded to that at which a single quantum of magnetic flux would be linking a single cell or cell pair. The periodicity observed as the field strength was increased corresponded to integer increments in the numbers of magnetic flux quanta linking the cells (Jafary-Asl et al., 1983).

Fröhlich was in close touch with this work throughout. On the 9th of June, 1980 he included the following paragraphs in a letter to the writer,

> ... Concerning interpretation you must realise (you may quote me on this if necessary) that magnetic flux is *always* quantised, but normally the respective energy states are so close together that the quantisation is

not observable (just as in angular momentum). It is not quite unique, however, to speak in such small regions of "superconductivity", other states might perform the same feature. In fact, you should not speak of the cell membrane as a superconductor; as a small ring over the membrane, with appropriate magnetic properties, would perform the quantisation.

Concerning Josephson effect, there exists a general theorem due to F. Bloch (Bloch, 1968) that given a superconducting ring (flux quantisation) containing one Josephson junction, the voltage induced by a time variation of magnetic field is exactly the Josephson voltage—the most general (model free) derivation of the Josephson formula. What you have done is to apply this theorem. But that does not prove that you actually have a Josephson junction. . .

Since 1980, we have had more instances of the applicability of the Josephson equation and of what appear to be magnetic flux quantization phenomena. So far, the most likely candidate for a weak-link Josephson junction is the septum forming between mother and daughter cells at cytokinesis. If nature is able to make use of magnetic flux quanta, then in principle it has the Josephson effect available for its use, since both derive from the quantized nature of a magnetic field.

NUCLEAR MAGNETIC RESONANCE CONDITIONS

At a Conference held in Nottingham University, England, to mark the retirement of Professor D. D. Eley in 1980, some raw data from dielectrophoresis measurements was shown by Professor H. Pohl. This contained an anomaly at about 2 kHz. Dielectrophoresis in this context involves the collection of biological cells from their suspension in highly de-ionized water or a nonionic isotonic medium or nutrient. The cells form as "pearl chains" at the tips of point electrodes that give the necessary nonuniform field. Since the effect depends on the square of the electric field, it can be observed with alternating fields as well as steady fields. The other important parameter involved is the difference between the permittivity of the cell and its surroundings (Pohl, 1978).

The appearance of an anomaly at a frequency of 2 kHz was very suggestive of a nuclear magnetic resonance interaction involving protons precessing in the geomagnetic field. This was confirmed and extended (Jafary-Asl et al., 1983; Aarholt et al. (to be published)). The simplest macroscopic demonstration of this phenomenon is obtained by setting up some "pearl chains" by dielectrophoresis and then with an externally applied magnet, slowly sweeping the ambient magnetic field through the proton NMR condition for the frequency being used to form the "pearl chains". Enough of the cells will repel each other to break up the chain formations. This can happen only if there has been a drop in the permittivity of these cells below that of their surround, and must be due to an increased lattice interaction of the magnetically resonant protons. That such a lattice interaction effect is possible may be demonstrated by measuring the permittivity of something as inorganic as the proton-conducting glass of a pH meter under proton NMR conditions.

The proton NMR condition represents a very sharply defined resonance condition whereby energy can be inserted into a living system in a very specific manner. When bacteria were grown under proton NMR conditions, twice as many cells of half the size were obtained, although the total cell mass remained unchanged. In experiments on bovine eye lenses, it was found that subcapsular cataracts located in the posterior cortex of the lens developed particularly well when the microwaves were modulated so that the modulation frequency satisfied proton NMR conditions in the ambient magnetic field. In this case the microwaves would be merely acting as a carrier able to deposit the NMR frequency as the microwave modulation signal within the tissue, where the nonlinearities would demodulate it. Since many microwave oscillators contain magnets, a technician moving about in the vicinity of such an oscillator would often be satisfying the necessary resonance conditions within the body tissues.

Once the possibility of magnetic resonance phenomena in living systems is appreciated, it is clear that such phenomena could also be a source of highly coherent radiation. If a magnetic field pulse is applied to a living system, as in stimulated bone healing, the betatron process will produce an increase in the proton population in the higher energy state; if the magnetic field is then reduced slowly, but in a time short compared with the spin relaxation times, the tissues will be subjected to the highly coherent spin relaxation frequencies containing all the chemical shift information, and will be able to undergo spin exchange processes.

ALLERGIC REACTIONS

Allergic responses have been found to occur so widely that allergy is now defined as "The failure of a regulatory system."

In 1982, we noted that persons exposed to environmental electromagnetic fields could be experiencing body currents, of the order of tens of microamperes, currents comparable to those known to be able to produce electroanaesthesia, which in turn is associated with the stimulation of endogenous opiates (Smith and Aarholt, 1982). The same year, endogenous opiates had been linked with allergies at two international conferences. This led Dr. Jean Monro (Allergy Unit, The Nightingale Hospital, London, U.K.) to contact the writer for help with the treatment of electrically sensitive multiple-allergy patients. Since then, about 60 patients have been tested and electrical sensitivities have been found to be critically dependent on frequency over a frequency range extending from millihertz to gigahertz, but less dependent upon the intensity so long as this exceeded a certain threshold.

One method of deriving a theoretical lower limit for the intensity of an electromagnetic field to which an allergic subject may acquire a sensitivity, is to take the energy volume density of the field and multiply this by the volume of the biological system that is cooperatively involved in the detection process; to assess the minimum value, this volume is assumed to be the total volume of the person's body. This is then equated to the thermal energy (kT), which is equivalent to assuming that the system is operating with a signal-to-noise ratio of unity (Smith et al., 1986a,b).

If the whole of a 70-kg man is an effective antenna for the detection of an electric field, this criterion would give a threshold electric field of 8 μV/m. It is of the order of the sensitivity of the most sensitive of the multiple-allergy patients tested. It occurred only while the patients were in a reacting allergic condition.

The lowest electric field that has been observed to evoke a response from certain fish is 1 μV/m (Bullock, 1977). Comparing this to the observed radiofrequency emission levels from yeast cells and assuming that there is nothing particularly unique about yeasts in this respect, we see that fish are going to have no problem in locating food represented by electrically noisy plankton.

One of the common features of living systems in general is the ability to sense minute changes in the environment, this in the presence of much larger irrelevant signals (Gamow and Harris, 1972). This is clearly an advantage to both predator and prey alike. Living systems have many biosensors that are clearly near to being quantum sensitive (Smith, 1986). To trigger a nerve impulse from a single quantum event requires an amplifier with a gain of the order of 10^9 to 10^{10}. The characteristics of such an amplifier would be determined by the feedback network. If this goes open-circuit, then any signal above a threshold will give saturation, that is, a "panic reaction". If the gain and phase shift are appropriate at some frequency, the system will oscillate. This is all consistent with the definition of allergy as "The failure of a regulatory system," and is an appropriate subject for the application of control theory.

A method of testing and a therapy has been devised for the treatment of electrically sensitive multiple-allergy patients (Smith et al., 1985). However, the testing of such patients must be regarded as a clinical procedure and must not be attempted without the immediate availability of facilities to treat anaphylaxis. It is based on the provocation-neutralization therapy of Miller (1972). It appears that increasing a coherent frequency has the same clinical effect on the electrically sensitive patients as increasing allergen dilution does on the chemically or nutritionally sensitive patients. It is the frequency and the coherence that are important; the field strength is less important so long as it is above a certain threshold.

CONCLUSIONS

The first conclusion must be that this is only the beginning. The ideas that Professor Fröhlich has begotten will work their way to fruition; whether the fruits are sweet or bitter does not depend only on science and its practitioners, but upon us all.

The following general conclusions apply.

Electric and magnetic fields and their frequency appear to be as important as chemical structure in living systems.

Living systems appear to operate optimally close to the quantum limits and probably use the whole electromagnetic spectrum extending from frequencies corresponding to the ultraviolet down to the frequency corresponding to the reciprocal of the lifetime of the organism.

Much of the apparent randomness in nature is because we are not measuring it with adequate precision. Trying to deduce anything about living systems from the electrical signals usually measured in respect of bioelectric phenomena is like trying to follow a television program from the sound channel alone.

A living system in homeostatic control behaves much like any other stable control system, but like a control system it can also go unstable under appropriate conditions. This offers the possibility of understanding why people become ill, and of stabilizing their regulatory systems before failure occurs. The diagnosis and treatment of what has gone wrong after an illness has become apparent represents, in general, the present state of medical science. This is neither necessary nor cost-effective in view of the possibilities of prophylactic treatment through the study of the body's regulatory systems.

There are far-reaching implications for investigating interactions between persons once it is realized that, when in a reacting condition, the most electrically sensitive allergic subjects can sense electromagnetic signals much weaker than those being emitted by other reacting allergic subjects. This implies that people should also be considered as coupled, coherent, nonlinear Fröhlich oscillators, both interpersonally and socially.

I gratefully acknowledge the many long discussions with a very patient Fröhlich, in Salford, Liverpool, and the many other parts of the world where we have chanced to meet. The experiments referred to in this paper are a part of the Fröhlich dialog and represent attempts by myself and others to determine whether Nature really orders its affairs in the manner that Fröhlich's theoretical model predicts and implies.

AD MULTOS ANNOS!

REFERENCES

E. Aarholt, E.A. Flinn, and C.W. Smith, Phys. Med. Biol., **26**, 613 (1981).

E. Aarholt, E.A. Flinn, and C.W. Smith, Phys. Med. Biol., **27**, 603 (1982).

E. Aarholt, M. Jaberansari, A.H. Jafary-Asl, P.N. Marsh, and C.W. Smith, NMR Conditions and Biological Systems, in *Handbook of Bioelectricity*, A.A. Marino, Ed., Marcel Dekker, New York, (to be published), Ch. 22.

N.A.G. Ahmed, J.H. Calderwood, H. Fröhlich, and C.W. Smith, Phys. Lett. **53**A, 129 (1975).

N.A.G. Ahmed, C.W. Smith, J.H. Calderwood, and H. Fröhlich, Coll. Phenom. 2, 155 (1976).

N.A.G. Ahmed and C.W. Smith, Coll. Phenom. **3**, 25 (1978).

F. Bloch, Phys. Rev. Lett. **21**, 1241 (1968).

J.G. Booth, J. Phys. Chem. Solids **27**, 1639 (1966).

T.H. Bullock, Neurosci. Res. Program Bull. **15**, 17 (1977).

G. Careri, L. De Angelis, E. Gratton, and C. Messana, Phys. Lett. **60**A, 491 (1977).

C.W. Chu, V.K.H. Chen, K. Sugawara, and C.Y. Huang, Solid State Comm. **19**, 357, (1976).

K.E. Davies and I.O. Walker, in *Chromatin Structure and Function*, C.A. Nicolini, Phenum, New York, 1979.

H. Fröhlich, in *Proc. 1st Int. Conf. on Theoretical Physics and Biology*, Versailles, 1967, M. Marois, Ed., North Holland, Amsterdam, 1969, p. 13-22.

H. Fröhlich, Proc. Nat. Acad. Sci. U.S.A. **72** 4211 (1975).

H. Fröhlich, I.E.E.E. Trans., **MTT26**, 613 (1978).

H. Fröhlich, Adv. Electron. Electron Phys. **53**, 85 (1980).

R.I. Gamow and J.F. Harris, I.E.E.E. Spectrum **9** 36 (1972).

I. Giaever, Phys. Rev. Lett. **14**, 904 (1965).

W. Grundler, F. Keilmann, V. Putterlik, L. Santo, D. Strube, and I. Zimmermann, in *Coherent Excitations in Biological Systems*, H. Fröhlich and F. Kremer, Eds., Berlin: Springer, 1983, p. 21.

A.H. Jafary-Asl and C.W. Smith, Ann. Rep. Conf. Electrical Insulation & Dielectric Phenomena, IEEE Publ. 83 **CH 1902-6**, 350 (1983).

A.H. Jafary-Asl, S.N. Solanki, E. Aarholt, and C.W. Smith, J. Biol. Phys. **11**, 15 (1983).

J.A. Kelly, A.R. Sielecki, B.D. Sykes, M.N.G. James, and D.C. Phillips, Nature (London) **283**, 875 (1979).

K.H. Li, F.A. Popp, W. Nagel, and H. Klima, in *Coherent Excitations in Biological Systems*, H. Fröhlich, and F. Kremer, Eds., Berlin: Springer, 1983, p. 117.

A. Mansingh and S.S. Bawa, Indian J. Pure Appl. Phys. **12**, 483 (1974).

J.B. Miller, *Food Allergy, Provocative Testing and Injection Therapy*, Springfield Ill.: C.C. Thomas (1972).

H. Pohl, *Dielectrophoresis: The Behaviour of Neutral Matter in Nonuniform Electric Fields.* Cambridge, U.K.: Cambridge University Press (1978).

H. Pohl, in *Coherent Excitations in Biological Systems*, H. Fröhlich and F. Kremer, Eds., Berlin: Springer, 1983, p. 199.

D. Rosen, Trans. Faraday Soc. **59**, 2178 (1963).

D. Schoenberg, Proc. R. Soc. London **175**A, 49 (1940).

T.M. Shaw, J. Chem. Phys. **10**, 609 (1942).

S.Y. Shaya and C.W. Smith, Coll. Phenom. **2**, 215 (1977).

C.W. Smith, Proc. 6th. Ann. Conf. IEEE Eng. in Med. & Biol. Soc., IEEE Publ. **CH 2058-6**, 176 (1984).

C.W. Smith, in *The Living State II*, R.K. Mishra, Ed., Singapore: World Scientific, 1985, p. 404.

C.W. Smith, *Industrial Biotechnology Wales*, (to be published, April/May 1986).

C.W. Smith and E. Aarholt, Health Phys. **43**, 929 (1982).

C.W. Smith, S.A.R. Al-Hashmi, A. Kushelevsky, M.A. Slifkin, R.Y.S. Choy, J.A. Monro, E.E. Cluclow, and M.J.C. Hewson, paper presented at 4th Int. Symp. on Man and His Environment in Health and Disease, Dallas, Feb. 27-Mar. 2, 1986 (submitted for publication).

C.W. Smith, A.H. Jafary-Asl, R.Y.S. Choy, and J.A. Monro, paper presented at Int. Symp. on Photon Emission from Biological Systems, Wroclaw, Poland, January 24-26, 1986, J. Slawinski, Ed. (to be published).

C.M. Sorensen, F.R. Fickett, R.C. Mockler, W.J. O'Sullivan, and J.F. Scott, J. Phys. C: Solid State Phys. **9**, L251 (1976).

S.J. Webb, paper presented at Fondazione Omeopatica Italiana Conference, 27 October, 1984, Medical and Surgical Faculty, University of Naples.

30. Dielectric Spectroscopy, Dielectrophoresis, and Field Interactions with Biological Materials

H. P. Schwan

With 4 Figures

It is an honor and a privilege to participate in this celebration of the 80th birthday of an eminently distinguished scientist. Fröhlich's contributions are many. His outstanding contributions to the theory of dielectrics and his important models that concern the effects of nonionizing electromagnetic fields on living matter have been of particular interest to me, as they relate to my research interests. Dielectric properties and field interactions are related, since interactions often are governed by properties. I concentrate on this point since the importance of dielectric work as it relates to interactions is not generally recognized.

A major interest of mine has been the dielectric properties of cells and tissues and their components, including cellular organelles, membranes, and proteins. We have reported on this previously in detail (Schwan, 1957; Schwan and Foster, 1980) and recently several reviews of the work have been published (Schwan 1983a, 1985b). I shall therefore limit myself to a brief summary and then turn to several examples of interactions and their dependence on dielectric properties.

DIELECTRIC PROPERTIES OF BIOLOGICAL SYSTEMS

The dielectric properties considered include the dielectric constant (permittivity) ϵ and conductivity σ. The complex dielectric constant is defined by

$$\epsilon^* = \epsilon' - j\epsilon'' \tag{1}$$

with $\epsilon' = \epsilon$ and $\omega\epsilon'' = \sigma$ (ω is angular frequency). The conductivity includes contributions from both dielectric losses and dc conductance.

The frequency response is well described by relaxation equations of the Debye type

$$\epsilon^* = \epsilon_\infty + \frac{\epsilon_0 - \epsilon_\infty}{1 + j\omega t} - j\frac{\sigma_0}{\omega} \tag{2}$$

or the more general Cole-Cole type

$$\epsilon^* = \epsilon_\infty + \frac{\epsilon_0 - \epsilon_\infty}{1 + (j\omega t)^\alpha} - j\frac{\sigma_0}{\omega}. \tag{3}$$

Here ϵ_0 and ϵ_∞ are limit values at low and high frequencies as compared to the relaxation frequency $f_0 = 1/2\pi t$. t represents a single or average time constant depending on whether we consider eq. 2 or 3 and α is a measure of the spread of relaxation times. A term σ_0 is included to represent low-frequency limit conduction. The dielectric response of biological matter is conveniently subdivided into three classes of effects, namely dispersions at low-frequency dispersions, labeled α, RF dispersions, termed β, and very high-frequency dispersions, labeled γ, as indicated in Fig. 1 (Schwan, 1957). In Table 1 we summarize these effects for water, biological macromolecules, and cells. In Table 2, we list the various relaxation mechanisms which have been identified so far. We demonstrated that the γ effect is largely caused by the polar water molecule and its rotation with the alternating field. The β effect results from the charging of membranes through intra- and extracellular conductive pathways. Finally the α effects can have several causes including counterion movement about the charged cell surface, subcellular membrane systems connecting with the outer cell membrane, membrane channel characteristics, and, perhaps, the boundary potentials caused by the diffuse ion distribution near the charged membrane surface (Schwan, 1981).

The identification of these different mechanisms is fairly easy with cell suspensions and protein solutions. It is more difficult for tissues. The reasons are: (1) the applicability of mixture theories suffers when the cell concentration increases, (2) there are difficulties in dealing with complex tissue cell geometry, and (3) the effects of gap junctions between cells must be considered, etc. Nevertheless, the general features of the dielectric response of tissues are similar to that of cell suspensions, strongly indicating that the same mechanisms are involved. Let us next consider known field interaction mechanisms and their dielectric response equivalent.

KNOWN SIGNIFICANT FIELD INTERACTIONS

First of all we consider the cells and their membranes as the site of excitation and contraction phenomena. Excitation of membranes is the basis of the nervous system's ability to transmit signals, and contraction is the basis of muscular performance. In both cases the response is caused by electrical fields that evoke membrane potential changes of the order of millivolts. Corresponding field strength values in the medium or tissue surrounding the cell under consideration are of the order of 1 V/cm, but can be larger or smaller depending on all sorts of qualifica-

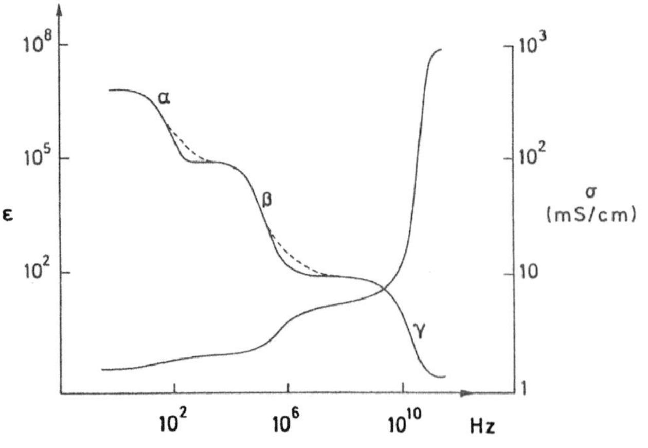

Fig. 1 — Dielectric constant ϵ and conductivity σ of muscle tissue as a function of frequency. The three major dispersions α, β, and σ are typical for all tissues and cell suspensions even though magnitudes and dispersion frequencies vary. Additional smaller relaxation effects contribute to the high-frequency tails of the α and β dispersions as indicated by the dashed curves.

Table 1

Water and electrolytes	γ
Biological macromolecules:	
Amino acids	$\delta + \gamma$
Proteins	$\beta_1 + \delta + \gamma$
Cell components	
Other than proteins, uncharged	$\beta + \gamma$
proteins	
with no surface charges	$\beta + \beta_1 + \delta + \gamma$
with surface charges	$\alpha_2 + \beta + \beta_1 + \delta + \gamma$
with subcellular organelles only	$\beta + \beta_1 + \beta_2 + \delta + \gamma$
with connecting membranes only	$\alpha_1 + \beta + \beta_1 + \delta + \gamma$
Total Cellular Response:	$\alpha_1 + \alpha_2 + \beta + \beta_1 + \beta_2 + \delta + \gamma$

Relaxation mechanism which determines the dielectric response of electrolytes, biological macromolecules, and cells. ELF, RF and MW responses are labeled α, β, and γ and are strong responses. γ stands for the water dispersion, β is the external cell membrane charging response, β_1, the weaker protein response and β_2 that from intracellular organelle membranes. α_1 and α_2 are contributions from membrane systems connecting with the outer cell membrane and counter ion displacement near fixed surface charges of the membrane glycocalix. The weaker γ-response at UHF-frequencies is caused by protein bound water and submacromolecular partial rotation. β_1, β_2, and γ are comparatively weak contributors to the cell's dielectric response as indicated in Fig. 1. The increasing participation of an increasing number of relaxation mechanisms as we proceed from simple water to the complex cell is apparent from the table.

tions. On the other hand the membrane thickness is only about 100 Å and, hence, the induced field strength change in the membrane about 10 kV/cm. In other words, an external field of about 1 V/cm is amplified in the membrane by a factor of 10^3 or 10^4; the amplification factor can be shown to be simply equal to R/d (R cell radius, d membrane thickness). It is this amplification that permits fairly low fields in the medium to become so effective in the membrane. However, this amplification occurs only at low frequencies, as we shall explain later. The membrane's electrical characteristics are highly nonlinear at these high field strength values. The well known Hodgkin-Huxley equations (1952) were developed to characterize these nonlinear properties. Briefly, membrane channels composed of membrane proteins control exchange of sodium and potassium across the membrane in a field-strength-dependent manner. Individual channels can be now observed with the voltage patch clamp technique of recent origin. However, kinetic details of the macromolecular conformation caused by the membrane poten-

Table 2. Relaxation effects of various kinds

Class of dispersion	Species or process concerned	Cause of relaxations	Contribution to dielectric properties
α_1	Counterions	I	M, S
α_2	Surface-connected membranes	I	S
α_3	Channels	?	M
α_4	Membrane surface and structure	?	?
β	Charging of membranes	I	S
β_1	Protein orientation	P	W
β_2	Organelles	I	W
γ	Normal Water	P	S
δ	Protein-bound water	P	W

I = mechanisms caused by orientation of induced dipoles (Maxwell-Wagner effects)

P = mechanisms caused by orientation of polar molecules

S, M, W = strong, medium, and weak contributions, respectively, to cellular dielectric properties

tial change remain to be resolved. In any case, this response is a "strong" one, requiring large field strength changes in the membrane.

Much dielectric work has been done with proteins and other biological macromolecules (Takashima and Minakata, 1975). The dielectric response for proteins is consistent with a molecular dipole moment of some hundred Debye units and the rotation of the molecule when an alternating field is applied. Relaxation frequencies are typically of the order of some megahertz. Dielectric saturation is approached with field strength values in the medium near 10 kV/cm. Experimental data are by and large in agreement with the Langevin criterion $L = 1$, where L is the Langevin function for significant orientation overcoming thermal motion. Thus a significant orientation of proteins in solution can be achieved only with very high fields. We note that the dielectric saturation response near 10 kV/cm occurs at fields which we stated earlier to cause membrane responses such as excitation. This agreement is probably not accidental.

ELECTRIC CIRCUIT ANALOGS OF CELL SUSPENSIONS AND MEMBRANE POTENTIALS

Much theoretical and experimental work has been concerned with the β-dispersion phenomena, yielding excellent agreement between theory and experiment. The analytical approach was to solve the Laplace equation for spherical or elliptical cells whose cytoplasm is surrounded by a poorly conducting membrane.

This solution in combination with the Maxwell mixture equation for spherical bodies in a medium yields the effective dielectric constant and conductivity for the cell suspension. This work resulted also in the recognition that the β-response can be modeled by the circuit displayed in Fig. 2. Here the membrane appears in series with the resistivities of cell interior and exterior. The time constant of the dielectric response must be then

$$t = RC_m\,(\rho_i + 0.5\,\rho_o) \tag{4}$$

where R is the radius of the cell, C_m is membrane capacitance, and ρ_i and ρ_o are the resistivities inside and outside the cell. At frequencies much lower than the characteristic frequency of the β dispersion, $f_\beta = 1/2\pi t$, the total potential applied to the cell will be delivered to the capacitor, thus resulting in the magnification of the external field in the membrane discussed above. At much higher frequencies, on the other hand, the membrane potential vanishes since the capacitor becomes short-circuited. At low frequencies the internal field is zero and at high frequencies it become comparable to the external field (identical to it if internal and external resistivities are equal). Figure 3 shows the frequency dependence of the membrane potential. It can be written as (Schwan, 1983b)

$$\Delta V_m = \frac{1.5\,ER}{\sqrt{1+(\omega t)^2}}\,. \tag{5}$$

On the other hand, the internal field strength approaches its peak value above the characteristic frequency f_β as demonstrated in Fig. 3. This suggests that any field effects taking place inside the cell or any chances to fuse cell nuclei are maximal at frequencies at or somewhat above the characteristic frequency of the β effects.

At high frequencies the membrane potential predicted by eq. 5 decreases proportionally with $1/f$. Therefore, induced membrane potentials in the high-frequency case $f > f_\beta$ are much smaller than at low frequencies. For the GHz-frequency range they are of the order of microvolts if the low-frequency limit is near millivolts. This corresponds to membrane field strength values of some

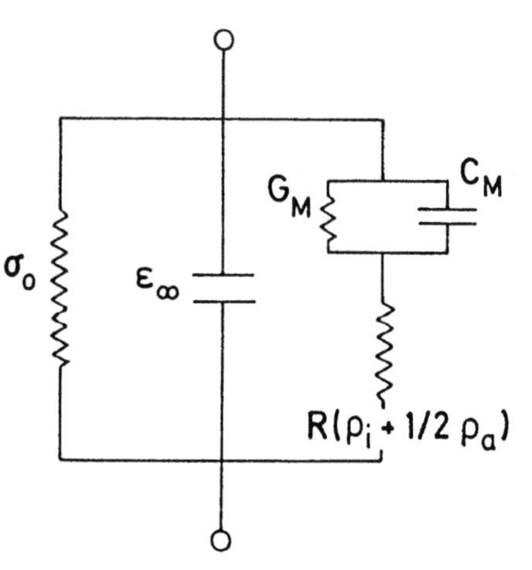

Fig. 2 — Dielectric properties of spherical biological cells represented by an electrical circuit as shown. ρ_i and ρ_o are resistivities of internal and external medium, R is the cell radius, and G_m and C_m represent membrane conductance and capacitance per unit area. σ_0 is the low-frequency limit of conductivity σ and ϵ_∞ is the high-frequency limit of dielectric constant ϵ. Usually the influence of G_m can be neglected, where- upon the value for t given in eq. 4 is obtained. The same circuit can be used to represent the dielectric properties of cell suspensions, provided that the series parameters G_m, C_m, ρ_i, and ρ_o are adjusted to include the volume fraction of the suspended cellular fraction.

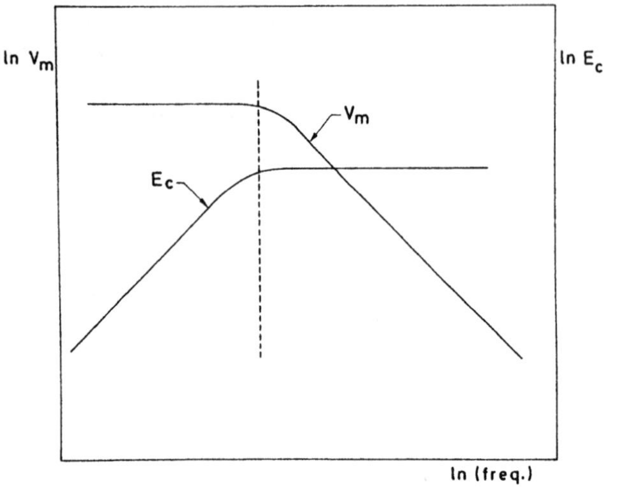

Fig. 3 — Frequency dependence of cellular membrane potential V_m and cytoplasmic field strength E_C. A frequency independent field is assumed in the medium outside the cell.

volts/centimeter or less. It is therefore much more difficult with high frequencies to apply membrane potentials that can cause significant bioeffects in the membrane. This is true as long as the Langevin criterion is valid, which appears to demand macromolecular exposure to fields of the order of 10 kV/cm as indicated above.

In recent years significant lateral diffusion of membrane proteins under the influence of external fields of some volts/centimeter has been reported. However, these movements are virtually observable only with dc-fields and not anticipated to occur under the influence of ac fields with frequencies larger than a fraction of 1 Hz. We want however, to complement the frequency dependence of the membrane potential eq. 5 by the corresponding radial field strength and by the lateral (tangential) field strength in the membrane as obtained from the Laplacian treatment of the cell aforementioned. In the low-frequency limit we obtain

$$E(\text{radial}) = 1.5\ ER/d \cos\delta \tag{6}$$

$$E(\text{tangential}) = 1.5\ E \sin\delta \tag{7}$$

where δ is the angle between field and radius vector to the point considered. The lateral field strength is therefore ctg$\delta \cdot R/d$ times smaller than the radial field. At higher frequencies it diminishes in proportion with $1/f$, as does the radial field as indicated in eq. 5. Therefore, the lateral field is always very small compared to the radial field except when δ is very close to 90° .

In conclusion, strong interactions of microwave fields with membranes are not expected from classical theory. If there are such interactions, then they must be much more sensitive and subtle than the excitation phenomena at low frequencies. The formulation of testable biophysical theories that can overcome this hurdle and that are quantitative or at least semiquantitative remains a challenge.

FURTHER COUPLING CONSIDERATIONS

Dielectric properties and knowledge of mechanism responsible for such properties aid our understanding of electromagnetic effects on biomaterials. We include, as an example, discussions of what we might call coupling considerations.

They may be subdivided into two subcategories. Thus, macroscopic dosimetry has been applied to (a) depth of penetration of microwaves and RF, (b) reflections (air-skin, skin-fat, fat-muscle), (c) heat production in skin, fat, and muscle, (d) man's radiation cross-section, (e) hot spot production by microwaves (mechanism, specific absorption rate (SAR), distribution), (f) thermoregulatory response to irradiation, and (g) exposure health standards. Microscopic "coupling" topics that have been studied include membrane potentials in cells and nuclei, and field strength in glycocalix, radial and lateral membranes, cytoplasm, organelles and nucleus, and inner membrane systems connecting with outer ones (tubule system, endoplasmic reticulum). On the macroscopic level absorption and refraction effects and spatial distributions of specific absorption rate (SAR) values in models of man have been extensively investigated (Schwan, 1965; Durney et al., 1980) and are generally considered of prime importance in the setting of standards of safety. Prerequisite to all these efforts is the knowledge of dielectric data for all body tissues over the total frequency range of interest.

Equally important are coupling considerations at the microscopic level (Schwan, 1983b; 1985a). An example was already presented above. Knowledge of the field strength values and their frequency dependence in all biological component systems such as cellular membranes, cytoplasma, membranes, and interior of organelles is necessary to apply any predictive theory of electromagnetic field interactions to biosystems. Do microwaves interact primarily with cell membranes, the cell membrane surface, the genetic apparatus in the cell nuclei, the cytoskeleton, or cytoplasmic proteins? For each case we can at least outline relevant field strength values and their frequency specificity utilizing the knowledge gained from the dielectric studies.

Consider for example the case that two cells have been fused to form a larger single one with two nuclei. What field parameters should be used first to join the two nuclei and then unite them by application of a nuclear membrane potential high enough to cause fusion at the point of contact? In Fig. 4 we present a simplified demonstration of the frequency dependence of the nuclear membrane potential and how it compares with outer cell membrane potential (Schwan, 1985a). Clearly,

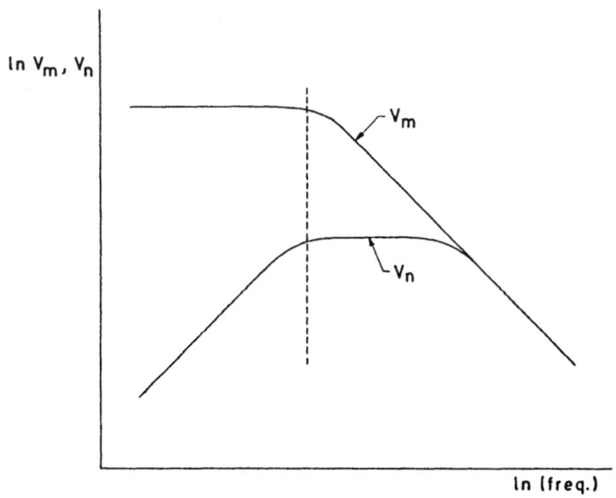

Fig. 4 — Frequency dependence of cellular membrane potential V_m and nuclear membrane potential V_n. A frequency-independent field is assumed outside the cell. V_n values may be higher or lower than V_m values at high frequencies, depending on electrical conductivities inside and outside the cell.

significant nuclear membrane potentials can be generated over only a limited frequency range. Thus dielectric breakdown of the nuclear membranes, if a prerequisite of fusion, can be accomplished only in this range. Similar considerations pertain to pulses, as the Fig. 4 can be readily transformed from the frequency domain into the time domain.

MANIPULATION OF CELLS BY ELECTROMAGNETIC FIELDS AND DIELECTRIC PROPERTIES

Alternating electrical fields exert forces on particles such as biological cells. These forces can manifest themselves in various ways (Schwan and Sher, 1969; Pohl, 1978; Schwan, 1985c;). Dielectrophoresis, i.e. the movement of cells in inhomogeneous fields, has been intensely studied over decades by Herbert Pohl and his associates (Pohl, 1978). Pearl chain formation and orientation of cells under the influence of alternating electromagnetic fields had been of interest to us also (Schwan and Sher, 1969). More recently, the fusion of cells by appropriate use of imposed fields and their rotation in a rotating field have been investigated by Zimmermann et al. and other groups (Zimmermann, 1982; Arnold and Zimmermann, 1984). Clearly the manipulation of cells by alternating fields can be used for all sorts of useful purposes. Earlier theoretical formulations provided a fair understanding of these effects, but a rigorous complete theory was lacking. Complete theoretical formulations have now been provided for the processes of pearl chain formation and rotation (Sauer, 1985; Sauer and Schloegl, 1985). In both cases threshold criteria can be formulated for the minimal field strength requirement to overcome thermal noise. The threshold equation for pearl chain formations is (Sauer, 1985)

$$E_{th}^2 \epsilon_2 R^3 \left| \frac{\epsilon_1^* - \epsilon_2^*}{\epsilon_1^* + 2\epsilon_2^*} \right|^2 = \frac{kT}{6p}. \tag{8}$$

Here E_{th} is the threshold field strength, R radius, k Boltzmann constant, T absolute temperature, p volume fraction of suspended particles, and the subscripts 1 and 2 are indicative of particle and medium. This equation, as well as earlier formulas, state that E_{th} is a constant which depends on the dielectric properties of the system. This means that threshold values decrease rapidly with increasing particle size, as has been demonstrated (Schwan, 1977). In eq. 8 the apparent dipole moment of the suspended particle or cell appears only as

$$u = \frac{\epsilon_1^* - \epsilon_2^*}{\epsilon_1^* + 2\epsilon_2^*}. \tag{9}$$

This expression fluctuates between the limits $-1/2$ and 1, both independent of ϵ_1^*. Thus we might anticipate that the particle property ϵ_1^* is but weakly related at best to the onset of pearl chain formation. This appears to be true from the data presented so far. However, if the medium dielectric constant ϵ_2 itself is directly related to ϵ_1^*, one would anticipate an effect. This may well explain some recently reported frequency dependencies of E_{th} (Takashima and Schwan, 1985).

A more direct demonstration of the effects of particle properties on a field effect is provided by the phenomena of cell rotation in a rotating field. Sauer and Schloegl (1985) have shown that the torque applied to the cell is given by

$$L = 4\pi R^3 \epsilon_2 E^2 \, Im \, \frac{\epsilon_1^* - \epsilon_2^*}{\epsilon_1^* + 2\epsilon_2^*} . \tag{10}$$

Again we note that the particle property appears through the u-transformation given in eq. 9. However the torque is now proportional to the imaginary part of this quantity. This imaginary value has been recently demonstrated to be often very sensitive to the particle property of biological cells (Sauer and Schloegl, 1985).

What about dielectrophoresis? Again, the speed of particle movement in the inhomogeneous field appears to be a function of the quantity u if multiplied with $E_{th} R$. But the dependence on u is now once more different as stated in eq. 11, which gives the force acting on the particle

$$F = \pi \epsilon_2 R^3 \nabla \mid E \mid^2 Re \, \frac{\epsilon_1^* - \epsilon_2^*}{\epsilon_1^* + 2\epsilon_2^*} . \tag{11}$$

The rotation technique is probably the leading competitor to dielectric spectroscopy for the evaluation of cellular dielectric data. This is because it registers the imaginary part of the quantity u, which displays the same frequency dependence as ϵ'' and can be readily observed, at least through the β-dispersion range (Schwan, 1985d). Furthermore pearl chain formation and dielectrophoresis are more cumbersome to execute and depend on the real part or magnitude of this quantity.

CONCLUDING REMARKS

I have chosen excitation, macro- and microdosimetry, and field-induced force effects as important examples in order to demonstrate the dependence of these interactions on dielectric properties. I have also restricted myself to these interactions since they are fairly well understood. This understanding is based on the mechanisms that have been recognized to account for observed dielectric properties. Only linear and dispersive properties need be discussed in this context. I do not maintain that the present excellent understanding of dispersive bioproperties excludes the possibility of resonant interaction. Indeed we know that at low frequencies "inductive" and "capacitive" channel responses may combine to resonant interactions, particularly in the nonlinear range. And, more recently, dielectric data at gigahertz frequencies indicate, apparently at the linear level, resonance features (Edwards, 1984). They may or may not relate to reported nonlinear millimeter-wave sensitivities of yeast cell growth (Grundler, 1985). But I suspect, if we fully understand the mechanism involved, we will again note that they are related to dielectric properties. Fröhlich has often stated that if there are nonthermal interactions aside from the cellular phenomena and field effects discussed above, then these interactions must be highly frequency-sensitive, nonlinear, and likely to occur at very high frequencies. They should reflect related dielectric properties, possibly in the nonlinear range not yet systematically examined.

REFERENCES

W.M. Arnold and U. Zimmermann, "Electric field-induced fusion and rotation of cells," in *Biological Membrances*, Vol. 5, Academic, London, 1984, p 389.

C.H. Durney, et al., "Radio Frequency Radiation Dosimetry Handbook," (1st ed., 1976; 2nd ed., 1978; 3rd ed., 1980), Reports SAM-TR-76-35, 78-22 and 80-32, prepared for U.S. Air Force, School of Aerospace Medicine, Brooks AFB, TX 78235. See also Proc. IEEE **68**, No. 1, various articles (1980).

G.S. Edwards, *Microwave-Field Driven Acoustic Modes in Selected DNA Molecules*, Ph.D. Dissertation, University of Maryland, 1984.

W. Grundler, "Frequency-dependent biological effects of low intensity microwaves," in *Interactions Between Electromagnetic Fields and Cells*, A. Chiabrera, C. Nicolini, and H.P. Schwan, Eds., NATO ASI series, Vol. 97, Plenum, New York, 1985, pp 459-481.

A.L. Hodgkin and A.F. Huxley, J. Physiol. **117**, 500 (1952).

H.A. Pohl, *Dielectrophoresis*, Cambridge University Press, Cambridge/New York, 1978.

F.A. Sauer, "Interaction-forces between microscopic particles in an external electromagnetic field," in *Interactions between Electromagnetic Fields and Cells*, A. Chiabrera, C. Nicolini, and H.P. Schwan, Eds., NATO ASI series, Vol. 97, Plenum, New York, 1985, p 181.

F.A. Sauer and R.W. Schloegl, "Torques exerted on cylinders and spheres by external electromagnetic fields: a contribution to the theory of field induced cell rotation," in *Interactions between Electromagnetic Fields and Cells*, A. Chiabrera, C. Nicolini, and H.P. Schwan, Eds., NATO ASI series, Vol. 97, Plenum, New York, 1985, p 203.

H.P. Schwan, "Electrical properties of tissues and cell suspension," in *Advances in Biological and Medical Physics,* vol. 5, J.H. Lawrence and C.A. Tobias, Eds., Academic, New York, 1957.

H.P. Schwan, "Biophysics of Diathermy," in *Therapeutic Heat*, 2nd Ed. Eliz. Licht Pub., New Haven, CT, 1965, pp 63.

H.P. Schwan, Ann. New York Acad. Sci. **303**, 198-213 (1977).

H.P. Schwan, "Electrical properties of cells: principles, some recent results, and some unresolved problems," in *The Biophysical Approach to Excitable Systems*, W.J. Adelman, Jr. and D.E. Goldman, Eds., Plenum, New York/London, 1981, pp 3-24.

H.P. Schwan, "Dielectric Properties of Biological Tissue and Cells at RF- and MW-Frequencies," in *Biological Effects and Dosimetry of Nonionizing Radiation—Radiofrequency and Microwave Energies*, M. Grandolfo, S.M. Michaelson, and A. Rindi, Eds., Plenum, New York/London, 1983a, pp 195-211.

H.P. Schwan, "Biophysics of the interaction of electromagnetic energy with cells and membrances," in *Biological Effects and Dosimetry of Nonionizing Radiation—Radiofrequency and Microwave Energies*, M. Grandolfo, S.M. Michaelson, and A. Rindi, Eds., Plenum, New York, 1983b, pp 213-231.

H.P. Schwan, "Biophysical principles of the interaction of ELF fields with living matter: 11. Coupling Considerations and Forces," in *Biological Effects and Dosimetry of Static and ELF Electromagnetic Fields,* M. Grandolfo, S.M. Michaelson, and A. Rindi, Eds., Plenum, New York, 1985a, pp 243-271.

H.P. Schwan, "Dielectric properties of cells and tissues," in *Interactions between Electromagnetic Fields and Cells*, A. Chiabrera, C. Nicolini, and H.P. Schwan, Eds., Plenum, New York/London, 1985b, pp 75-97.

H.P. Schwan, "EMF-induced force effects," in *Interactions between Electromagnetic Fields and Cells*, A. Chiabrera, C. Nicolini, and H.P. Schwan, Eds., Pleunum, New York, 1985c, pp 371-389.

H.P. Schwan, Studia Biophys. **110**, 13 (1985d).

H.P. Schwan and K.R. Foster, Proc. IEEE **68**, 104 (1980).

H.P. Schwan and L.D. Sher, J. Electrochem. Soc. **116**, 22C (1969).

S. Takashima and A. Minakata, "Dielectric behavior of biological macromolecules," 37:602 in *Digest of Dielectric Literature*, National Research Council, Washington, DC, 1975.

S. Takashima and H.P. Schwan, Biophys. J. **467**, 513 (1985).

U. Zimmermann, Biochim. Biophys. Acta, **694**, 227 (1982).

31. Condensed Matter Physics and the Biology of the Future

S. Rowlands*

In a modern textbook of histology or cell physiology, electron micrographs illustrate the complexity of living matter. As the resolution of electron microscopes improves, more and more details appear. A nerve fiber is usually regarded as a simple structure, yet at high resolution we can recognize a reticulum of fibers named microtubules, microfilaments, and neurofilaments. Moreover, the fibers are regularly interconnected and in the connections are little rings. What is the function of this network? It looks like a microprocessor.

To take such photographs we have to poison, kill, and embalm the living tissue. We can also mince it up while it is still alive and pick up some of the bits. Biochemistry tells us that there are over 30,000 different proteins in the human body, not to mention all the other molecules. Biophysics uncovers thousands of intricate processes. In intact life these processes occur in an orderly manner. The acinar cell of the pancreas is a factory that manufactures digestive enzymes. It cuts, shapes, inspects, and packages them in a way that prevents the digestion of the cell itself; and as it does so it moves them the length of the cell and then delivers them through the cell wall into the pancreatic duct outside the cell, and so to the gut where the now free enzymes digest the food of the organism.

The task of studying the individual biochemical reactions in minced-up tissue has been so enormous and so interesting that molecular biologists have had little time to devote to how the reactions are organized and how the reactants and products are moved. Thermal diffusion (Brownian motion) is always present but there are no thermal gradients as in a heat engine. The cell is isothermic, yet efficient in the use of energy. Most biochemical reactions involve at least three molecules, two or more reactants and an enzyme that is the catalyst. The chance collision of three entities in Brownian motion is an unlikely occurrence. Moreover, the collision must be made with the correct orientation for the active parts on the several molecules to mesh. A further oriented collision may have to follow the first. After a three-body collision transfer RNA has its coded-for amino acid attached. It is now a transporter that finds its correct code-word on messenger RNA as this "computer tape" speeds through a ribosome. This is the machine that joins together hundreds of the twenty or so different amino acids, on instructions from messenger RNA, into a specific protein. It is inconceivable that the cell has to wait for the chaotic Brownian movement to bring about all these things by chance. An automated factory would be quite inefficient if it operated by means of a random movement of parts from floor to ceiling. It requires coordination and control, conveyer belts and robots.

*Present address: RR1, Site 160, Box 31, Bowser, British Columbia, Canada, VOR 1GO

As cellular structure is elucidated in more and more detail by electron microscopy, the cell appears less and less like a bag containing molecules dissolved in water, which implicitly was the classical view, and more and more like a solid-state structure. A cell 16 μm in diameter has an internal surface area of perhaps 100,000 μm^2 (Clegg, 1983). Such a structure will impose a spacial order of sorts but gives no clue of how motion is ordered in relation to such a structure. Some years ago London (1950) speculated that the behavior of superfluids might in some way help in understanding the dynamics of the immensely complicated macromolecular systems of biology. He suggested that a coherent fluid state of zero entropy combines the stability of quantum states with the possibility of motion without the dissipation of energy.

The suggestion of coherent quantum phenomena in biology usually evokes contemptuous dismissal because these phenomena require bizarre, extreme conditions such as extremely low temperatures (for superconductivity and superfluidity) or extremely high ones (in the laser). There is, however, a bizarre extreme electric field across all the membranes in and around a living cell, and the purpose of this field is unknown except in a nerve fiber, where a transient interruption of field is the nerve action potential. If the modest transmembrane potential of the order of 100 mV is divided by the thickness of the membrane (10 nm), the field works out to be enormous ($\approx 10^7 V.m^{-1}$), enormous in comparison even with the lethal field imposed in the electric chair! Fröhlich seems to have been the first to publish this simple but vital calculation (1968) and he went on (1980) to develop a theory of coherent excitations in biological systems based on this elementary observation.

Briefly the theory is this: All membrane molecules have the thermal vibrations appropriate to the organism's temperature; but, because the membrane molecules are in an intense electric field, they will be electrically polarized and will therefore interact, giving rise to modes of vibration. These phonon modes will exchange nonlinearly with the random phonons of the surrounding cell water. If energy above a threshold level is supplied to these modes, Fröhlich's theory predicts that the lowest mode will condense into the coherence of a giant vibration with a frequency in the millimeter wave band. One of the important consequences of this theory is that it predicts the existence of ultra-long-range interactions between macromolecular systems, of much longer range than that of chemical forces. Moreover, the forces will be specific in that a cell membrane could recognize a macromolecule by its inherent molecular vibrations and either attract or repel it, depending on the relative phase of the coupled vibrations. The interacting systems may well have other systems in between them. And so we have the beginnings of an explanation of the orderly motion of cellular structures and of macromolecules. There is evidence of such specific action-at-a-distance between living mammalian red blood cells, such action mediated by extended macromolecules of molecular weight greater than 40,000 daltons. This interaction is experimentally consistent with the requirements of the Fröhlich theory (Rowlands et al. 1982; Fritz, 1984).

The energy that an organism gets from food, sunlight, etc., is converted by complex metabolic reactions into adenosine triphosphate (ATP), which is the "fuel" for many if not all cellular reactions. Until ten years ago a most baffling problem in biology was how the 0.5-eV energy from the decomposition of ATP is conserved. It should quickly be degraded into heat by bond vibrations, whereas actually most of the energy appears as an intact quantum quite a long way, on a microscopic scale, from where it was produced. The work of Davydov (1979) and of Scott and his colleagues (1981) now makes it look as though the α-helix, so characteristic of proteins, is uniquely designed to carry the quantum of energy of ATP over considerable biological distances with little loss, in the form of solitons (solitary waves). These are well known in theoretical and experimental physics and there is experimental evidence of their existence in crystalline acetanilide, which has structural features in common with the much more complicated α-helix of protein molecules. Moreover, there is now a theory (Del Giudice et al., 1982) that combines those of Fröhlich and Davydov into the beginnings of a comprehensive theory of cell dynamics in which the Fröhlich vibrations are fed with energy by solitons in molecular chains.

So there is plenty of scope in the years to come for interpreting new experiments and for reinterpreting the old ones in terms of mechanisms already familiar to condensed matter physics. But the problem of greatest difficulty (and interest) is the growth and development of organisms. The fertilized ovum of man is about a tenth of a millimeter in diameter and yet it contains all the information for building an adult. If you argue that in 40 weeks gestation much information may come from the mother, then consider a fish, a highly complex animal but one completely lacking in motherly care. Inside the fertilized ovum are all the instructions for building man, given energy, simple chemicals, a few amino acids not synthesized in the body, and vitamins. A human being is much more complex than an airliner. Yet an airliner does not build itself from energy and raw materials; nor could we conceive of packaging the "building instruction book" within a one-tenth-millimeter sphere.

How is it done? There are some clues but it is difficult to see where to make a start experimentally. Fanchon Fröhlich (1977) thinks that there is much more to DNA than instructions for building macromolecules. As we understand the genetic code, it is like the Morse code in which a group of up to four binary digits corresponds to a letter of the alphabet. In the DNA genetic code a group of three nucleic acids corresponds to the position of a particular amino acid in the primary structure of a protein. In terms of linguistics we are reading DNA in a very primitive fashion. In a language, letters form words and the order becomes important. Sentences bring another level of order. Sentence follows sentence and contradictions have to be avoided if there is to be an unequivocal buildup of ideas. When you think of a sonnet, fourteen ten-syllable lines, and all the subtleties that Shakespeare can convey in one, maybe we are reading the genetic code at the level of first-year junior school. In linguistics the distinction is between letter-by-letter coding and the collective complexity of language; so we may be dealing with a collective cooperative phenomena as in Fröhlich's theory of coherent excitations in biological systems.

In any living creature there is more DNA than is needed for building proteins. The DNA content per cell for various organisms varies over a wide range from 0.005 to 200 pg. *Homo sapiens* weighs in at about 7 pg per cell. The record seems to be held by a marine organism consisting of just a single cell. Less than 2% of its DNA is needed for protein synthesis; the rest is surplus and has been called "junk"

DNA. Life is very efficient and I cannot accept that even a single species wastes 98% of its DNA. This marine creature, a dinoflagellate, being a single cell, has no brain. But it has behavior, so perhaps the "junk" DNA serves as a brain which instructs it how to behave, to react to stimuli, and to reproduce. The code may be different or interpreted in a different way or there may be an entirely different code lodged in the intramolecular vibrations of DNA or elsewhere. Popp et al. (1984) suggest that excimer-laser-like excitations in DNA give rise to the so-called ultraweak photon emission from living cells. They base this suggestion partly on experimental evidence for coherence in such radiation. If this view is correct, DNA vibrations could constitute an enormous reservoir of information. Most other workers in this field, however, believe that the ultraweak photon emission arises from the decomposition of lipid peroxides; that is, it has its origin in well-known chemical reactions (Quickenden et al., 1985).

In higher organisms there are other possibilities. The support for Fröhlich's theory shows that cells are communicating with organelles, and cells with cells, by mechanisms additional to simple electrical and chemical methods. These excitations have existed since life began, and since life seems to use every available mechanism, it may well have developed the use of Fröhlich excitations for very subtle communication. At frequencies of 10^{11} Hz quite a narrow waveband could transmit a vast quantity of information. For instance, a tenfold band could carry 150,000 television programs. There may be a nervous system of higher order than the one we know, a second nervous system (Rowlands, 1983) and the action potential may be just a crude fast transmitter of urgent messages to make us move and to control whole organs. Neurotrophism, which has been defined as "interaction between nerve and other cells which initiate or control molecular modifications in the other cells" and which has not been much studied in comparison with what might be called "action potential" neurology, may turn out to be much the more interesting.

The vibrations might be transmitted along nerve fiber membranes but they would be interrupted by action potentials. It is more likely that the microtubules, microfilaments, and neurofilaments mentioned in the first paragraph of this essay are used. Nonlinear integrated optics is beginning to replace electronics in communication and an analogous process for Fröhlich excitations and/or solitons could transmit masses of information, without loss, in these structures. The cross connections and the associated ring proteins may be switches and parts of a "microprocessor". Already the Naval Research Laboratory in Washington D.C. is considering computer elements made of macromolecules and molecular switches in which information is transmitted by solitons. Some patents have been taken out!

In the central nervous system itself there are myriads of neurons (nerve cells) but lots more of neuroglia; they outnumber neurons by about ten to one. They have a complex architecture but we do not really know what they do. Hypotheses of their function range from their being merely a mechanical support structure to the seat of memory! The membranes of the glial cells invest those of the neurons extensively and the membranes come closer together (10 nm) than any cells that are not actually in contact. This gap is small enough for the tunnelling of quantum particles between the two moieties of specialized cells in the central nervous system. Glial cells could be processing information from "junk" DNA.

Evidence is very slender for such thoughts but in science we should keep open minds— though not of course so open that our brains fall out! Wilfred Trotter wrote: "The mind likes a strange idea as little as the body likes a strange protein and resists it with similar energy. It would not be not fanciful to say that a new idea is the most quickly-acting antigen known to science". We must beware of orthodoxy; it is a way of going wrong with confidence.

This essay, written to honor Fröhlich's 80th birthday, is a brief account of ideas that were inspired by reading, in 1980, his theory of coherent excitations in biological systems. Another purpose of the essay is to persuade young scientists, not yet "weighed down with the burden of inert ideas" (Whitehead), that a rigorous training in condensed matter physics would be a major asset for a successful career as a 21st-century life scientist.

ACKNOWLEDGEMENT

This paper is based on a lecture given at the University of Toronto on 25th February 1984 during the tenure of a Medical Research Council of Canada Visiting Professorship.

REFERENCES

Clegg, J.S. (1983), *Coherent Excitations in Biological Systems.* (Fröhlich, H. and Kremer, F., eds.) Springer-Verlag, Berlin, pp. 162-177.

Davydov, A.S. (1979), *Physica Scripta*, **20**, 387-394.

Del Guidice, E., Doglia, S. and Milani, M. (1982), *Physica Scripta*, **26**, 232-238.

Fritz, O.G. (1984), *Biophys. J.* **46**, 219-228.

Fröhlich, H. (1968), *Int. J. Quantum Chem.* **2**, 641-649.

Fröhlich, H. (1977), *Synergetics.* (Haken, H., ed.) Springer-Verlag, Berlin, pp. 241-246.

Fröhlich, H. (1980), *Adv. Electron. Electron Phys.* **53**, 85-152.

London, F. (1980), *Superfluids.* Vol. 1, Wiley, New York.

Nagele, R.G. and Roisen, F.J. (1982), *Brain Res.* **253**, 31-37.

Popp, F.A., Nagl, W., Li, K.H., Scholz, W., Weingartner, O. and Wolf, R. (1964), *Cell Biophys.* **6**, 33-52.

Quickenden, T.I., Comarmond, M.J. and Tilbury, R.N. (1985) *Photochem. Photobiol.* **41**, 611-615.

Rowlands, S., Sewchand, L.S. and Enns, E.G. (1982), *Can. J. Physiol. Pharm.* **60**, 52-59.

Rowlands, S. (1983), *Coherent Excitations in Biological Systems.* (Fröhlich, H. and Kremer, F., eds.) Springer-Verlag, Berlin, pp. 145-161.

Scott, A.C. (1981), *Nonlinear Phenomena in Physics and Biology.* (Enns, R.H., Jones, B.L., Miura, R.M. and Rangnekar, S.S., eds.) Plenum, New York, pp. 7-82.

Publications of Herbert Fröhlich from 1930–1985

Among the many significant contributions he has made, there are broad themes discernible in the long and continuing distinguished career of Herbert Fröhlich. These broad themes address the following major topics:

1. Dielectric theory: dielectric breakdown; high-field electronic conductivity; spatial heterogeneity; nonlocal field theory.
2. Polaron theory: electron-lattice coupling; electron-phonon interaction, coherent lattice polarization; collective states.
3. Superconductivity: collective electron-phonon model; order and cooperativity.
4. Biological energy transfer: off-diagonal long-range order; cooperative phenomena; long-range vibrational energy transfer; polar metastable states; coherent lattice polarization.

However, in order to sample the intellectual leadership offered by Professor Fröhlich, one may choose among the following list of his publications.

Books

1. *Elektronentheorie der Metalle.* Springer, Berlin, 1936; reprinted, Ann Arbor 1943, republished by Springer, 1969 [in German].

2. *Theory of Dielectrics.* Clarendon Press, Oxford, 1949; 2nd ed., 1958. Translations in Russian and Japanese.

3. *Coherent Excitations in Biological Systems.* H. Fröhlich and F. Kremer, Eds. Springer, Berlin, 1983.

Original Papers and Review Articles

4. Photoelectric effect in metals. Ann. Phys. **7**, 103-128 (1930) [in German].

5. Shot effect in quantum mechanics. Z. Phys. **71**, 715-719 (1931) [in German].

6. Theory of secondary emission from metals. Ann. Phys. **13**, 229-248 (1932) [in German].

7. Light absorption and selective photoelectric effect. Z. Phys. **75**, 539-543 (1932) [in German].

8. Determination of energy levels of metallic electrons from their optical constants. Naturwissenschaften **20**, 906 (1932) [in German].

9. Position of the absorption spectra of photochemically coloured alkali halide crystals. Z. Phys. **80**, 819-821 (1933) [in German].

10. Absorption of metals in the visible and ultraviolet. Z. Phys. **81**, 297-312 (1933) [in German].

11. Magnetic interaction of metallic electrons. Criticism of Frenkel's theory of superconductivity (with H.A. Bethe). Z. Phys. **85**, 389-397 (1933) [in German].

12. The calculation of the work function in Sommerfeld's model of a metal. Z. Phys. Sov. **7**, 509-510 (1935) [in German].

13. The number of free electrons in a metal. Proc. Cambridge Phil. Soc. **31**, 277-280 (1935).

14. Inner photoelectric effect in semiconductors. Z. Phys. Sov. **8**, 501-510 (1935) [in German].

15. Time orientation of nuclear spins in a magnetic field (with W. Heitler). Z. Phys. Sov. **10**, 847-848 (1936) [in German].

16. Time effects in the magnetic cooling method (with W. Heitler). Proc. R. Soc. London, A**155**, 640-652 (1936); Proc. VII Int. Congr. Refrigeration (with W. Heitler and E. Teller), p. 1-2 (1936).

17. Quantum mechanical discussion of the cohesive forces and thermal expansion coefficients of the alkali metals. Proc. R. Soc. London, A**158**, 97-110 (1937).

18. Electronic specific heat of small metal particles at low temperatures. Physica **4**, 406-412 (1937).

19. Theory of electrical breakdown in ionic crystals, I. Proc. R. Soc. London, A**160**, 230-241 (1937).

20. Theory of the λ-point of helium. Physica **8**, 639-644 (1937).

21. Magnetic moments of the proton and the neutron (with W. Heitler). Nature **141**, 37 (1938).

22. Nuclear forces and the magnetic moments of the neutron and proton (with W. Heitler and N. Kemmer). Proc. R. Soc. London, A**166**, 154-177 (1938).

23. Solution of the Schrödinger equation by perturbation of the boundary conditions. Phys. Rev. **54**, 945-947 (1938).

24. Deviation from the Coulomb law for the proton (with W. Heitler and B. Kahn). Proc. R. Soc. London, A**171**, 269-280 (1939).

25. Mean free path of electrons in polar crystals (with N.F. Mott). Proc. R. Soc. London, A**171**, 496-504 (1939).

26. Theory of electrical breakdown in ionic crystals, II. Proc. R. Soc. London, A**172**, 94-106 (1939).

27. Dielectric breakdown in ionic crystals, III. Phys. Rev. **56**, 349-352 (1939).

28. Dielectric breakdown in solids. Rept. Progr. Phys. **6**, 411-430 (1939).

29. Photodisintegration of deuteron in meson theory (with W. Heitler and B. Kahn). Proc. R. Soc. London, A**174**, 85-102 (1940).

30. Orientation of nuclear spins in metals (with F.R.N. Nabarro). Proc. R. Soc. London, A**175**, 382-391 (1940).

31. On the dielectric strength of mixed crystals. Proc. R. Soc. London, **A178**, 493-505 (1941).

32. Dielectric loss in dipolar solids. Solution of dipolar molecules in solid paraffin. Rep. Brit. Elec. Allied Indust. Res. Assoc.*, Ref. L/T 121, 12 pp (1941).

33. Theory of the dielectric properties of dipolar solids. E.R.A. report, Ref. L/T 124, 55 pp (1941).

34. Electric breakdown in ionic crystals. Phys. Rev. **61**, 200-201 (1942).

35. Meson theory and the magnetic moments of protons and neutrons. Phys. Rev. **62**, 180 (1942).

36. Dielectric loss in paraffin wax solutions. Proc. Phys. Soc. London **54**, 422-428 (1942).

37. Theory of the dielectric properties of dipolar solids. Calculation of dielectric losses. E.R.A. report, Ref. L/T 132, 8 pp (1942).

38. Theory of dielectric breakdown. Nature **151**, 339-340 (1943).

39. Theory of the dielectric properties of dipolar solids. Influence of dipolar interaction. E.R.A. report, Ref. L/T 142, 12 pp (1942).

40. Dielectric properties of dipolar substances (with R.A. Sack). Proc. R. Soc. London, A**182**, 388-403 (1944).

*Hereinafter denoted by E.R.A. report

41. Theory of the dielectric constant and energy loss in solids and liquids. J. Inst. Elec. Eng. London **91**, 456-463 (1944).

42. Phase transitions of solid paraffins and the flexibility of hydrocarbon chains. Trans. Faraday Soc. **40**, 498-502 (1944).

43. Theory of dielectric breakdown in amorphous solids. E.R.A. report, Ref. L/T 153, 7 pp (1945).

44. Dielectric and thermal properties of long-chain substances. E.R.A. report, Ref. L/T 156, 7 pp (1945).

45. Dielectric properties of solids at very high frequencies. E.R.A. report, Ref. L/T 157, 6 pp (1945).

46. Dielectric properties of solids at ultra-high frequencies. E.R.A. report, Ref. L/T 163, 5 pp (1946).

47. Dielectric properties of dipolar solids. Proc. R. Soc. London, Ser. A**185**, 399-414 (1946).

48. Theory of rheological properties of dispersions (with R.A. Sack). Proc. R. Soc. London, A**185**, 415-430 (1946).

49. Shape of collision-broadened spectral lines. Nature **157**, 478 (1946).

50. Theoretical physics in industry (Roy. Institution Lecture). Nature **158**, 332-334 (1946).

51. Dipolar interaction. Trans. Faraday Soc. **42**A, 3-7, disc. 36-39 (1946).

52. Light absorption and selective photo-effect in adsorbed layers (with R.A. Sack). Proc. Phys. Soc. London **59**, 30-33 (1947).

53. On the theory of dielectric breakdown in solids. Proc. R. Soc. London, A**188**, 521-532 (1947).

54. Energy distribution and stability of electrons in electric fields. Proc. R. Soc. London, A**188**, 532-541 (1947).

55. Decay of negative mesons in matter. Nature **160**, 255 (1947).

56. The binding energies of very light nuclei (with K. Huang and I.N. Sneddon). Proc. R. Soc. London, A**191**, 61-82 (1947).

57. General theory of the static dielectric constant. Trans. Faraday Soc. **44**, 238-243 (1948).

58. A quantitative discussion of the interaction between nuclear particles (with W.H. Ramsay and I.N. Sneddon). Rep. Int. Conf. on *Fundamental Particles* (1946), in Phys. Soc., London **1**, 166-175 (1948).

59. Polarization of dielectrics by slow particles (with H. Pelzer). E.R.A. report, Ref. L/T 184, 8 pp (1948).

60. Decay and capture of slow mesons in dielectrics (with R. Huby, R. Kolodziejski and R.L. Rosenberg). Nature **162**, 450-451 (1948).

61. Nuclear ferromagnetism (with F.R.N. Nabarro). Rep. Int. Conf. *Low Temperatures*, Cambridge (1946), in Phys. Soc., London **2**, 130-134 (1948).

62. Time dependence of electronic processes in dielectrics (with J. O'Dwyer). E.R.A. report, Ref. L/T 219, 5 pp (1949); Proc. Phys. Soc. London A**63**, 81-85, and 299 (1950).

63. Properties of slow electrons in polar materials (with H. Pelzer and S. Zienau). E.R.A. report, Ref. L/T 221, 12 pp (1950); Phil. Mag. **41**, 221-242 (1950).

64. Conduction electrons in non-metallic solids. Research London **3**, 202-207 (1950).

65. Isotope effect in superconductivity. Proc. Phys. Soc. London A**63**, 778 (1950).

66. Notes on the theory of dielectric breakdown in ionic crystals (with F. Seitz). Phys. Rev. **79**, 526-527 (1950).

67. Intrinsic dielectric breakdown in solids (with J.H. Simpson). Advan. Electron. **2**, 187-217 (1950).

68. Theory of the superconducting state, I. The ground state at the absolute zero of temperature. Phys. Rev. **79**, 845-856 (1950).

69. Theory of the superconducting state, II. Magnetic properties at the absolute zero of temperature. Proc. Phys. Soc. London A**64**, 129-134 (1951).

70. Crystal structure and superconductivity. Nature **168**, 280 (1951).

71. Superconductivity and the effective mass of electrons. Nature **168**, 280-281 (1951).

72. On the theory of dielectric breakdown in ionic crystals. E.R.A. report, Ref. L/T 277, 14 pp (1952).

73. Interaction of electrons with lattice vibrations. Proc. R. Soc. London, A**215**, 291-298 (1952).

74. Theory of the superconductive state. Proc. Washington Conf. (1952).

75. Superconductivity and lattice vibrations. Proc. Lorentz-Kamerlingh Onnes Conf., in Physica **19**, 755-764 (1953).

76. Energy loss of moving electrons to dipolar relaxation (with R.L. Platzman). Phys. Rev. **92**, 1152-1154 (1953).

77. Rotational transitions in solids. 2^{e} Reunion de Chimie Physique Paris, p. 231-234 (1953) [in French].

78. Electrons in dielectrics. *Proc. Int. Conf. of Theoretical Physics, Kyoto and Tokyo, 1953*, p. 805-809, publ. by Science Council of Japan 1954.

79. Remarks on the theory of superconductivity. *Proc. Int. Conf. of Theoretical Physics, Kyoto and Tokyo, 1953*, p. 909-915, publ. by Science Council of Japan, 1954.

80. On the theory of superconductivity: the one dimensional case: Proc. R. Soc. London, A**223**, 296-305 (1954).

81. Electrons in lattice fields. Adv. Phys. **3**, 325-361 (1954).

82. Dielectric polarization in polar substances. Remark on a paper by F.E. Harris and B.J. Alder. J. Chem. Phys. **22**, 1804-1806 (1954).

83. Heat conduction in semiconductors. Remark on a paper by G. Busch (with C. Kittel). Proc. Amsterdam Conf. on Semiconductors, in Physica **20**, 1086 (1954).

84. Plasma oscillations and energy loss of charged particles in solids. E.R.A. report, Ref. L/T 322, 14 pp (1955); Proc. Phys. Soc. London **A68**, 525-529 (1955).

85. Theory of secondary electron emission from solids. Proc. Phys. Soc. London **B68**, 657-660 (1955).

86. Dielectric breakdown in solids (with B.V. Paranjape). Proc. Phys. Soc. London **B69**, 21-32, 866 (1956).

87. The influence of interelectronic collisions and of surfaces on electronic conductivity (with B.V. Paranjape, C.G. Kuper and S. Nakajima). Proc. Phys. Soc. London **B69**, 842-845 (1956).

88. Plasma interaction and conduction in semiconductors (with S. Doniach). Proc. Phys. Soc. London **B69**, 961 (1956).

89. Remark on the calculation of the static dielectric constant. Physica **22**, 898-904 (1956).

90. Debye loss in ionic solids. Arch. Sci. (Geneva) **10**, 5-6 (1957) [in French].

91. Speculations on the masses of particles. Nucl. Phys. **7**, 148-149 (1958).

92. A survey of the theory of dielectrics. Proc. 8th Colloque Ampere, in Arch. Sci. (Geneva) **12** (special number), 5-8 (1959).

93. Electric conduction in semiconductors (with G.L. Sewell). Proc. Phys. Soc. London **74**, 643-647 (1959).

94. Phenomenological theory of the energy loss of fast particles in solids. In *Max Planck Festschrift, 1958*, W. Frank, Ed., p. 277-284. VEB Deutscher Verlag der Wissenschaften, Berlin, 1959.

95. Space-time reflexions, isobaric spin and the mass ratio of bosons. Proc. R. Soc. London, A**257**, 147-164 (1960).

96. Light quanta and heavy bosons. Proc. R. Soc. London, A**257**, 283-290 (1960).

97. Space-time reflexions, light quanta and heavy bosons. Helv. Phys. Acta **33**, 803-828 (W. Pauli Memorial Issue) (1960).

98. The theory of the superconductive state. Rept. Progr. Phys. **24**, 1-23 (1961).

99. Breakdown in non-polar semiconductors. Soviet Phys. Solid State **3**, 491 (1961).

100. New heavy bosons. Proc. Phys. Soc. London **77**, 1223 (1961).

101. The structure of momentum space, the neutrino and the Pauli principle. Nucl. Phys. **26**, 324-337 (1961).

102. Hot electrons. E.R.A. report, Ref. L/T 414, 11 pp (1961).

103. Isobaric spin algebra (with C.M. Terreaux). Nucl. Phys. **42**, 21-26 (1963).

104. On isobaric spin space. Nucl. Phys. **45**, 609-613 (1963).

105. Low mobility materials and Debye dielectric loss due to electrons. E.R.A. report 5003, 8 pp (1963); with S. Machlup and T.K. Mitra. Phys. Kondens. Mat. **1**, 359-366 (1963).

106. Electron-phonon interaction, superconductivity and the third law of thermodynamics. Phys. Lett. **7**, 346-347 (1963).

107. Introduction to the theory of the polaron. In *Polarons and Excitons*, C.G. Kuper and G.D. Whitfield, Eds., Oliver and Boyd, Edinburgh and London, 1963 (Scottish Universities' Summer School 1962), pp 1-32.

108. The Boltzmann Equations in electron-phonon systems (with A.W.B. Taylor). Proc. Phys. Soc. London **83**, 739-748 (1964).

109. Quasi-superconductive transitions in strong magnetic fields (with C.M. Terreaux). Proc. Phys. Soc. London **86**, 233-236 (1965).

110. Geometrical interpretation of electrodynamics. Progr. Theor. Phys. Suppl. dedicated to H. Yukawa on the 30th anniversary of meson theory, p 1-13 (1965).

111. Macroscopic wave-functions in superconductors. Proc. Phys. Soc. London **87**, 330-332 (1966).

112. Generation of dual transformations through fields in electrodynamics. Progr. Theor. Phys. **36**, 636-647 (1966).

113. Superconductivity and the many body problem. In *Perspectives in Modern Physics*, R.E. Marshak, Ed., Interscience, NY, 1966 (Essays in honor of Hans A. Bethe), pp 539-552.

114. Limits of the band model and transitions to the metallic state. In *Quantum Theory of Atoms, Molecules and the Solid State*, P.O. Löwdin, Ed., Academic Press, NY, 1966 (A tribute to John C. Slater), pp. 465-468.

115. A contradiction between quantum hydrodynamics and the existence of particles. Physica **34**, 47-48 (1967).

116. Microscopic derivation of the equations of hydrodynamics. Physica **37**, 215-226 (1967).

117. Dielectric Instabilities. In *Ferroelectricity*, E.F. Weller, Ed., Elsevier Publ. Co., Amsterdam, 1967 (Dedicated to P.J.W. Debye), pp 9-15.

118. Proposal of crucial experiments for superconductors with incomplete inner bands. Phys. Lett. **26**A, 169-170 (1968).

119. Localized versus band model of electrons in solids. Helv. Phys. Acta **41**, 838-839 (1968) (G. Busch Festschrift).

120. Superconductivity in metals with incomplete inner shells. J. Phys. C **1**, 544-548 (1968).

121. Superconductivity and the magnitude of the electron-phonon interaction (with T.K. Mitra). J. Phys. C **1**, 548-549 (1968).

122. Theoretical problems in superconductivity. Phys. Bull. **19**, 209-212 (1968) (Jubilee article).

123. Bose condensation of strongly excited longitudinal electric modes. Phys. Lett. **26**A, 402-403 (1968).

124. Long-range coherence and energy storage in biological systems. Int. J. Quantum Chem. **2**, 641-649 (1968).

125. Storage of light energy and photosynthesis. Nature **219**, 743-744 (1968).

126. General remarks on the connection of the laws of micro and macro physics: Proc. Int. Conf. on Statistical Mechanics, Kyoto 1968, in J. Phys. Soc. Jpn. **26** (Suppl.), 189-195 (1969).

127. Macroscopic wave functions and wave equations. In *Problems of Theoretical Physics,* Nauka, Moscow, 1969 (Essays in honor of N.N. Bogoliubov), pp. 373-378.

128. Quantum mechanical concepts in biology. *Proc. 1st Int. Conf. Theoretical Physics and Biology, Versailles, 1967*, M. Marois, Ed., North Holland, Amsterdam, NL, 1969, p. 13-22.

129. Proposed model experiments on the storage of light energy in photosynthesis. Nature **221**, 976 (1969).

130. The macroscopic wave equations of superfluids. Phys. Kondens. Mat. **9**, 350-358 (1969).

131. Theoretische Phys. und Biologie. In *Wohin führt die Biologie*, M. Lohmann, Ed., Verlag Karl Hasner, München, 1970 (Transcript of a German radio broadcast) [in German], pp 147-173.

132. Possibility of a second acoustic branch in transition metals. Mat. Res. Bull. **5**, 607-609 (1970) (N.F. Mott Festschrift).

133. Long-range coherence and the action of enzymes. Nature **228**, 1093 (1970).

134. What future for superconductivity. Nature **228**, 1145-1146 (1970).

135. Superconductivity, lattice stability and phonon frequencies. Phys. Lett. **35**A, 325-326 (1971).

136. From theoretical physics to biology. *Proc. 2nd Int. Conf. Theoretical Physics*

and Biology, Versailles, 1969, M. Marois, Ed., Edition du C.N.R.S., Paris, 1971, pp 17-19.

137. Wohin führt die Physik. In *Bild der Wissenschaften*, Vol. 8, Heft **1**, p. 64-70 (1971) [in German].

138. Selective long-range dispersive forces between large systems. Phys. Lett. **39**A, 153-154 (1972).

139. General introduction. *Proc. 3rd Int. Conf. From Theoretical Physics to Biology, Versailles, 1971*, M. Marois, Ed., S. Karger, Basel, 1973, pp 2-6.

140. The connection between macro- and microphysics, Riv. Nuovo Cimento **3**, 490-534 (1973).

141. Collective behaviour of non-linearly coupled oscillating fields with applications to biological systems, Collect. Phenom. **1**, 101-109 (1973).

142. Organisation and long-range selective interaction in biological and other pumped systems. In *Synergetics*, H. Haken, Ed., Teubner, Stuttgart, 1973, pp 241-245.

143. Low frequency vibrations of macro-molecules, Phys. Lett. **44**A, 385 only (1973).

144. Theory of superconductivity without solution of many-body problems, Collect. Phenom, **1**, 173-184 (1974).

145. Collective phenomena in biological systems. In *Progress in Synergetics*, H. Haken, Ed., North Holland, Amsterdam, 1974, pp 263-267.

146. Evidence for bose condensation-like excitation of coherent modes in biological systems, Phys. Lett. **51**A, 21-22 (1975).

147. Evidence for collective magnetic effects in an enzyme: likelihood of room temperature superconductive regions (with N.A.G. Ahmed, J.H. Calderwood and C.W. Smith) Phys. Lett. **53**A, 129-130 (1975).

148. Dielectric properties and measurement techniques for biomolecules, (with N.A.G. Ahmed, J.H. Calderwood and C.W. Smith) *Proc. 1975 Conference on Dielectric Materials, Measurement and Applications, Cambridge,* I.E.E. Conf. Publ. No. 129, 1975, pp 44-47.

149. The extraordinary dielectric properties of biological materials and the action of enzymes, Proc. Nat. Acad. Sci. U.S.A. **72**, 4211-4215 (1975).

150. Electric and magnetic properties of lysozyme and other biomolecules, (with

N.A.G. Ahmed, J.H. Calderwood and C.W. Smith), Collect. Phenom. **2**, 155-156 (1976).

151. Low frequency Raman spectra of lysozyme (with L. Genzel, F. Keilmann, M.W. Makinen, T.P. Martin, G. Winterling and Y. Yacoby) Biopolymers **15**, 219-225 (1976).

152. Makroskopische Quanteneffekte in Phys. und Biologie. In *75 Jahre Quantentheorie*, W. Brauer, H-W. Streitwolf and K. Werner, Eds., Akademie-Verlag, Berlin, 1977, pp 45-51.

153. Possibilities of long- and short-range electric interactions of biological systems, in *Brain Interactions with Weak Electric and Magnetic Fields*, S.M. Bawin, and W.R. Adey, Eds., Neuroscience Res. Progr. Bull. **15**, 67-72 (1977).

154. Dielectric theory and ion channels in nerve membrane, Biosystems **8**, 193-194 (1977) (Dedicated to L. Onsager).

155. Coherent excitations in biological systems (reply to M.A. Livshits), Biophysics (USSR) **22**, 777-778 (1977).

156. Long-range coherence in biological systems, Riv. Nuovo Cimento **7**, 399-418 (1977).

157. Biological control through long-range coherence. In *Synergetics*, H. Haken, Ed., Springer, Berlin, 1977, pp 241-246.

158. Resonant growth rate response of yeast cells irradiated by weak microwaves (with W. Grundler and F. Keilmann), Phys. Lett. **62**A, 463-466 (1977).

159. Evidence for non-thermal excitation of energy levels in active biological systems (with S.J. Webb and M.E. Stoneham), Phys. Lett. **63**A, 407-408 (1977).

160. Coherent electric vibrations in biological systems and the cancer problem, IEEE Trans. Microwave Theory and Techniques MTT **26**, 613-617 (1978).

161. Collective electron conduction, J. Less Common Metals **62**, 367-368 (1978) (B.T. Matthias Festschrift).

162. The biological effects of microwaves and related questions, Adv. Electron. Electron Phys. **53**, 85-152 (1980).

163. Recollections of the development of solid state physics, Proc. R. Soc. London **A371**, 102-103 (1980).

164. Coherence in biological systems, Coll. Phenom. **3**, 139-146 (1981).

165. Electret model for the collective behaviour of biological systems (with H. Bilz and H. Buttner) Z. Naturforsch. **36**b, 208-212 (1981).

166. Conditions for coherent excitations in biological systems (reply to M.Y. Yushina), Phys. Lett. **93**A, 105-106 (1982).

167. What are non-thermal electric biological effects? Bioelectromagnetics **3**, 45-46 (1982).

168. Types of coherent electrical vibrations in biological systems. In *Physics as Natural Philosophy*, A. Shimony and H. Feshbach, Eds., MIT Press, Cambridge, 1982 (Essays in honour of L. Tisza), pp 287-293.

169. Coherence in biology. In *Coherent Excitations in Biological Systems*, H. Fröhlich and F. Kremer, Eds., Springer, Berlin, 1983, p 1-5.

170. History of the theory of superconductivity. In *Advances in Superconductivity*, B. Deaver and J. Ruvalds, Eds., Plenum Press, New York, 1983, pp 7-11.

171. Evidence for coherent excitation in biological systems, Int. J. Quantum Chem. **23**, 1589-1595 (1983).

172. Biophysics, Nature **308**, 8 only (1984).

173. *Nonlinear Electrodynamics in Biological Systems*, W.R. Adey and A.F. Lawrence, Eds., Plenum, NY, 1984, pp 491-496.

174. Further evidence for coherent excitations in biological systems, Phys. Lett. **110A**, 480-481 (1985).

175. Coherent excitation in active biological systems. In *Modern Bioelectrochemistry*, F. Gutmann and H. Keyzer, Eds., Plenum Press, New York, 1985, pp 241-261.

Subject Index

GPSR Compliance

The European Union's (EU) General Product Safety Regulation (GPSR) is a set of rules that requires consumer products to be safe and our obligations to ensure this.

If you have any concerns about our products, you can contact us on ProductSafety@springernature.com

In case Publisher is established outside the EU, the EU authorized representative is:

Springer Nature Customer Service Center GmbH
Europaplatz 3
69115 Heidelberg, Germany

Batch number: 10584556

Printed by Printforce, the Netherlands